Agglomeration of Iron Ores

Agglomeration of Iron Ores

Authored by

Ram Pravesh Bhagat

CRC Press
Taylor & Francis Group
Boca Raton London New York

CRC Press is an imprint of the
Taylor & Francis Group, an **informa** business

CRC Press
Taylor & Francis Group
6000 Broken Sound Parkway NW, Suite 300
Boca Raton, FL 33487-2742

Printed on acid-free paper

International Standard Book Number-13: 978-1-138-03508-9 (Hardback)

Library of Congress Cataloging-in-Publication Data

Names: Bhagat, Ram Pravesh, author.
Title: Agglomeration of iron ores / Ram Pravesh Bhagat.
Description: Boca Raton : CRC Press, Taylor & Francis, 2019. | Includes bibliographical references and index.
Identifiers: LCCN 2018060465 | ISBN 9781138035089 (hardback : alk. paper) | ISBN 9781315269504 (ebook)
Subjects: LCSH: Iron ores. | Ore-dressing. | Agglomeration.
Classification: LCC TN538.I7 B44 2019 | DDC 622/.7—dc23
LC record available at https://lccn.loc.gov/2018060465

Visit the Taylor & Francis Web site at
http://www.taylorandfrancis.com

and the CRC Press Web site at
http://www.crcpress.com

Dedication

Dedicated to my parents
The late Ganesh Prasad Bhagat
The late Janki Devi

Table of Contents

Preface

The book titled *Agglomeration of Iron Ores* is about the size enlargement with reference to iron ores and consists of mainly the sintering and pelletization processes. The iron ore industry is going through an unprecedented period of expansion to meet the increasing demand of feed materials for iron making. Simultaneously, the quality parameters for raw materials have become more stringent. This requirement coupled with increasing regulations for environmental protection and the compulsion before the industries to utilize lean ores (around 45 wt. % Fe) has become a challenging task for the iron and steel industries. We need to develop technology that produces high-quality product from relatively low–grade ore, through beneficiation at finer size ranges followed by their preparation as feed to the blast furnace and other iron-making processes through agglomeration.

Importantly, the process adopted in this respect is material specific as the characteristics of ores change significantly, depending upon their sources, genesis, etc., which influence the agglomeration practice. This means that while the basic underlying principles in agglomeration remain the same, we cannot have a generalized approach for the process in designing a plant. The optimization of the process in this respect should be not only based on the "transfer of best practice elsewhere" but also requires conceptual understanding of the subject—why things happen so, besides how—and the research and development (R&D) remains unavoidable in this respect. Interestingly, much of the information on the latest technological developments in the field of agglomeration is yet to be covered in a text/reference book published so far. The present book underlines the importance of this aspect and has been framed to have a collected knowledge base contributed through the research in the field of agglomeration by various workers including the author. Since R&D is a continuous effort, one has to look for the current journals. Nevertheless, this book will serve as a base/starting material to initiate the research.

The aim of writing this book is to share adequate and updated information on this important topic, which is occasionally gathered on a piecemeal basis. The author has contributed a Chapter on Agglomeration in the *Encyclopaedia of Iron, Steel, and their Alloys* (Eds. Rafael Colas, George E. Totten) published by the CRC Press in 2016. The Chapter provides a broader picture of the subject but cannot make the subject complete in view of space limitations. Hence, the idea of writing a book on the subject came about to cover the subject in a more exhaustive manner. The framework of theoretical and practicing knowledge to which one is exposed today requires a broader perspective.

This book is designated to provide integrated and comprehensive coverage on the subject that should be helpful to graduate students in their curriculum on Mineral Processing and Ferrous Extractive Metallurgy pertaining to Agglomeration of Iron Ores. This book will also serve as reference material on the technical details to provide updated information on the subject to the teachers helping in preparing the lectures and also to the researchers and practicing professionals, working in the agglomeration and iron-making areas, assisting to enhance their knowledge base.

There are not many books with adequate and broader coverage on the subject; and hence, I do hope this volume will fill the gap.

The chapters and topics are arranged in a systematic way so that the readers get thorough knowledge and understanding of the subject. The book comprises 16 chapters. The first Chapter comprises an introduction to agglomeration including the necessity and scope. The processes of agglomeration requires raw materials of specified characteristics in terms of their purity (chemical compositions), size parameters, as well as mineralogical, thermal, and assimilation behaviors. This requires their detailed characterization and preparation (enrichment) before these are used. These are covered in Chapter **2**. The **third** and **fourth** Chapters cover topics on iron-making processes and agglomerates in iron making. These topics are important and relevant to explore the required physical-chemical characteristics of the agglomerates in order to extract iron metal in an efficient manner. The topics of sintering and pelletization are covered in Chapters **5** to **10** and Chapters **11** to **16**, respectively. The processes, fundamentals, and quality parameters with respect to each of these agglomeration processes are detailed. In keeping with the existing trends, topics on newer processes and techniques of agglomeration in the case of difficult ores, including high alumina ores that prevail in India and other Asia-Pacific countries, have been included. Lastly, common topics related to the transport parameters for sintering and pellet induration, and the test of quality parameters with respect to sinter and pellets are described as Annexures.

I do hope that the present book, updated through the current topics on its theoretical and practical aspects in the iron ore agglomeration, will be a significant addition to the metallurgical literature in this area.

Author Biography

Ram Pravesh Bhagat (b. 1954) served the Indian Institute of Technology (IIT) - (ISM) Dhanbad, India, as Professor (SAIL-NMDC Chair) on two years contract which he joined in November 2014 after superannuation from the service of CSIR-National Metallurgical Laboratory (NML), Jamshedpur, as Chief Scientist. He has over 38 years of research experience in iron ore beneficiation and agglomeration.

During his initial research career (1978-'89) at R&D Centre for Iron & Steel - Steel Authority of India Limited (SAIL), Ranchi, he worked with Soviet experts to implement R&D Projects in the sintering plants of Bhilai Steel Plant (BSP)–SAIL basically aimed at to improve the sinter productivity and its quality.

In the 1990s, Dr. Bhagat pursued research projects in the areas of beneficiation and dewatering in collaboration with IIT Kharagpur, RAS Russia, and with support from private industries. His subsequent research career at NML (2000–2014) concentrated on agglomeration, and he initiated and executed several R&D projects sponsored by the Ministry of Steel, Government of India, as well as private organizations, such as Tata Steel, MECON, Tata Metaliks, etc. Development of pilot-scale pelletization technology for Indian goethitic/hematite ores, Agglomeration of iron ore fines following nonconventional routes, Improved granulometry, and Cooling characteristics of sinter had been some of his recently concluded projects.

As a faculty member of Academy of Scientific & Industrial Research -New Delhi, also of ISM in the recent past, Prof. Bhagat developed courses and delivered lectures in agglomeration technology. He has supervised two Ph.D. scholars at IIT Kharagpur and several M. Tech students at ISM.

He has contributed three book chapters including "Agglomeration" in the *Encyclopedia of Iron, Steel, and Their Alloys* (CRC Press). He has published 45 research papers in SCI journals and 56 in conference proceeding volumes, He has also delivered several invited lectures. Prof. Bhagat is a reviewer of national and international journals. He is also an examiner of Indian Universities.

Prof. Bhagat received his M. Tech. from IIT Kanpur and Ph.D. from IIT Kharagpur. He is a life fellow and member of several professional bodies. He is also the recipient of various awards and fellowships, including CSIR-DAAD.

Author Biography

Acknowledgment

Numerous professionals have made significant and important contributions to the field of agglomeration and related literature. The author has used information from these sources and gratefully acknowledges the authors, journals, publishers, and also the societies for granting their kind permission to reproduce their data in the book:

a. Association for Iron & Steel Technology (AIST) | AIST Foundation (AIME), USA
b. Australasian Institute of Mining and Metallurgy (AusIMM), Australia
c. BHP, Australia
d. Bauverlag BV GmBH (*Aufbereitungs Tech.*)
e. Elsevier, UK
f. Indian Institute of Metals, India
g. Iron and Steel Institute of Japan, Japan
h. Institute of Chemical Engineers, London
i. JFE Steel (NKK), Japan
j. Kobe Steel Limited, Japan
k. Metso Minerals, India
l. Springer, US
m. Springer-Verlag, Berlin
n. Society for Mining, Metallurgy & Exploration Inc., US
o. Taylor & Francis Group, UK
p. Tata Steel Limited, India
q. Wiley-VCH, Weinheim, Germany

The author has greatly benefited from his work and discussions with the Russian researchers, Dr. B. M. Boranbaev and Dr. G. A. Grebenkin and acknowledge some of their input reflected in the book. He is grateful to R&D Centre for Iron and Steel- Steel Authority of India Limited (RDCIS-SAIL) and Council of Scientific & Industrial Research – National Metallurgical Laboratory (CSIR-NML) for having given him the opportunity to be associated the various R&D projects on agglomeration, and the sponsors, Tata Steel Limited, Ministry of Steel, and Department of Science & Technology, Govt. of India, in particular. He acknowledges CSIR-NML for extending library facility.

The author has had a long academic and research association with his doctoral guides, Prof. S. K. Gupta and Prof. H. S. Ray, his students at Indian Institute of Technology (ISM) and Academy of Scientific and Industrial Research, also several professionals and he records his appreciation to them. Thanks to his fellow colleagues associated with the author in various research programs at RDCIS-SAIL and CSIR-NML. Various professionals: Prof. Kal. V. S. Sastry (Univ. of California), Prof. S. Komarov (Tohoku University), Mr. S. K. Sarna (Ispat Guru), Prof. F. Matsuno

(Japan), my past colleagues, Dr. Nayak and Mr. Hembram have provided their input in the form of Figure(s) - the author thanks them.

My contribution, as a Chapter on Agglomeration, in the Encyclopaedia of Iron, Steel and their Alloys (Eds. Rafael Colas, George E. Totten) has given me opportunity to expend the work in the present form. I thank the publisher, CRC Press/Taylor & Francis, their team of editing and production for their cooperation at various stages of this project.

The cooperation extended by my wife, Manju, enabled me to devote time to prepare the book. Thanks to my daughters, Puja, Prabha, and Pragya, for their encouragement.

Ram Pravesh Bhagat

1 Introduction

ABBREVIATIONS

BF	*Blast furnace*
CCS	*Cold crushing strength*
DRI	*Direct reduced iron*
LOI	*Loss on ignition*
OSM	*Oxygen steel making*
NMDC	*National Mineral Development Corporation*
R&D	*Research and Development*
SAIL	*Steel Authority of India Limited*

1.1 PREAMBLE

Agglomeration is an important process in the extraction of iron ore to produce iron. The process, being the interface between the mineral processing and extractive metallurgy, is encountered by the professionals in these fields. Optimization of the agglomeration process is one of the major contributory factors in deciding techno-economics of the entire iron-making process following blast furnace and direct reduction route(s). The process has undergone through many innovations over the recent period. The effect of quality parameters of the agglomerates on the performance of iron-making practices is required to be studied for the process optimization.

The sintering and pelletization are the pertinent agglomeration processes with respect to iron making. In addition to knowledge of the basic and fundamental principles of these processes, insights on correlating the quality parameters of agglomerate with its mineralogy vis-à-vis the process variables are important for the technology development in these areas.

The technological innovations on agglomeration in phases has led to the development of newer practices and their assimilation in the plants. These basically aim to (a) increase productivity, (b) reduce raw material consumption and other operating costs, (c) improve the product quality, and lastly, (d) reduce environmental pollution. The influencing parameters can be categorized into those that are (a) raw material based and (b) process based.

1.2 DEFINITION AND CATEGORY

1.2.1 AGGLOMERATION

Agglomeration, according to *Merriam-Webster's Dictionary*, is the process of making larger-sized particles from smaller ones. The process has varied applications in (a) pharmaceutical industries, (b) chemical industries, and (c) metallurgical industries. With reference to the metallurgical usage of agglomerates in iron and steel industries, the process can be classified into (a) sintering, (b) pelletizing, and (c) briquetting (Figure 1.1)

FIGURE 1.1 Photographs of (a) sinter, (b) pellets, and (c) briquette

The agglomeration process has a much wider application in the iron and steel industries. It is the common method of burden preparation in the iron making following blast furnace and direct reduction processes. Growth of the steel industry with depleting resources of high-grade ores has led to a very strong demand for both pelletizing and sintering of iron ores.

1.2.2 Sintering

The process of sintering involves the agglomeration of the ore fines admixed with fluxes and other additives into a porous mass by the reactions of ingredient minerals through solid state diffusion at higher temperature as well as incipient fusion and bonding by slag, which are caused by the heat produced through combustion (of coke) within the mass itself.

Important **steps** in the sintering process are:

- Mix preparation,
- Charging the mix into a sintering strand,
- Ignition (of the top layer)
- Sintering
- Cooling

1.2.3 Pelletizing

Pelletization is the process of forming nearly spherical balls by tumbling moist particulate fines (approximately below 72 micron size) with/without the addition of binders and additives in a balling drum or disc. The green balls, thus formed, undergo an induration cycle consisting of drying, preheating, firing at approximately 1350°C (1623 K) temperature. and subsequent cooling.

Important steps in the pelletization process are:

- Making green pellets,
- Heat hardening the green pellets,
- Cooling the fired pellets

1.2.4 Briquetting

Briquetting is the process of making compressed blocks of particles admixed with binder such as coal tar and other additives. It is a cold process, and compression is imparted by the pressure between two blocks.

1.3 SCOPE OF AGGLOMERATION

The integrated blast furnace and oxygen steelmaking (BF-OSM) route is a dominant process for worldwide steel production, currently producing approximately 60% of the world's total output of crude steel. The rest of the (iron-making) processes are contributed by (a) coal- and gas-based DRI processes for production of non-liquid iron and (b) smelting–reduction processes for production of liquid iron, like Corex - a VAI Iron-making Technology.

These processes of iron making essentially require a prepared feedstock: iron ore in lump form (below 30 mm and above 10/6.3 mm size) and agglomerates in the form of sinter and pellets. These are the size enlarged products resulting from the process of *agglomeration*. In most cases, sinter finds its application in blast furnace iron making. The pellets also find their usage in the iron-making processes where agglomerates are charged. These include (a) blast furnaces, (b) direct reduction plants (coal- and gas-based), and also in the (c) Corex process.

One important aspect of the usage of agglomerates in the ferrous extraction process is that the agglomerates could be *tailor-made*. That is, while the quality parameters of raw materials are guided by natural constraints, the quality parameters of agglomerates, sinter in particular, can be regulated, which makes them important for their usage in a BF as burden. For instance, lump ore is the product of nature with inherent variability in its physical and chemical properties, whereas the agglomerates of desired quality specifications could be produced by modifying the process conditions and the chemical composition of agglomerates.

1.4 NEED FOR AGGLOMERATION

The sintering process is required to utilize the iron fines that are generated in the mines along with the lump ore. The fines can't be used as blast furnace burden, even if they meet the chemical specification, as their charging in the blast furnace will reduce the permeability of bed and hence productivity of the blast furnace. A sinter plant requires ore fines of lesser purity (Fe approximately 62 wt. %). In several cases, the fines generated in the mines through the dry and wet processing of ores meet the specification and, hence, do not require elaborate treatment.

In addition to ore fines, large amounts of lean-grade ores are excavated during the mining operation. In addition, the wet processing of ores generates slimes having the size below 150 micron. It is essential to have Fe value recovered from these materials for material conservation, also for the profitability of plant on one hand and for the sustainable development on the other. The agglomeration process provides an important avenue to utilize these materials, but only after these are enriched in terms of Fe value in the concentrate product.

The concentrate generated from lean-grade iron ores and slimes following their intense beneficiation process are in the sub-sieve range (approximately below 150 micron size) and hence could not find its place in conventional sintering for the obvious reason of unfavorable particle granulometry. Incorporation of the concentrate in the sinter mix will reduce the bed permeability during the sintering process and hence the strand productivity. Therefore, pelletization of the generated concentrate becomes an obvious choice. Alternatively, *pellet-sintering* technology, which combines pre-balling and the sintering process, has the potential to take care of the slimes and fines in the sub-sieve range. The cost of induration is eliminated through the incorporation of this technology. The technology, which mostly prevails in China, has drawn attention in other countries, like India, these days. Chapter 5 covers salient feature of the technology

Lime, which is required to neutralize the acidic constituents (SiO_2 and Al_2O_3) of the furnace burden, can be incorporated through fluxed sinter. In this way, the calcination of limestone inside the furnace which is highly endothermic, is eliminated (through the incorporation of fluxed sinter). The process of sintering provides an excellent way for the diversification of the fuel sources by incorporating cheaper fuel, that is, coke breeze. Besides, volatilities, moisture, CO_2 and most of the sulfur and arsenic in ore being eliminated during the sintering process.

1.5 RAW MATERIALS FOR AGGLOMERATION

The agglomeration processes require prepared raw materials, most importantly, ferruginous ones. The characteristics of the raw materials, in terms of their granulometry and chemistry, do have a significant effect on the process. The iron ores that are excavated from the mines are never pure and contain impurities, such as silica, alumina, and volatilities. This requires beneficiation. The process of beneficiation may adopt different routes depending on the nature and complexity of the iron ores being treated and giving rise to different flow-sheets. These topics are covered in Chapter 2.

Along with the lump ore, ore fines are generated in large quantities during the mining operation. The size of lump ore, as well as that of fines, is decided by the blast furnace requirement. Normally, sized iron ore (above 10/6.3 mm) in lump form, is directly used in the furnace; accordingly, undersized fractions (below 10/6.3 mm) of ore are considered ore fines and become feedstock in the agglomeration processes.

The ore fines, if rich in Fe (Fe greater than 62 wt. %) and low in silica and alumina content, can be directly used in sinter making, however, after wet scrubbing and washing. The sintering process requires iron ore of a coarser size. Preferred size of the fines is below 6.3/10 mm depending on the size of lump ore being used in the BF. Finer size fraction of ore fines (below 150 micron) should not exceed 20 wt. % since excessive quantities of very fine material affects the sinter bed permeability and, hence, productivity.

In addition to the ore fines of acceptable grade (by the sinter plant), large quantities of lean-grade ores are excavated during the mining operation. Besides, the wet processing of ores generates slimes having the size -150 micron. These materials, in their present form, can't be used directly in the iron-making processes. The problems are with respect to their purity in terms of high contents of silica and alumina as well as their granulometry.

The silica and alumina that constitute gangue minerals are associated with the hematite and magnetite minerals in iron ore. These gangues need to be removed/reduced to enrich Fe value in the product, called, *concentrate* while removing the undesired portion referred as *tailing* through the process of beneficiation. The process could be a prerequisite in case the deposit is composed of low-grade ores constituting high silica and/or high alumina also in cases of complex ores and slimes having high alumina and/or high content of loss on ignition (LOI). These ores are difficult to upgrade in view of their poor liberation characteristics and hence require specific beneficiation treatment to make suitable concentrates. In some cases, it is essential to have pre-concentration of high LOI ores using reduction roasting, which transforms goethite to hematite mineral and eventually maghemite. These are recovered by magnetic separation, allowing enrichment of the ore.

Utilization of slimes happens to be an important issue where modern agglomeration techniques need to find an inroad. The slimes are highly heterogeneous material having Fe content that may vary from 50 wt. % to sub-60 wt. %. Their chemistry and beneficiation-ability depend on the iron ore resource: whether the ore is hard massive, friable, compact laminated, or soft lateritic. Technically, because the slimes material is in the sub-sieve range, it need not be further ground, so the cost involved in size reduction that is usually practised in case of ores is absent. However, it is difficult to beneficiate because of the fact that alumina minerals are not liberated even in a finer size, and some are present in chemical combination of iron oxide.

Growth of steel industries in recent years coupled with depleting resources of high-grade ores have made a strong case toward the utilization of lean-grade iron ores and slimes after their intense beneficiation. Many of the integrated iron and steel plants have now adopted advanced beneficiation techniques in their mines to produce the concentrate having stringent chemical specification. In order to have adequate liberation of the gangue minerals from the Fe-containing minerals, the processing of the lean grade ores is required to be carried out in their sub-sieve sizes to produce the concentrate which is used in the pelletization process.

1.6 THE AGGLOMERATION PROCESSES

1.6.1 Sintering

A continuously moving bed of the moistened mix constituting ore fines (normally below 10/6.3 mm size) admixed with fine particles (normally below 3 mm size) of fluxes, limestone, and dolomite is agglomerated by partial fusion of the quiescent mass. Heat is generated by combustion of admixed solid fuel (normally 3 mm size). Air is infiltrated through the top of the bed. The resulting high temperature zone moves through the bed by an induced draft applied at the bottom of the bed. The bonding is affected by a combination of fusion, grain growth, and slag liquidation. The product of sintering consists of a cellular mass of ore bonded in a slag matrix. The sinter (cake), thus produced, is subsequently crushed and sized. Sized sinter is used in iron making. Normally, a +6.3 mm or +10 mm size fraction of sinter is accepted in the blast furnace as its feedstock. Fractions finer in size (than this) are recycled to the sintering plant as 'return sinter fines' and make up part of the sinter mix material. The return fines have a lower fusion temperature than the raw mix. The return fines serve as seeds for initiation of the sintering process. Chapter 5 describes the sintering process in detail.

The sintering is an extremely efficient heat transfer process: The air entering the bed is heated by the finished sinter until it attains its peak temperature in the combustion zone. The gases leaving this zone yield their heat to the feed below. The physicochemical reactions in the sintering process occur in the solid stage at different stages of temperature. The process of sintering is split into various zones, and it is unlikely that an equilibrium is established during its short duration. The phenomena of gas dynamics, heat transfer, and the movement of reaction front across the bed are crucial in deciding the sintering process. The sinter plant productivity is influenced by the bed permeability before ignition and during the sintering process. Improved granulation of the sinter mix has significant potential in this respect and is more important in case of the complex ores that prevail today. Chapter 6 describes fundamentals related to the sintering process. The influencing process and operational variables that do affect the productivity of sinter and its quality are outlined in Chapter 7 and Chapter 9, respectively.

It is generally accepted that the quality of iron ore sinter is decided by the sinter structure formed during the sintering process as well as by the nature, abundance, and mutual interactions of the mineral phases present in the sinter. In addition to the physical phenomena and chemical reactions during sintering, the cooling of sinter, does affect the sinter mineralogy. The topic related to sinter mineralogy is described in Chapter 8.

The quality parameters of sinter are important from the perspective of blast furnace operation. Cold strength of sinter also affects the sinter productivity since a stronger sinter has a higher yield and consequently the amount of recirculated return fines is low. Therefore, the variables that do influence the quality parameters of sinter are also interinfluenced with respect to the sinter productivity.

1.6.2 Pelletization

The concentrate of finer size (normally -100 micron, following beneficiation of iron ore, is pelletized into 10 to 15-mm-diameter sphere shaped ball by using binders like bentonite clay, water, and fluxes in a palletizing disk or drum. The green pellets, thus obtained, are subsequently indurated following the stages of ldrying, preheating, and heat hardening at around 1350°C (1623 K). These pellets can be charged into the blast furnaces or used for production of DRI as they contribute to faster reduction and high metallization rates. The induration process is outlined in Chapter 14.

Chapter 12 describes the basics pertaining to green pelletization. The process of green pelletization is divided into several steps and involves various mechanisms. The role of binder as well as that of moisture is significant in affecting the productivity of green pellets and their qualities. The process of green pelletization and qualities of green pellets, depend on many variables including the type of ore: whether it is hematite or magnetite. The ore characteristics affect the mineralogical characteristics of indurated pellets also. The quality parameters of these pellets, like those of sinter, are influenced by their mineralogical characteristics. Topics pertaining to quality parameters of green- and indurated pellets are outlined in Chapters 13 and 16, respectively.

1.7 IRON MAKING

Chapters 3 and 4 include brief description of the iron-making processes and correlations related to the quality parameters of agglomerates and the BF productivity and coke rate. The iron-making process and that through the BF route in particular, is highly capital intensive; Hence, any improvement in productivity increases its profitability. The BF iron-making process uses coke through the carbonization of coking coal. Coke serves both as fuel and as reducing agent during the smelting reduction in the BF. The coking coal is a very costly commodity, and its sources are limited. Therefore, much attention is paid by the furnace operators toward savings in coke rate. Hot metal is a feedstock of steel making, and its quality is equally important to produce quality and specialty steel at cheaper costs.

The countercurrent principle on which the blast furnace operates makes it heavily reliant on the quality of its burden to maintain a sufficient level of gas permeability in its upper shaft and of liquid and gas permeability in its lower part. Therefore, the quality parameters of ferrous burdens—namely, lump ore, sinter, and pellets—play an important role in this respect.

In an integrated iron and steel plant, the performance of sintering and/or pelletizing plants should not been seen in isolation. More important is the overall operation of these plants and that of iron making (combined together). It is not only the quantity of agglomerated products but their quality criteria that matter.

Enhancing the quality parameters of sinter and pellet and their control is required for smooth and efficient furnace operation. These parameters, therefore, are measured in the *laboratories* on a routine basis. Various standards have been formulated and subjected to revision from time to time in this respect. Important quality parameters that need routine and/or periodic measurement include: size (range), strength under cold condition and during reduction, reducibility, softening and meltdown temperature, swelling index, porosity. Standard methods to measure the agglomerate qualities are briefly outlined in Annexure 2.

1.8 R&D

Laboratory studies are critically important to understand the agglomeration fundamentals and to derive the dependence of process and operational variables on plant productivity and quality parameters of agglomerates. It is prudent to mention that the process of sinter and pellet making is material specific. Therefore, it requires detailed R&D studies pertaining to the ores in question. The ores constituting goethite or blue dust, for instance, do have difficult sinter-ability. In case of pelletization, special induration treatment is required for goethitic ore since combined water is required to be removed during its preheating stage. Certain groups of goethite ore contain higher LOI (approximately 10 wt. %) 0%) but much lower Al_2O_3 (in between 1 wt. % and 3 wt. %). Complexity arises with those ores that contain high alumina in addition to high LOI: these are typical and difficult to be agglomerated in the normal course of operation.

One of the major problems, that many steel plants in the Asia-Pacific region (including India) face, is that they will have to use high alumina iron ores in their sinter blends for either economic reasons or due to a shortage of high-grade iron ores. The effect of alumina loading into the blast furnace on increased furnace slag volume leading to an increased coke consumption is well established. A small increase in the alumina content of sinter blends can have a significant adverse impact on the strength and reduction degradation characteristics of the sinter., thereby leading to deterioration in gas permeability in the upper part of the blast furnace. Hence, it is essential to evolve a sintering methodology such that the adverse effect of Al_2O_3 in the sinter is neutralized. At the same time, we need to evolve an intense beneficiation process to yield the concentrate having over 65 wt. % Fe and Al_2O_3 between the ranges of 1 wt. % and 1.5 wt. % followed by its pelletization.

Agglomeration research with regard to difficult ores, that has been recently carried out in many countries (including Asia-Pacific), is important in this regard. In addition, various plants have adopted current practices of agglomeration based on process- and operation improvements. These topics are mentioned in the book.

1.9 TECHNO-ECONOMICS

1.9.1 Sinter (Pellet) Plant Operation

The economics of a sintering (pellet) plant operation are decided by how much sinter (pellet) is produced per unit of time per unit of area/volume with respect to the sintering (pelletizing) operations. The objectives of a sintering (pellet) plant operation, in this regard, are:

- To maximize production, and
- To produce sinter (pellet) with the best possible qualities with reference to its:
  a. Mean size and its size range: The size should be compatible with the lump ore.
  b. Cold strength: The sinter (pellet) should have higher resistance to tumbling.
  c. Reducibility: The sinter (pellet) should be reduce by the reactant gases with ease and at a higher rate.
  d. Low temperature breakdown during reduction: The sinter (pellet) should withstand the stresses during its reduction so that lesser fines are generated in the furnace stack.
  e. High and uniform softening and melt-down temperature of sinter (pellet) to have conducive cohesive zone inside the furnace: In addition, the pellets should not swell much during the reduction inside the furnace to ease better gas flow.
- Decrease in solid fuel (in case of sinter)/gaseous (in case of pellet) consumption.

  To reduce the cost of sintering (pellet production) operation:
  a. Utilizations of waste gas and sensible heat.
  b. Value added substitution: Raw materials with metallurgical/industrial wastes.

- Reduction of environment pollution. The emission of solid- and gaseous effluents from the plants is required to be minimized by adopting advanced technology. Recycling of hot gases to the plant also reduces the energy requirement in plant operation.

  A successful sintering/pelletizing technology incorporates the possible ways to exploit all these goals simultaneously. The economics is favored significantly when the process and operational parameters are optimized to improve upon the performance in terms of the product quality and the plant productivity. In the case of sintering, in particular, the quality of sinter product including its chemistry (the sinter basicity) can be modified by changing the composition of the mix materials: ore fines vis-à-vis the fluxes. Importantly, while the quality of lump ore is guided by natural constraints, the agglomerates, and sinter in particular, could be tailor-made for the aforementioned reasons.

A sintering (or a pellet) plant is a highly capital-intensive installation; hence, any improvement in its productivity has a direct bearing on its profitability. In addition to productivity, the strength and reduction properties of agglomerates (sinter and pellets) are important parameters that do influence the iron-making process. The quality of agglomerates is influenced by their mineralogical and morphological characteristics, which are decided by the influencing variables. It is, therefore, important to understand the interrelationship, among the process and operational variables, the mineralogy, and the quality parameters of agglomerates. These topics are mentioned in Chapters 8 and 9 (in case of sintering) and Chapters 15 and 16 (in case of pelletization).

1.9.2 Usage of Agglomerates

The techno-economics of the usage of pellets versus sinter in BF iron making has been a subject of discussion and has gone through several changes. Iron ore sinter constitutes a major proportion of BF burden in most countries in the Asia-Pacific region, (including India) and the use of pellets was practically missing/less in BF iron making. With abundant availability of high-grade ores, direct sintering of the generated fines had been considered a viable and economic option.

With the fast-changing techno-economic scenario across the globe, the integrated steel plants are now considering incorporation of some portion of pellets in the BF burden in view of the fact that the reserves of high-grade ore would not last long enough to sustain the long-term growth of the industry. Complex beneficiation of low-grade ore and slimes followed by pelletization has become a perquisite. The fact that gaseous fuel is available now at cheaper prices and the pellets do have superior chemistry and reduction properties over lump ore also add to the reasons why the pellets should replace calibrated lump ore.

1.9.3 Reducing Operational Cost

The cost of raw materials and energy is an important factor in deciding the economics of sinter plant operation, which need to be reduced without compromising on the product quality. The consumption of solid fuel in many of the sintering plant installations, particularly those in Southeast Asia, is very high. This adds to the cost of the sintering operation besides increasing the pollution level through the emission of more SOx and NOx gases. The sinter plants will have to face stricter environmental regulations in days to come, which adds to the cost of operation. Hence, the scope of reducing the operational cost through the development of newer technologies and their implementation is enormous. In this regard, the improved granulation of sinter mix and the sintering at low temperature are important for enhancing the sintering rate with the production of quality sinter and for reduced solid fuel consumption. The substitution of solid fuel (coke breeze), recirculation of off gases, and usage of industrial wastes are some of the other means to reduce the solid fuel requirement as well as the pollutant emissions. Besides, the process and operational variables need to be optimized suiting the existing local technology. These are described in Chapter 10.

In the case of pelletization, there exists a scope for the reduction of gaseous fuel requirements by optimally selecting the induration cycle, depending on the ore characteristics and the manner the green pellets are formed. The optimization of influencing variables is important in this regard. Besides, the gaseous fuel, required for induration, can be partially substituted by solid fuels either through direct firing or through incorporating it in green pellets. This is important in case the pellets serve as feedstock in the direct reduction processes where one does not require the pellets of higher cold crushing strength. In general, the selection of process and additives depends on the end use. For instance, bentonite could be partially replaced by organic binders in cases where the end use is in non-BF iron making. Topics pertaining to the induration process and pellet quality are described in Chapters 14 and 16, respectively.

1.9.4 Utilization of Slimes and Metallurgical Wastes

The techno-economics of agglomerate production are also influenced by the extent the mines and metallurgical wastes are utilized through the process.

a. The slimes generated during wet processing of iron ore are important waste materials, and this is where the modern beneficiation and agglomeration techniques need to find an inroad.
b. Several waste materials: basic oxygen steel making slag, flue dust, mill scale that are rich in iron, flux, and fuel values, are generated by an integrated iron and steel plants. Recycling of these metallurgical wastes to the sinter plant decreases the input of fresh material, and hence decreases the cost of sinter production. Besides, this facilitates the sinter plant to meet the statutory environmental regulations and make an effective utilization of natural resources.

1.10 EVOLUTIONARY PHASES

The agglomeration technologies have undergone radical changes, both in operation and process, throughout the world during the recent years. Important inventions relating to blast–furnace burden preparation are well summarized by Eu. F. Wegmann [1]

1.10.1 Sintering

The technological developments in sintering have undergone evolutionary phases in stages incorporating several innovations since the early 20th century. The technological innovations relate to the development of equipment and processes to utilize iron ore fines of different forms[1]. A. S. Dwight and R. L. Lloyd, both of the United States, invented a belt-type sintering machine for vacuum sintering (US patent Nos. 882517 and 916396) in the year 1906. Von Schlippenbach invented a rotary-type sintering machine (German patent No. 226033) in 1909. In the year 1987, F. Heberlein and T. Huntington, of England, invented the process of agglomeration sintering of sulphide ores (with the sintering bed being blown with air from bottom upwards). W. Job invented the process of sintering of pyrite cinder and iron ore fines admixed with coal and air being blown through the bed from bottom upwards in 1902 (German patent No. 137436). Eu. F. Wegmann, A. G. Mikhalevich and A. A. Buyarov designed a sintering machine for pressure sintering in 1967 (USSR Inventor's Certificate No. 508647). The first sintering machine was put into operation in the year 1930. Since then, sinter has become the widely accepted and preferred BF burden material.

1.10.2 Pelletization

The essentiality of the pelletization process arose from the need to utilize the lean-grade iron ore fines and slimes generated in the mines along with lump ore. These materials, being subgrade (low in Fe), need deep beneficiation at their finer size to improve Fe grade. The concentrate, thus generated, cannot be incorporated in the sinter mix as it will reduce the bed permeability in the sintering process (also the sinter productivity). Therefore, the concentrate is required to be pelletized.

A.G. Anderson, a Swede, invented the process of pelletizing of fine concentrate in 1912 (Swedish Patent No. 35124, 1912). The commercial use of pellets, however, began in the United States after World War II to facilitate the utilization of finely ground iron ore concentrates in steel production. The world's first pelletization plant went into operation in 1955 at Silver Bay, Minnesota [2]. At the beginning of the 1980s, 50% of the blast furnaces in the United States, Canada, and Mexico employed pellets. They constituted over 55% of the iron component. (For 15% of blast furnaces, pellets amounted to over 85% of the iron-ore component.)

Various studies were conducted with the aim of developing the vast reserves of taconite in the area around the Great Lakes. The U.S. iron making has historically been based to a large extent on pellets primarily because all the local iron ore needed beneficiation (upgrading) by grinding it to fine particles (size less than 0.1 mm) and agglomerating these fines into pellets but also because sinter plants are not/lesser used due to environmental reasons. In the mid-1980s, a number of pelletizing and blast furnace trials were conducted to evaluate the benefits of limestone dolomite fluxed pellets. Toward the end of the decade, fluxed pellets had been firmly established as a major product, accounting for about 30% of North America's pellet production.

REFERENCES

1. Eu. F. Wegmann, *A reference book for blast furnace operators*, Mir Publishers, Moscow, 1984, 78-79.
2. S. G. Savel'ev, G. V. Gubin, and Y. A. Stoikova, "Prospects for the production of iron ore pellets," *Steel in Translation*, 2013, 43, No. 8, 499–502.

2 Raw Materials: Characterization and Preparation

ABBREVIATIONS

BF	*Blast furnace*
BHJ	*Banded hematite jespar*
BHP	*(Formerly known as BHP Billiton)*
BHQ	*Banded hematite quartzite*
BIF	*Banded iron formation*
CC	*Concentration criterion*
CV	*Calorific value*
DCM	*Dry circuit material*
DR	*Direct reduction*
FC	*Fixed carbon*
HMS	*Heavy media separation*
LD	*Linz-Donawitz (basic oxygen steelmaking process)*
LOI	*Loss on ignition*
RDCIS	*Research & Development Centre for Iron & Steel*
RMBBP	*Raw material bedding & bedding plant*
SAIL	*Steel Authority of India Limited*
SEM	*Scanning electron microscopy*
TEM	*Transmission election microscopy*
VM	*Volatile matter*
WHIMS	*Wet high intensity magnetic separation*
XRD	*X-ray diffraction*

2.1 CATEGORIES AND SPECIFICATIONS

The raw materials' characteristics, in terms of their chemical constituents and size, are important in affecting the agglomeration process and the quality of agglomerates. The specifications of raw materials in sinter making and those in the pelletization process are different and are mentioned as follows.

2.1.1 Iron Ore

Iron ore is the principal raw material in the agglomeration process that needs to be prepared following the methods, depending on the type and nature of ore and its end use. The chemical composition of iron ore plays an important role in determining the primary melt formed during agglomeration and, consequently, the structure of agglomerates and quality. The process of agglomeration remains smooth in case the chemical composition of the ore has a higher degree of consistency (does not fluctuate).

Sinter Making: The sintering process requires ore fines of coarser granulometry and relatively lower Fe content (Fe more than 62 wt. %) than in the case of pelletization. Hence, the ore may not require elaborate and intensive treatment of beneficiation. Iron ore fines having the size below 10/6.3 mm and up to 0.15 mm, are generated in the mines after wet processing of iron ore raw material, is used in producing sinter. In case the chemical purity of the ore fines are not up to the norm, that is, wt.% Fe below 63, the material may need the beneficiation process to be adopted. The ore is required to meet certain specifications in terms of quality parameters, and these are:[1a, 1b]

a. **Size:** Below 8 mm; with micro fines (below 150 micron) should be maximum 25 wt. %. The micro fines should be as low as possible since sintering with more micro fines means more fuel consumption and decreases the productivity. Above 8 mm size fraction should be maximum 5 wt. %. These fractions should be as low as possible, since these will not participate in the sintering process. Fractions of 2 mm to 5 mm should be as high as possible for a better sintering process.
b. **Chemistry:** Fe in ore fines should be above 62 wt. %, SiO_2 and Al_2O_3 is needed to be as low as possible.

Pellet Making: Pellet feed is the iron ore concentrate with a minimum of 64 wt.% Fe and combined contents of SiO_2 and Al_2O_3 should be below 3.5 wt.%. In view of this, the ore needs to be enriched intensively, especially with respect to lean ore and slimes. Given here is the specification of iron ore (concentrate) in pellet making.[2]

Size		Chemical Analysis				
Below 150 micron	Blain number	Fe	$(SiO_2 + Al_2O_3)$	LOI	P	Alkalis
100 wt.%	more than 1800	63% (min.)	3.5% (max.)	~ 3.5%	0.09%	0.15%

2.1.2 Fluxes

Limestone and dolomite are the principal resources of CaO and MgO, respectively. Magnesium silicate in the form of dunite or pyroxenite is also used, which substitutes dolomite. Dunite is comprised of pure olivine (2 $MgO.SiO_2$). This rock is preferred because of its high MgO content and favorable $MgO:SiO_2$ ratio. Theoretical values of MgO and silica are 57 wt. % and 43 wt. %, respectively. Pyroxenite comprises the enstatite variety (an orthorhombic mineral) of the pyroxene ($MgO.SiO_2$) group consisting of magnesium silicate and usually occurs in a massive form. Its color varies from grayish white to olive green and brown. It is the second-best

TABLE 2.1
Magnesium Silicate Minerals: Chemical Formula and Theoretical MgO Content

Mineral	Chemical Composition	Theoretical Values of MgO
Pyroxene	$MgO.SiO_2$	40
Olivine	$2MgO.SiO_2$	57
Serpentine	$3MgO.2SiO_2.2H_2O$	43

alternative as an MgO source. Theoretical values of MgO and silica are 40 wt. % and 60 wt. %, respectively (**Table 2.1**).

Sinter Making: The incorporation of fluxes in the sinter mix provides lime (CaO) and MgO to make the sinter basic, which is desired by the slag chemistry in blast furnace iron making.

The fluxes, those that are used in the sintering plants, in general, do not require beneficiation. Limestone, which is generated in mines, with normal size below 10 mm, is further crushed in a hammer mill to below 3 mm in size and used in sinter making. Usage of flux in a closer grain size is important since limestone with a grain size higher than 3 mm does not calcine and assimilate fully during the short duration of sintering. On the other hand, the finer sized fluxes (below 0.25 mm) are not preferred in the process as they adversely affect the sintering efficiency (**Sec. 7.7.2, 9.6.2, 10.6.1**).

Major impurities in limestone (and dolomite) are the contents of silica and alumina. Besides, the alkalis in fluxes affect the blast furnace performance adversely when added through sinter. These impurities should be within the limits as follows.

SiO_2	<3.6%	Al_2O_3	<0.6%	Na_2O	<0.08%	K_2O	<0.3%

Pellet Making: Fluxes are normally used to tailor the chemistry of pellets and to produce the desired metallurgical and physical properties in the fired pellets. Limestone levels of 0.5 wt. % to about 2 wt. % of pelletizing mix are common for production of direct reduction (DR) - grade pellets. Blast furnace (BF) - grade pellets have (higher) levels of carbonate flux addition in the green balls, which may exceed 10 wt. % in some cases.

The specification of flux in the production of pellet, with respect to its chemical composition, is similar to that used in sintering. Finer-sized flux, ground to below 150 micron size, is used as an additive in pellet making.

If the pellet chemistry calls for magnesium content, dolomite could be incorporated as a substitute for the limestone. The chemistry of pellet is customer driven. As an alternate to dolomite, dunite and pyroxenite also are used as an Mg source in order to improve the metallurgical performance of fired pellets (**Sec. 10.5.4**).[2]

	Size	Chemical Analysis					
	Below 74 micron	CaO	MgO	LOI	SiO_2	Fe	Al_2O_3
Limestone	100%	50%	3%	~ 42%	2.5-3.5%		
Pyroxenite		~3%	25-30%	~10%	45-50%	6-10%	1%

2.1.3 Solid Fuel

Sinter Making: Coke breeze is used as solid fuel in sintering. The fractions of coke sized below 10 mm that are generated in the coke ovens are known as coke breeze. It is crushed to less than 3.15 mm in size for its use in sinter making. The coke breeze of a finer size (below 0.25 mm) as well as of a bigger size (above 3.15 mm) is not good for sinter making as incorporation of these fractions ,in the sinter mix causes a mismatch of flame front and heat front (Sec. **10.7.2**); 1 mm to 3-mm-sized coke breeze should be 85 to 90% with a greater amount of fixed carbon (min. 76 wt. %).

Pellet Making: Solid fuel in the form of coal/coke breeze is added to green balls to improve furnace productivity and with respect to overall fuel economy. This is, however, in a limited quantity, which ranges from 1 to 2 wt. % of the pellet mix. The aim is to partially substitute the gaseous (or liquid) fuel requirement during the induration (heat hardening) process of green pellet. Both proximate analysis and calorific value (CV) of coal/coke breeze decide the quantity that can be added to the mixer. Anthracite is a substitution for the coke breeze (Sec. **10.7.6**).

The specification of coke for its use in pelletization is as follows:[2]

Size-74 micron	Fixed Carbon	CVKcal/kg	Moisture	Ash%	Sulfur	V M coal	Ash fusion point
100%	65%	~ 6,500	7%	15 max.	1% max.	10% max.	1330°C

2.1.4 Revert Materials/Metallurgical Wastes

The by-/waste products of the metallurgical plant namely, mill scale, blast furnace flue dust, LD slag, LD sludge, etc.—are recycled through the sintering route, however in limited quantity, so that the performance of the sintering plant and product quality should not be impaired. These provide iron, CaO/MgO values (Sec **5.5**).

LD Slag: This is a source of lime, MgO, and Fe. Recycling it through the sintering process saves fluxes and iron but at the cost of plant productivity, fuel rate, and product quality.

LD Sludge: It consists of iron and CaO materials. The sludge is profitable metallurgical waste for recycling. Harmful materials in the sludge are alkali, Zn, Pb, etc.

Mill Scale: This is very high iron material. Almost 100% of the mill scale, non-oily, is recycled through sinter making.

BF Flue Dust: High iron and carbon content make it very attractive for recycling. It consists of a high alkali content.

2.1.5 Binders

Binders are incorporated to provide sufficient strength of the green balls following pelletization. The binders ensure sufficient strength to the pellets during their heat hardening and thereafter. These include mainly bentonite—a type of clay mineral and lime. Organic binders can be/are also used as a partial substitute for bentonite.

Bentonite is used as a binder in production of BF-grade pellets or DR-grade pellets. Bentonite is formed by hydrothermal alteration of volcanic ash deposits. It is a mixture of clay minerals, with the primary component being the montmorillonite mineral, which has the composition: (Na, Ca) (Al Mg) $Si_4O_{10}(OH)_2$. The acidic constituents of bentonite are considered as impurities, especially for iron ores with high acidic ($SiO_2 + Al_2O_3$) content. Bentonite adds up to approximately 0.3% by weight to the silica content of fired pellets.

Hydrate lime [$Ca(OH)_2$)] acts as both a binder and a basic additive in the pelletization process. It is used with ores that already have some porosity for self-agglomeration. Hydrate lime contributes almost no silica to the final product. It acts as a silica flux when the pellet is reduced in the blast furnace and improves the metallurgical performances of the pellet. The use of hydrate lime as a binder is most attractive when fluxed pellets are being produced.

It is preferable not to incorporate lime in its original form in the pellet mix. This is because it gets slaked with water that is injected into the pellet mix, and this generates calcium hydrate, which is not evenly distributed. Lime is readily available and is also less expensive than bentonite.

The size, consistency, and chemical composition of these binders are as follows.

Binders	Size	Chemical Analysis					
	-74 micron	CaO	MgO	LOI	SiO_2	Fe	Al_2O_3
Bentonite	100%	1-2%	2-3%	~ 8%	~60%	~2%	~16%
Lime		90%		~10%	2%		

Organic polymers are also used as binders to partially substitute for the bentonite. Their advantage lies in the fact that they do not add to impurities in the pellet mix. However, the organic binders alone cannot provide adequate strength to the pellets after firing. Besides, they are costlier than bentonite.

2.2 CHARACTERIZATION OF RAW MATERIALS

The raw materials need characterization prior to their end use in sinter and pellet making. Iron ore, especially, prior to its beneficiation (the procedure of enrichment) is characterized in order to get information about its grain size, types of minerals locking, liberation sizes, etc. The group of certain experimental procedures needed to characterize the ore sample are mentioned here.

Experimental Procedure	Purpose	Remarks
Sieve analysis	Suitable size distribution	Estimation of top size Content of below 125 micron size is determined
Chemical analysis of each sized fines	Confirmation of the size range of the fines with rich iron content	Scope of upgrading the fines (higher Fe) by their processing
Identification of mineral phase by X-ray diffraction and other techniques	Mineralogical identification of iron oxides and gangue minerals	Granulation of fines, breakdown and melting by heating. Estimation of minerals' reactivity during sintering and the possible phases in the sinter product
Microscopic observation of coarse iron ore	Mineralogical identification composed of dense or porous structure	

2.3 GENESIS OF IRON ORE

The metal iron is one of the predominantly available elements that constitutes 4.6% of the earth's crust. There are a number of commonly occurring iron minerals, and the deposits of iron ores are found at the earth's surface. The three most common iron ore minerals are hematite, magnetite, and goethite, which together account for an estimated more than 99% of the total Fe constituting minerals.[3a]

Iron ore is formed through the geological process of crystallization from silicate solution. Heavier ferrous materials precipitated to the bottom of the ocean. This resulted in the formation of a final rock mass confined to certain areas that are richer in iron than others. Iron-bearing minerals were deposited and gangue minerals, especially silica, could be leached out by water solution when surface water descended and underground water ascended through rock strata. Besides, the iron-bearing minerals were also concentrated by precipitation from water reservoirs, lakes, seas, rivers, and so on following chemical and/or bacteriological actions. Goethite mineral is enriched along with hematite ore during the process of beneficiation, which results in an increased percentage of loss on ignition (LOI) in the concentrate. Oxidation of lower-grade oxides, such as magnetite to hematite, takes place in the upper zone of ore bodies. These minerals deposited on the sea floor in great abundance forming what are known as the "banded iron formations". The rocks are "banded" because the iron minerals deposited in alternating bands with silica and sometimes shale.

Typical banded iron formation consists of repeated, thin layers (a few millimeters to a few centimeters in thickness) of silver to black iron oxides, either magnetite (Fe_3O_4)

(a)

(b)

FIGURE 2.1 (a) Banded Iron Formation at the Fortescue Falls, Karijini National Park, Western Australia (after Ref. [3b]) (b) Photograph of a black-band ironstone (after Ref. [3c])

or hematite (Fe_2O_3), alternating with bands shales and cherts, often red in color, of similar thickness. Figure 2.1 shows some photographs of banded iron formation. Banded iron formations account for more than 60% of global iron reserves, and can be found in Australia, Brazil, Canada, India, Russia, South Africa, Ukraine, and the United States.

Although the geological formation of iron-bearing minerals in the earth's crust remains similar across the globe, there happens to be minor variation from place to place within the globe.

a. Indian hematite ore deposits belong to the Precambrian Metasedimentary Sequence, known in Indian geology as "Iron Ore Series." The iron ore body was formed by a secondary process of leaching and enrichment of iron-bearing rocks (banded hematite quartzite/banded hematite jasper, or

BHJ/BHQ) under certain structural and meteorological controls. This process had also produced different ore types of varying physical, textural, and chemical compositions. The hematite ores occur in massive, laminated, friable, and powdery forms. Pockets of unreplaced or partly replaced parent rocks are present in the ore body. BHJ occurs as "horses" in the zone, indicating the original unaltered rock.

The major economic deposits of iron ore in India are distributed across five broad zones:[3d]

Zone A: Singhbhum in Jharkhand and Cuttak in Odisha

Zone B: Dantewada and Durg in Chattisgarh and Chandrapur and Gadchiroli in Maharashtra

Zone C: Bellary-Hospet iron belt in Karnataka

Zone D: Goa and Ratnagiri in Maharashtra and North Karnataka

Zone E: Metamorphosed BIF along the west coast in Karnataka and Kerala.

b. The deposits in Brazil/United States are primarily due to a high level of metamorphism with the subsequent generation of layered hematite. These are known as itabirite/taconite.

   Iron ore mineral deposits are widely dispersed in the United States. These are formed in a wide variety of geological environments that include sedimentary, metamorphic, and igneous rock formations.[3e]

c. The deposit in Krivoy-Rog, Ukraine (erstwhile USSR), is less metamorphosed; and hence, it consists of a higher amount of jasper.

d. Iron ore from the Sheini Hill ore deposit (Ghana) is basically stone and contains 47 wt. % Fe and 28 wt. % SiO_2.

e. Australia has large reserves of pisolitic limonite. The ore contains about 10 wt. % combined water. Iron ore in Australia is mainly found in rocks that are more than 600 million years old.[3f]

f. South Africa has approximately 9 billion tons of iron ore reserves, of which 45% is located in the Northern Cape. Kumba Resources produces iron ore from its two mining operations at Sishen in the Northern Cape and Thabazimbi in Limpopo.

2.4 CLASSIFICATION OF IRON ORES

2.4.1 Hematite Ores

Hematite is an iron oxide (Fe_2O_3) mineral and widespread in the earth's crust. The mineral has the crystal structure similar to corundum. Hematite is colored black to steel or silver-gray, brown to reddish brown, or red. Huge deposits of hematite are found in banded iron formations. While the forms of hematite vary, they all have a rust-red streak.

These ores, depending on their chemical purity, hardness, and density, are classified as follows[4]

a. **Hard Ores:** These are hard, compact, and steel gray colored. These are lumpy having specific gravity around 4.5. The ore generally contains high iron (above 63 wt. %) and less alumina (below 2 wt. %). These ores, if processed individually, generate around 80% calibrated lump ore having 1.2 to 1.6% Al_2O_3. There remains hardly any scope for improvement of lump quality as alumina in this is mostly occupied in hematite crystal structure and is difficult to separate.

The hard ores are of the massive and laminated type. The massive ores show nonplanar structure and occur as massive bodies, whereas the laminated ores have planar structures.

b. **Lateritic Ore:** This occurs in top weathered profile and weathered products of other types of ores. Its specific gravity is much lower—around 2.3. The ore is of an extremely lean grade having Fe as low as 55% and Al_2O_3 as high as 7.0%. Lateritic ore contributes alumina in ore fines.
c. **Flakey Ore:** The ore occurs as flakes with alternate bands of iron ore and gangue. These ores are friable, inherently soft, and contain a very high percentage of natural fines. This type of ore is also leaner in iron content and richer in alumina.

   Blue dust, however, contains high iron and less alumina and is not suitable for wet processing as the majority of finer fractions are carried away with slime.
d. **Soft Ore:** This type of ore is soft and non-crystalline. The ore is light gray in color and has low specific gravity. Typically, it contains mixtures of oxy-hydroxides of iron and silicate gangue material. The ore has a high content of gangue (alumina and silica) minerals and is low in iron.

2.4.2 Magnetite Ore

The mineral magnetite is an oxide of iron having the chemical formula Fe_3O_4. It is ferromagnetic and attracted to a magnet. Magnetite contains both ferrous and ferric iron. Magnetite differs from most other iron oxides in that it contains both divalent and trivalent iron. Magnetite is black or brownish-black with a metallic luster.[5]

2.4.3 Goethite Ore

Goethite ore is one of the most common and the third most abundant source of iron. Goethite, whose composition corresponds to α–FeOOH, is an oxyhydroxide mineral and is formed in the orthorhombic crystallographic system. This is formed under oxidizing conditions as a weathering product of iron-bearing minerals. It has a multitude of shapes and sizes but occurs commonly with a basic acicular morphology.[6]

The ores may contain Al_2O_3, MnO, CaO, and SiO_2 as impurities within their matrix and do have varying chemical composition. This is due to fact that Fe in the crystal lattice of goethite is normally replaced by Al and Si with varying proportions due to weathering effects.[7] A case of higher replacement results in above 10 wt. % Al_2O_3 content compared to 2 to 3 wt. % in others. Earthy-like varieties frequently have higher Al_2O_3 and SiO_2.

The pisolitic limonite deposits of Australia as well as iron ores from the Brazilian Quadrilatero Ferrifero and Minas Gerais deposits do contain higher LOI (around 10 wt.%) but much lower Al_2O_3 (1-3 wt.%). Whereas the ores that prevail

TABLE 2.2
Chemical Analysis of the Kiriburu high LOI (Goethitic) Ore

Radical	Fe	CaO	MgO	SiO_2	Al_2O_3	LOI	S	P	Mn
Wt.%	56.3	0.33	0.04	2.14	6.32	10.31	0.013	0.056	0.01

in the Asia Pacific regions of India and China consist of higher LOI coupled with high alumina content.

The chemical composition of typical high LOI and high alumina goethite ore from the Kiriburu mines of Steel Authority of India Limited (SAIL) in Jharkhand State, India, is given in **Table 2.2**.[8] The ore is of high LOI (10.3 wt. %) content and very high Al_2O_3 content (6.3 wt. %), which necessitates the beneficiation before it is used as feedstock for iron making.

2.4.4 Banded Hematite Quartzite/Jasper (BHQ/BHJ)

Banded Hematite Quartzite/Jasper (BHQ/BHJ) are of a BIF series and considered as polycrystalline materials that have undergone complex geological processes—namely metamorphism, recrystallization, and weathering. These low-grade ores contain 28 to 40 wt. % Fe, 40 to 55 wt. % SiO_2, and 1 to 2 wt. % Al_2O_3. Silica is the major impurity present in the BHQ sample. The percentages of other impurities are comparatively small. The minerals are liberated at extremely fine sizes. In addition, these ores typically consist of bands of iron-bearing hematite or magnetite intergrown with quartz. For this reason, the nature of BHQ appears quite complex.

BHJ is unevenly banded with alternating layers of hematite and jasper, and the thickness of individual bands may vary from a few millimeters to about 2 cm while the average thickness is observed to be 1 to 2 mm. The bands also vary in color from gray or white to red, brown, or black. The typical chemical composition of BHJ is Fe: 34 to 36 wt. %, SiO_2: 47 to 48 wt. %, Al_2O_3: 1 to 2 wt. %, LOI: 1 to 2 wt. %.

The liberation analysis says that in BHJ samples, clay content is negligible, but silica content is very high. Hematite and quartz are medium to very fine grained. About 80% of hematite grains carry extremely fine-grained inclusions of quartz.

Figure 2.2 shows photographs of typical iron ores.

I. Hematite ore with laterite: The ore has a distinct earthy luster and is granular in form. A weathering effect is visible.
II. Kiriburu goethitic ore: The colors of the ore vary from shades of yellows to reddish brown.
III. Magnetite from Bolivia: Spinel structure group. Black, gray with brownish tint in reflected sun.
IV. Banded hematite quartzite (BHQ): Alternating layers of hematite and quartz.

(I) Hematite ore

(II) Goethite ore

(III) Magnetite ore

(IV) BHQ

FIGURE 2.2 Photographs of typical iron ore samples

2.5 MINERALOGY OF IRON ORES

2.5.1 Iron Ore-Bearing Minerals

Iron ore constitutes different iron-bearing minerals, each having definite characteristics in terms of chemical constituents, structural arrangement, density, and hardness. Iron ores that are excavated from the mines, like any other ore, are never pure. These constitute the "gangue" in the ore, which is basically nonferrous minerals. Alumina in hematitic iron ore exists as hydroxides of aluminum. Important minerals are: gibbsite [$Al(OH)_3$], kaolinite [$Al_2O_3.SiO_2.(OH)_4$], and diaspore [$Al_2O_3.H_2O$]. Also, silica minerals in the form of quartz having the chemical formula SiO_2 are present in the iron ore. The types of minerals and their abundance depend on their geological formation which are as follows:

a. Hematite (α–Fe_2O_3): This is a major mineral constituent of iron ore. It has hexagonal, close-packed corundum structure anion packing. It is weakly ferromagnetic at room temperature. Its specific gravity is approximately 5.1.

b. Magnetite (Fe_3O_4): This is a major mineral constituent of iron ore. It has an inverse spinel structure and is ferromagnetic at room temperature. Its specific gravity is approximately 5.17.
c. Maghemite (γ–Fe_2O_3): It has an inverse spinel structure like magnetite but with cation-deficient sites. It is ferromagnetic at room temperature.
d. Martite (Fe_2O_3): It is hematite that has replaced primary magnetite. Its occurrence is in Australian ore.
e. Goethite (α–FeOOH): It has a diaspore structure (hexagonal close packing of anions). Goethite-bearing ores have a low magnetic susceptibility. It is antiferromagnetic at room temperature. Its specific gravity is around 4.2. The goethite ore has three subtypes, namely,
   i. Yellow ochreous: having excess water.
   ii. Brown: the most common variant with stoichiometric.
   iii. Vitreous: glassy and contains silica and alumina, either as an impurity or as submicron inclusions.
f. Gibbsite ($Al[OH]_3$)/diaspore ($Al_2O_3.H_2O$): These are aluminum-bearing minerals.
g. Kaolinite ($Al_2\ O_3.SiO_2.[OH]_4$): This is aluminum and silicon-bearing clay.
h. Quartz (SiO_2): It is a common oxide of silicon.

2.5.2 Ores and Associated Minerals

Important minerals associated with different types of ores from important resources are given here.

Ore Type	Main Minerals	Subordinate/Accessory Minerals
Hematite ore	hematite and martite	magnetite, limonite, and goethite
Itabirite ore	hematite, martite, magnetite, and quartz	chlorite, muscovite, Fe-bearing dolomite, kaolinite, kyanite, manganese oxides, gibbsite, etc.
Taconite ore	magnetite, hematite, siderite, and quartz	pyrite, calcite, and pyrrhotite
Hard ore	hematite followed by goethite; traces of martite, magnetite, and rutile	quartz, jasper followed by clay minerals
Soft ore	goethite, limonite, and hematite	quartz, clay minerals, gibbsite, and chert in varied proportions
Flaky ore	hematite and martite	lateritic clay in traces
Blue dust	traces of goethite	quartzite
BHJ	hematite and martite	quartz, chert, and jasper

2.5.3 Mineralogical Characterization

It is important to have information about the mineralogical characteristics of iron ore before the ore is beneficiated. The following studies are important in this regard.

Macroscopic Examination The study is carried out by the unaided eye, with a hand lens and/or with the stereomicroscopy. Considerable information can be drawn by macroscopic study, such as whether the ore is crystalline or amorphous, soft or hard.

Microscopic Study Preliminary information about the mineralogical constituents of iron ore and other raw materials can be obtained through the X-ray diffraction (XRD) analysis of the sample. The minerals associated with the ores are compounds in crystal form. The peaks are obtained having certain intensities at different d-values through XRD, which are characteristics of these compounds. XRD data files are available to scan through the peaks. The mineralogical constituents of the ore sample (polished) are studied using an optical petrological microscope having both reflected and transmitted light facilities at different magnifications.

The microscopic examination of typical iron ores from India reflects the following observations.

Hematite Ores A certain group of iron ores from Barsua Mines of SAIL in Eastern India is aluminous and presents typical problems associated with fine-grained alumina within a hematite matrix. This needs the exhaustive characterization of ore to elucidate the nature of occurrence of aluminum in it. Detailed characterization through the application of electron microscopy, microprobe, and electron diffraction techniques should be attempted to study the fine-grained mineral samples for which petrography and X-ray diffraction techniques are inadequate.[9] Electronic microscopy, namely transmission electron microscopy (TEM) and scanning electron microscopy (SEM), is applied for the characterization of minerals in the micron level sizes.[9] The size at which the gangue constituent minerals are liberated from the Fe-bearing minerals (hematite) may go below a few microns. The ore, however, consists of a scattered amount of the minerals that are loosely bound and happen to be fairly liberated at a coarser size.[9a,9b]

Goethite Ores The goethite ore from Kiriburu Mines, SAIL in Eastern India is high in LOI content and also highly aluminous. The goethite mineral occurs in various forms such as vitreous, colloform bands, fine crystals, spongy, earthy, and so on. The identified minerals are:

a. Major iron-bearing minerals: hematite and goethite
b. Gangue minerals: kaolinitic clay, quartz, gibbsite, and limonitic matter

The mineralogical characteristics of typical goethitic iron ore are demonstrated in **Figure 2.3**.[8] The figure shows the presence of goethite crystals and an alternate form of hematite–goethite that is trapped in clay minerals (Figure 2.3, I). The silica was mainly composed of quartz of various grain sizes. These fine quartz crystals were of the order of 10 to 20 microns. Conversely, the silica also contained iron minerals. Cases of scattering of very fine crystals of iron-bearing minerals along

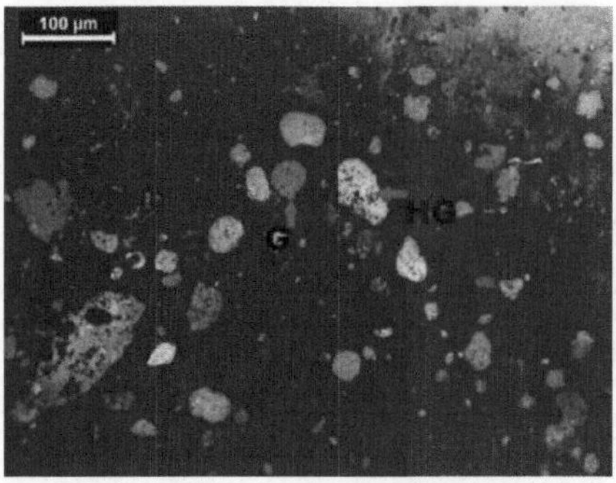

I Finer to coarser different phases of goethitic crystals are scattered within the ground mass of clay materials.

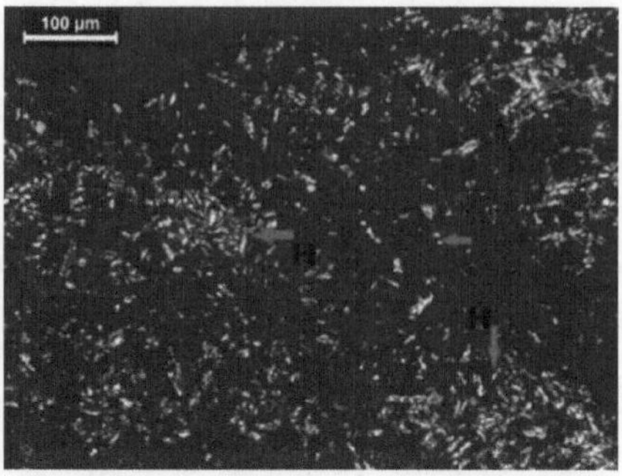

II Very fine crystals of iron ore are scattered along with clay materials.

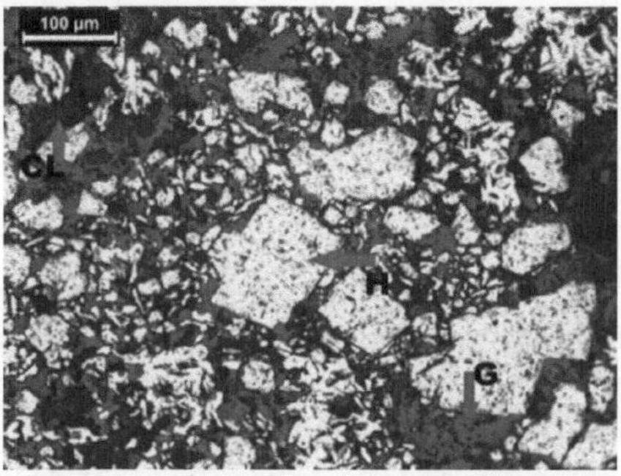

III Hard massive iron ore shows coarser to fine crystals of hematite scattered in ground mass of goethite.

FIGURE 2.3 Mineralogy of typical Indian goethite ore (after Ref. [8])

with clay minerals (Figure 2.3, II) and the appearance of hematite crystals within the hard laminated goethite were observed (Figure 2.3, III) through the present microscopic study.

2.5.4 Liberation Characteristic of Minerals

The liberation characteristic of the minerals associated with ore is important in deciding the beneficiation process. It determines the size at which the value-constituting minerals dissociate themselves from the gangue-constituting minerals. In the case of iron ore, the value-constituting minerals are hematite, magnetite, and goethite (limonite), while the gangue-constituent minerals are silica and alumina.

The liberation characteristic of ore is determined using a standard procedure. Typically, the ore, ground to below 2 mm in size, is sieved to different micron sizes. Individual sieve fractions are studied under a microscope to delineate the liberation characteristics of the ore in terms of free iron minerals, free gangue minerals, and interlocked particles by counting of grains.

Results of the macroscopic examination in the case of the ore from Kiriburu mines of SAIL, India, is shown in **Figure 2.4**. Chemical composition of the ore is given

FIGURE 2.4 Photograph of the liberated and interlocked grains of different minerals in typical Indian goethite ore

in Table 2.2. The ore is of high LOI (10.3 wt. %) content and very high Al_2O_3 content (6.3 wt. %), which necessitates beneficiation before it is used as feedstock for iron making.[8]

Table 2.3 shows the liberation characteristics of the ore. The amount of interlocked particles was maximum in the ore at its coarsest size (2 to 1.4 mm) range and decreased as the sample size decreased.[8]

The optimum of the size reduction and classification of ores for its processing can be decided through the liberation characteristics data. The proportion of interlocked minerals decreases as the ore is ground to its finer size (Table 2.3). The grinding operation involves cost; obviously, one has to make a call about the extent to which the sample is required to be size-reduced for effective separation between the valuable and gangue minerals.

TABLE 2.3
Data Showing Weight Percent Distribution of Liberated and Interlocked Minerals within the Ore Sample at its Different Size Fractions

Size fraction in microns	Weight %				
	Goethite	Hematite	Interlocked	Gangue	(Go+He)
(-2000+1410)	15.92	5.10	77.07	1.91	21.02
(-1410+841)	19.76	6.85	67.74	5.65	26.61
(-841+595)	20.04	7.84	62.75	9.37	27.89
(-595+320)	21.14	9.19	57.71	11.96	30.33
(-320+230)	23.80	10.43	52.94	12.83	34.22
(-230+150)	25.62	12.30	46.60	15.48	37.92
(-150+100)	26.02	13.78	42.34	17.86	39.80
(-100+74)	27.04	14.27	38.09	20.60	41.31
(-74+57)	27.71	15.05	35.60	21.65	42.75
(-57+48)	30.44	19.08	28.32	22.16	49.52

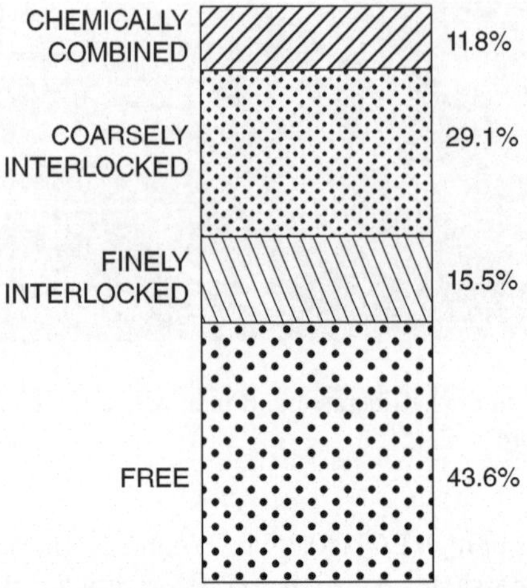

FIGURE 2.5 Various modes of alumina occurrence in typical Indian iron ore (hematite) (after Ref. [10])

During the beneficiation process, the interlocked particles report to the middling, while the valuable minerals report to the product (called the concentrate), and the gangue minerals report to "tailings."

Certain groups of iron ores are very complex in view of their poor liberation characteristics. The gangue minerals constituting silica, alumina, and combined water (in case of high LOI goethite ore) are intimately associated with the iron-bearing minerals. These ores are encountered with unfavorable liberation characteristics. The liberation size of the constituting minerals may go below a few microns. A typical example is the ore from the Barsua Mines of SAIL, situated in Eastern India. The ore, however, consists of scattered amounts of the minerals that are loosely bound and happen to be fairly liberated at a coarser size. In view of this, the differential or preferential grinding is not ruled out in beneficiation.[9a,9b]

The occurrence of iron ore remains in different forms. Less than half of the total alumina is free in typical Indian iron ore (**Figure 2.5**), which could be eliminated following the beneficiation. Also, one-third of alumina in the ore is coarsely interlocked, which could be liberated from iron-constituting minerals following the intensive size reduction. The remaining portion of alumina, constituting one-fourth of the total, remains as chemically combined and finely interlocked.[10]

2.6 PREPARATION OF ORES AND OTHER RAW MATERIALS

The preparation of iron ore includes steps that can be grouped into three as follows.

a. Pre-beneficiation operations
    - Size reduction (crushing and grinding) of ores that are explored from the mines
    - Sizing (screening and classification), scrubbing

b. Physical beneficiation: the removal of gangue mineral matter from the ferruginous minerals of ore to generate concentrate
c. Post-beneficiation treatments
    - De-watering: the removal of water from the beneficiated ore products, the concentrate, as much water is added during sizing and cleaning as is economically feasible.
    - Blending and homogenization of the ore products: lump ore and ore fines.

In addition to the aforementioned, the material handling, that is, the transport of processed ore from mines to sinter/pellet plants is important.

2.6.1 Size Reduction and Size Classification

Comminution The size reduction, termed comminution, of iron ore involves its crushing and grinding.

The grinding methods are roughly categorized as follows. These grinding methods are used in combination depending on the types and characteristics of the ore and the mixing ratio, taking into account the economic feasibility.[11a]

a. Dry grinding/wet grinding: The discharge from the dry grinding unit (ball mill) is processed through an air separator to get the product (**Figure 2.6**). The product must be pre-wetted with adequate water in green pelletization. The medium is generally water in cases of wet grinding. A wet grinding system (Figure 2.7) accompanies a dewatering unit.
b. Open-circuit grinding/closed-circuit grinding: In the case of a closed circuit, the discharge from the ball/rod mill in slurry form is fed to the classifier to separate coarser particles from the finer ones (slime). A cut point is

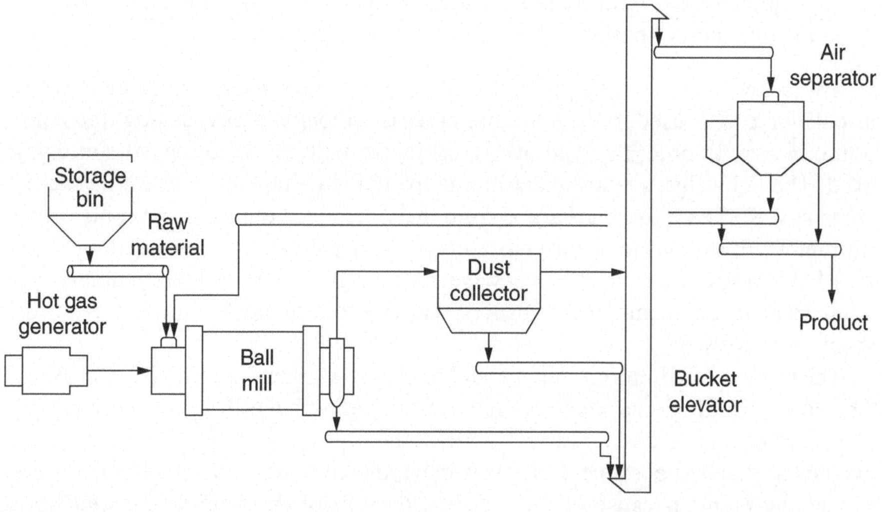

FIGURE 2.6 A closed-circuit dry-grinding system

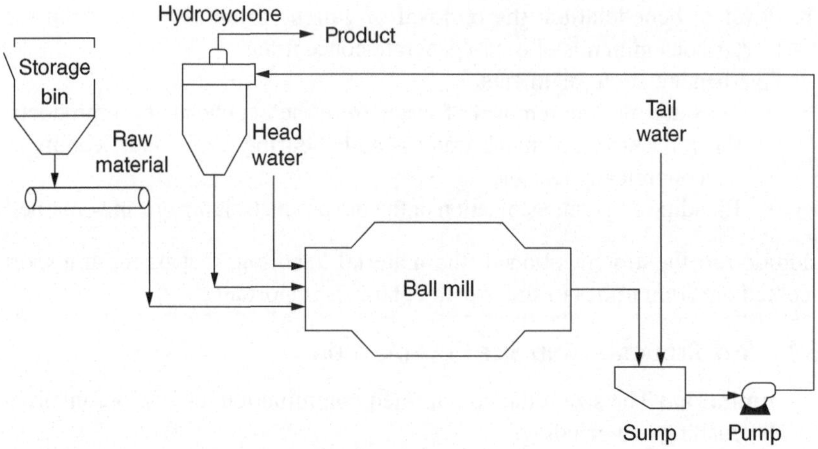

FIGURE 2.7 A closed-circuit wet-grinding system

set in the classifier, which is attached to the grinding circuit. Accordingly, the coarser particles are recirculated back to the mill along with the fresh feed for further grinding (**Figure 2.7**).

Closed-circuit grinding has the ability, over the open circuit, to control the product size distribution and when the size distribution of feed (ore) varies considerably. This is important when the product of beneficiation (concentrate) is feedstock to the pelletizing unit.

c. Single stage grinding/multiple stage grinding: In the latter case, the coarser size fraction of the ground product from one grinding unit is fed to another similar unit for further comminution. The product, through multiple stage comminution, has the granulometry in closed size; hence, the over-comminution of the material is avoided/reduced.

The below 10-mm-sized product of the coke oven, which is generally termed as coke breeze, is used in the sintering process—though after crushing it to below 3 mm in size. In order to avoid over-crushing, a multiple-stage crushing circuit is used. The Schwelgern sinter plant has adopted a two-stage coke crushing facility (**Figure 2.8**). The coke, discharged from the bin, passes through a scalping screen (to remove fines) and then through a magnetic separator (to remove tramp iron). It is fed at a given rate to the two-stage roller mills equipped with hydraulic gap adjustment mechanisms. This ensures that the rollers are fed uniformly across their entire width.[11b]

The process of comminution, in plant practice, is generally termed as "milling." The cost of milling equipment is high. Additionally, the milling is highly capital- as well as energy-intensive. In view of this, the milling operation is required to be avoided as much as possible. But this is inevitable in most of the mineral (iron-ore) processing plants because of the fact that the liberation of minerals occurs at a finer size.[11a]

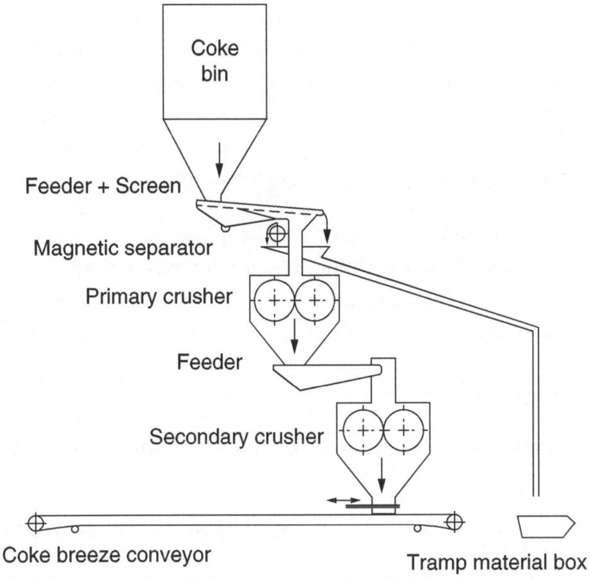

FIGURE 2.8 Two-stage coke grinding facility at Schwelgern (after Ref. [11b])

Size Classification The function of a classifier is to separate particles of various sizes and density in fluid (usually water) under the influence of gravitational or centrifugal forces. In relation to iron-ore processing, a classifier has various applications, though the basic principle (of separation based on size and density) remains the same. These include:

a. *Closed circuit wet grinding* A classifier is generally attached with the closed circuit grinding unit. The discharge from the ball/rod mill in slurry form is fed to the classifier to separate coarser particles from the finer ones (slime). The coarser particles are recirculated back to the mill along with the fresh feed for further grinding.
b. *De-sliming of ore fines* The slimes are removed from the ore for further treatment using a classifier.
c. *Dewatering of slurry* The ore slurry from the wet circuit is dewatered. Sand–slime separation may be achieved in one or more stages of classification.

The principle of classification is based on the differences in particle settling velocity in fluid. The nature of particle movement in fluid medium depends on whether the flow is laminar or turbulent, depending on the Reynold number.

d. **Hydrocyclone** is the most common classifier and uses centrifugal force in separation. It is widely applied in mineral processing for fine classification (mostly less than 100 micron). The feed is injected tangentially into the hydrocyclone, which causes swirling and thus generates centrifugal force within the cyclone. The feed is accelerated and rotates at a high speed within the vessel, transporting the coarse particles by centrifugal forces to the inner wall; from there, it moves down along the conical part and leaves the vessel through the underflow spigot. The slower settling fine particles stay in the center of the fluid, which forms an inner upstream current and leaves the vessel through the central upper discharge opening.

The cyclone performance is decided by the yield and cleanliness of the underflow and overflow products and is influenced by the parameters: (a) solid concentration of the pulp, (b) the suspension pH value, (c) diameter of the vortex finder, and (d) diameter of the apex.

2.6.2 Beneficiation Process

Most beneficiation operations will result in the production of three materials:

a. Concentrate (the valuables)
b. Middling or low-grade concentrate, which is either reprocessed, or stockpiled
c. Tailing (waste), which is discarded

The beneficiation process comprises various separation techniques that are basically unit physical operations. The separation of valuables from the gangue is based on the differences in their physical characteristics, as follows.

Unit Operation	Difference in physical property being exploited
Comminution	Breakage characteristics of different constituents of ore. Though it is bulk operation, some sort of concentration can be achieved by exploiting the phenomenon through selective grinding.
Gravity	Difference in the specific gravity of the iron-bearing and gangue minerals. Hematite and magnetite minerals are heavier than the minerals constituting silica and alumina (quartz, bauxite, corundum, etc.).
Magnetic	Difference in the magnetic susceptibility of hematite/magnetite and gangue minerals. Magnetite is ferromagnetic, and hematite is paramagnetic; whereas silica and alumina are nonmagnetic.
Flotation	The process is based on the floatability of one type of mineral along with air in an aqueous medium while rejecting another type(s) of minerals. Though the oxidic minerals constituting iron ore are hydrophilic (water loving) with different intensities. Surface active agents can be induced in such a manner that either hematite or the gauge oxidic minerals are floated.
Dispersion	Utilizes the differences in surface chemical properties of various particles in the suspension. Selective dispersion is based on the preferential adsorption of a selective dispersant on the particular minerals to be dispersed, leaving the remainder of the particles in suspension, which is settled under gravity.
Selective Flocculation	The process is based on the agglomeration of fine-sized particles into larger ones by polymer bridging so that these are settled quickly

Iron ores are beneficiated by using one or more of the following methods:

a. Gravity separation: spiral, hydrocyclone, and jigs
b. Flotation: mechanical (conventional) cell and column
c. Magnetic separation
d. Selective flocculation

Latest technologies include Batac jigs, packed flotation column, centrifugal concentrator—such as the Falcon concentrator, and the Knelson concentrator for the beneficiation of iron ore slimes.

Final products of the beneficiation processes are:

a. Lump ore (40 to 10/6.3 mm)
b. Fines/Concentrate (10/6.3 to 0.15 mm)
c. Slimes (below 150 micron)

Lump ore is used in blast furnaces as feed material, also in direct reduction (DR) application. The fines are used the sintering plants. Besides, the slimes are generated from the wet processing of ores, which are traditionally discarded and dumped in tailing ponds present in many of the mines, and need intensive beneficiation to yield concentrate for pellet making.

Gravity Separation This is a physical process and exploits the differences in densities of minerals to bring about a separation. The gravity separation of the associated minerals that are present in iron ore is carried out by their relative movement in fluid (normally water) under gravity. The forces affecting their relative movement are:

a. Drag force: Resistance to motion of particles in a fluid due to its viscous effect.
b. Specific gravity of the particles: Iron ore constitutes of ferruginous (hematite/magnetite) and gangue (silica and alumina) constituting minerals. Each group has a different density.

The gravity concentration process finds immense application in the processing of iron ores. The major limitation with the gravity concentration is the treatment of fines and ultrafines. In the fine size ranges, the fluid and viscous forces become dominant relative to the gravity, and this in turn affects the separation efficiency.

The concentration criterion (CC) in relation to gravity separation is defined as:

$$CC = (dH - dF) / (dL - dF) \tag{2.1}$$

where,

dH, dF, and dL = density of heavy mineral, fluid, and light mineral ($kg*m^{-3}$), respectively.

The ease of gravity separation depends on:

a. Concentration criterion: the higher the CC, the more effective the separation.
b. Particle size: The separation is more effective in the case of coarser ores.

More effective separation can be achieved in relation to the ore that has higher CC.

a. **Jigging** is normally used to concentrate relatively coarser material. The separation of minerals having different specific gravities is accomplished in a bed that is rendered fluid by a pulsating current of water so as to generate stratification. The aim is to dilate the bed of material being treated so that the heavier, smaller particles penetrate the interstices of the bed, and the larger, high specific gravity particles fall under a condition similar to hinder settling.
b. **Spiral** is composed of a helices conduct of a modified semicircular cross section. Feed pulp is introduced at the top of the spiral and, as it moves spirally downward, the particles stratify due to the combined effects of centrifugal force, thus affecting the separation.
c. **Hydrocyclone** is basically a size classification unit (Sec. **2.7.2**). Nevertheless, this also serves as a beneficiating unit since the separation of particles gets affected based on their density difference. The hydrocyclone is widely used in beneficiating iron ore slimes.

Magnetic Separation The magnetic separation equipment is categorized according its field intensity, as follows:

a. Low-intensity magnetic separators with field intensity up to 0.15 Tesla have been in use in iron-ore processing plants to treat iron ore that are rich in magnetite.
b. Wet medium-intensity magnetic separators consist of a revolving drum made of stainless steel. Permanent magnet bars, NiFeB, are positioned horizontally inside the drum. A field strength up to 0.7 Tesla is provided on the drum surface. These are used in magnetic separation of low-grade martitic iron ore.
c. Wet high-intensity magnetic separator (WHIMS): These separators are based on Jones' design and are available in batch and continuous models for their laboratory scale and plant operations, respectively. The separator consists of a separation chamber packed with ferrous matrix. Electromagnetics of high intensity are provided to the matrix through the supply of electricity (DC).

The process is completed in two cycles:

Cycle 1: Iron ore slurry is fed from the top of the separator chamber. The iron constituting minerals (hematite, etc.) of high magnetic susceptibility get attached to the matrix, while the nonmagnetic gangue minerals (such as silica and alumina) get discharged.

Cycle 2: The electricity is turned off, which causes the discharge of electromagnetics in the chamber. The compartment with the ferrous matrix is rinsed with water, and the magnetic particles are recovered in another container placed beneath the separator.

Flotation Froth flotation utilizes the differences in physical-chemical surface properties of the minerals. In the process, air bubbles attach themselves to particles and lift these to the water surface. The air bubbles displace water from the mineral surface. This is possible when the mineral is water repellent or hydrophobic: the minerals are selectively made hydrophobic through the addition of a flotation reagent, called a collector. In the flotation concentration, the mineral is usually transferred to the froth or float fraction, leaving the gangue in the pulp or tailing. This is called direct flotation. In the reverse flotation, the gangue is separated into the float fraction.

The process of flotation for concentrating iron ores received a big impetus in the United States after World War II. This was due to the depleting resources of direct shipping iron ores in the Lake Superior District. Flotation offers a very efficient, low-cost treatment method for beneficiating of low-grade iron ore.

Typically, in relation to a hematite quartz system, either hematite or quartz can be floated depending on the nature of surfactant and aqueous medium used:

Reagents	pH	Mineral floated
Alkyl amine salt	6-7	Quartz
Sodium oleate + starch + Ca^{++} ions	11-12	Quartz
Sodium alkyl sulfonate	2-4	Hematite
Sodium oleate	6-8	Hematite

The flotation recovery can be enhanced by modifying the adsorption of the collector at the mineral water interface.[12] It is also possible to suppress the flotation of a particular mineral by reducing the adsorption of the collector onto its surface. In the case of an iron ore quartz system, the separation has been made commercially viable when starch is used to depress hematite[13] and calcium ions to activate quartz, which is then floated at alkaline pH with an anionic collector. Hematite-rich ores have greater hydrophobicity, which makes them easier to separate. Hence, they may not require as much comminution as goethite-rich iron ore, which has lesser hydrophobicity.

Dispersion The fine particles in the suspension undergo hetero-coagulation, which makes the beneficiation difficult. The dispersant helps in reducing the hetero-coagulation, making the separation cleaner. Depending on its applicability in mineral processing, the dispersion is categorized into two groups.

a. *Non-Selective Dispersion:* The suspension of fine particles, irrespective of its nature, is destabilized. In the case of an iron ore system, the separation of heavy iron-bearing minerals and lighter gangue constituents are affected through density differences following dispersion.

b. *Selective Dispersion:* The reagent (selective dispersant) is preferentially adsorbed on the particular minerals, which are stabilized, while the remainder of the particles in suspension settle under gravity.[14] In relation to iron ore, hematite minerals have a different ability for dispersion than that of clay minerals. This results in separation by selective dispersion.[10]

Selective Flocculation The process of selective flocculation, in general, consists of following four major steps:

a. General dispersion of mineral particles: The particles are made stable and uniformly distributed in the suspension.
b. Selective adsorption of the flocculants and floc formation: The natural and synthetic polymer-based flocculants do have/are provided specific sites onto their chains, which are adsorbed on the iron-bearing minerals.
c. Floc conditioning: This aims at obtaining flocs of desired properties for their subsequent separation and with minimum entrapment of dispersed particles.
d. Floc separation from the suspension.

The beneficiation plant at the Tilden concentrator processes finely disseminated large deposits of oxidized taconites containing predominantly hematite and goethite.[15,16] The process involves selective flocculation and desliming followed by cationic flotation of coarser silica. The flow-sheet involves dispersion of minerals using sodium hydroxide followed by selective flocculation of iron minerals using starches (e.g., tapioca flour starch). The settled (flocculated) concentrate is then subjected to reverse flotation with cationic amine reagents in order to remove coarse silicates. Starch thus works both as a selective flocculant and as a dispersant for iron mineral.

2.6.3 Dewatering of Iron Ore Fines and Slimes

The wet processing of iron ore requires a considerable amount of water, which needs to be recovered through the process called *dewatering*. The purpose of dewatering is to enable handling of ore and concentrates and allow further processing including the disposal of gangue. The water extracted from the ore by dewatering, following its treatment, is recirculated for plant operations.

The main processes that are used in dewatering of ore products include: (a) dewatering screens, (b) sedimentation, and (c) filtration. The process of dewatering through screens is only viable for coarse ores that have a close size distribution. The process is applicable when the product is sinter-grade ore fines. The dewatering of suspension of fine iron ore (slimes) is carried out in two stages, namely, (a) thickening, and (b) filtration.

a. Thickening: Most of the fine grains settle out of the water in a thickener. It serves the dual functions of clarifying the suspension to recover a clear water for its recycling from the top of the equipment and also thickening the suspension from the bottom to recover a thick pulp of high solids content for its further processing—both occur simultaneously.

The capacity enhancement of the thickeners draws a considerable benefit to the iron ore processing plant in view of the fact that:

- Life of the existing slime pond can be increased significantly.
- More water can be recycled to the process from the thickener itself.

This can be achieved by:

i. Mechanical means: Trouble-free operation by avoiding jamming, etc.

ii. Enhanced settling: The sub-sieve particles can be subjected to the enhanced settling in the thickener by agglomerating them to bigger ones through the use of flocculants.

b. Filtration

The suspension of fine ores is further dewatered in filters. The cake coming off the filters has around 60 to 65 volume percent solids and could be handled as a *solid* on a conveyor belt. With regard to pellet making, moisture content of the magnetite ore concentrate is one of the most important parameters to influence the green pelletization (**Secs. 13.3 and 13.4**). Obviously, a real economic advantage accrues due to the ability to control the moisture content of the filter cake within the range required for optimum pellet production.

Filtration is effected by application of capillary, vacuum, pressure, or centrifugal force. Pressure filters are normally used in the filtration of iron ore concentrate because of the high pressure differential compared to vacuum filters.

Effect of Flocculation An acceptable solid–liquid separation efficiency frequently demands a pre-agglomeration of the fine particles within the suspension in order to obtain particle clusters that then either possess a sufficient mass for a rapid sedimentation (in a thickener) or can form coarse pore structures possessing an adequate permeability (in the case of filtration).

The aim of agglomeration is to convert the suspended fines to mechanically separable flocs. The main criterion for the desired floc structure is the nature of the subsequently applied separation technique. (i) The filtration and sedimentation in a centrifugal field demand compact agglomerates with high mechanical stability. Whereas, (ii) the sedimentation by means of earth's gravity is facilitated by large, high-density flocs.

2.7 INDUSTRIAL PRACTICE ON IRON ORE BENEFICIATION AND PROCESS FLOW SHEETS

2.7.1 Industrial Practice

a. Hydrocyclone

Paraburdoo Fines Processing Plant treats the fines produced at Paraburdoo and Channar. The ultrafine fraction (under 10 micron in size) is removed using a combination of wet screens and hydrocyclones. The plant is designed to process 13 million ton annually and achieve a mass recovery of 85% of the feed.[17]

b. Spirals

Essar Steel produces a concentrate of 68% Fe with a weighted yield of 90% from a feed having 64.5% Fe using gravity concentration by means of spirals. The spiral tails are further processed through WHIMS.[18]

c. Jigging Operation

- The upgrading of ferrous ores by jigging is a recent practice in South Africa as well as in countries, namely, Australia, India, and in South America. In cases where the cut density exceeds 4.0 g/cm^3, jigs offer satisfactory performance. Jigs, due to their simplicity and low running costs, are now the preferred route as opposed to other density–separation techniques. Beeshoek Mine in South Africa, using the jigging operation in their plant, is able to treat low-grade ore having 50% Fe compared to an earlier level of 58% while maintaining the desired product grade.
- Coarse jigging is a proven technology for lump ore. The coarse jig (capacity 120 ton per hour operates in South Australia for the size range below 32 and above 8 mm and produces a concentrate: Fe: 62.5 to 66 wt.%, Al_2O_3: 0.8 to 2.0 wt.%, with a weighted yield of 75 to 85% from feed having Fe: 54 to 58 wt.%, Al_2O_3: 2.5 to 4.0 wt.%.
- The Batac Jigs of KHD Humboldt Wedag operating in Golds Worthy Mining Ltd., Hedland, West Australia; Ferteco Minerco SA, Brazil; and Beeshock SA produce a concentrate with an increase in Fe value by 3% and alumina reduction greater than 40%. The material yield is around 70%.
- The all-mineral fine jig operating in Southern Australia produces a concentrate grade of 66.6 to 67.9% Fe from 64 to 65.5% Fe with 50% reduction in alumina.

d. Jigging and Magnetic Separation[17]

- **Australia** Crushing followed by heavy media separation (HMS)/lump ore jigging for coarse size (below 40 to above 8 mm), fine jigging (below 8 to above 0.15 mm) for finer size, and WHIMS for ultra-fines (below 0.15 mm) is the most common beneficiation sequence used in iron ore beneficiation plants in Australia.

  A beneficiation plant, BHP Billiton at Newman, utilizes heavy media drums, cones, and cyclones to process low-grade ore at around 57 wt. % Fe to produce marketable high-grade iron ore. Newman iron ore fines are beneficiated in the concentrator consisting of jigs, spirals, and WHIMS circuits to produce high-grade feedstock.
- A beneficiation plant at Finucane Island employs jigs, spirals, and WHIMS to treat low-grade lump and fines at around 57 wt. % Fe from the Goldsworthy operations. Batac jigs operating at Goldsworthy mines in Port Hedland improve the iron value from 60.5 wt. % in the feed (below 8 mm size) to 63.4 wt. % in the product with 66% yield.
- The concentrator at Tom Price in Western Australia beneficiates low-grade ore by HMS of the coarse fraction (0.5 mm to 31.5 mm) and WHIMS for the finer fraction (0.063 mm to 0.5 mm) to produce lump and fines products.

e. Magnetic Separation and Flotation
United States: According to the American Iron Ore Association, milling and magnetic separation are the most common methods used in iron ore beneficiation. Gravity concentration is seldom used in existing U.S. facilities. Flotation is primarily used to upgrade concentrates arising out of magnetic separation by reducing the silica content in the concentrate.

f. Flotation
The iron ore industry in Minnesota and Michigan in the United States uses cationic flotation of silica from magnetic taconites. Column flotation technology has been applied for rejecting fine silica using a variety of cationic amines. This technology has been commercialized in several installations including that at Kudremukh in India.

g. Selective Dispersion
Groups of phenol-based and lignite-based low molecular weight polymeric dispersants have been developed for the preferential removal of alumina in iron ore systems. These dispersants are specific in action on hydrous minerals such as clay and laterite. The dispersion is associated with an increase in hydrophilic character and stability of suspension is maintained by prohibiting particle-particle contact.

Several industrial trials at the Barsua iron ore mines of Rourkela Steel Plant of SAIL, in association with Research and Development Centre for Iron and Steel (RDCIS), have indicated partial success in selective removal of alumina from iron ore fines when the dispersant was added to the washing circuit compared to plain water washing.[19, 20] The ratio of ore to water was maintained at approximately one in all the trials. Additive concentration between 0.08% and 0.05% (W/V) was maintained. An increased removal of Al_2O_3 by about 17% was observed due to additive treatment compared to normal water washing in the case of Barsua ore.

2.7.2 Process Flow-sheet

The process flow-sheet is developed involving the beneficiation stages applicable to a particular ore under consideration. The flow-sheet also comprises the flow of material, mass balance with respect to different constituents of the ore, and water balance. A flow-sheet may be simpler in case the quality of product is achieved through a shorter/simpler processing operation. Conversely, a greater number of unit operations makes the flow-sheet more complex, which demands a larger capital investment in setting the plant and higher outflow through operation. The simplicity (or in other words, the complexity) of the process flow-sheet is guided by the physical and mineralogical characteristics of the ore to be processed. It also depends on the end usage of the beneficiation products: whether it is for sinter making or in the production of pellets that require the ore concentrate in a finer size. As each and every step in the ore processing involves cost, the objective/strategy should be to make the flow-sheet as simple as possible without compromising the quality of products and their productivity while designing the process plant.

High-Grade Ores High-grade ores and/or the ores with favorable liberation characteristics will require a simpler beneficiation circuit involving crushing, scrubbing and screening to get sized products. In the case of hematite bearing ores, these operations are accomplished by using the conventional gravity separations, flotation, and/or a wet-type high/medium magnetic separation. In case of magnetite-bearing ore, the beneficiation is accomplished mainly by magnetic separation. The process has been employed in North America.

Lean Ores/Complex Ores High alumina-content lean ore and complex ores constituting silica, alumina, and combined water are difficult to beneficiate as these ores are encountered with unfavorable liberation characteristics. In the case of high LOI (goethite) ore, the alumina present in the goethite matrix appear to be less amenable to physical separation. These ores require comminution to their finer sizes and multistage operation for their enrichment.

Exploitation of these ores from mines and their processing has drawn considerable attention in recent years:[21]

a. For sustainable development as the high-grade ores are being steadily depleted.
b. To produce (high) value-added concentrate. The competitiveness in the market demands for good-grade iron ores/concentrates.
c. For additional material recovery (from these complex ores). As a consequence, opening of new mines that require deforestation could be delayed.

These ores do not meet specifications in terms of their chemical constituents and are upgraded through the beneficiation process, which is rather complex. An assessment for better technological development is required to set the appropriate beneficiation strategy in terms of:

a. The mode of occurrence of gangue-constituting minerals, especially alumina. Their association with hematite and magnetite. Their liberation characteristics.
b. Comparative evaluation of the separation efficiency of different beneficiation processes (unit operations) in terms of the recovery-grade plots for the mineral constituents (hematite/magnetite vis-à-vis kaolinite/gibbsite) of ores at different size ranges.
c. Techno-economic evaluation of the various technology options for a typical iron ore mine.

Categories of Flow-Sheets The flow-sheet comprising the process of iron ore beneficiation may be categorized as:

a. **Dry Circuit** When the ore is massive to hard laminated and constitutes a sizeable amount of blue dust, also during the dry season, it is prudent to only process the ores through a dry circuit since the dewatering operation following the wet circuit involves costs and incorporates slime disposal.

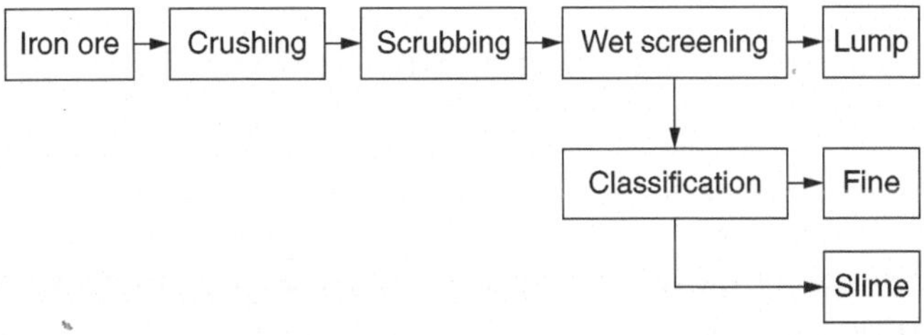

FIGURE 2.9 Simple wet circuit for the enrichment of iron ore (schematic) (ROM: run of mines)

b. **Wet Circuit** These flow-sheets are based on

Only Water-Washing If the ore is richer in Fe (approximately 60 wt.%), its processing may be limited to crushing, scrubbing, and simple wet screening (**Figure 2.9**) to remove adhering impurities. These techniques are traditionally practiced in many of the Indian plants but have their own inherent limitations. These are not effective for reducing alumina below particular limits.

The rejection of silica (and, to some extent, alumina in the form of clays) occurs during washing and size classification of the crushed ore involving wet screening. The products: lump ore and fines are dispatched to the blast furnace, direct reduction plant and to the sinter plant, respectively. Another coproduct is slime (Ref, **Sec. 2.7.6**), which is generated following the scrubbing of ore and wet screening. This is dumped in the tailing pond or can be further processed. Also, the tailings, constituting gangue, are dumped.

Simple Circuit It is easier to beneficiate the ores that consist of magnetite/hematite and silica-bearing minerals and that have favorable liberation characteristics. For example, itabirite ore is poorer in terms of iron value and consists of a lower amount of Fe. with the rest being silica-bearing minerals. The ore is more amenable for beneficiation: A reasonable good quality concentrate can be produced from the ore by following the flow-sheet involving wet magnetic separation and dewatering.

Complex Circuit Low-grade ores with complicated mineral assemblages and slimes would require an intensive beneficiation process, which becomes complex if the ore consists of:

- High amount of alumina (the ore is more lateritic)
- High LOI (the ore contains high amount of goethite or/and limonite

The alumina minerals, unlike the silica-constituting minerals, are intricately (complicated) associated with the hematite mineral in these ores.

The beneficiation amenability of high LOI ore (containing a high amount of combined water) is also poor in view of its complex mineralogy. The ore may require pyro-metallurgical pretreatment, such as reduction–roasting, before the physical methods of separation including the magnetic one.

Typical hematitic iron ores from the eastern part of India, though rich in Fe value, are not excellent feed to process in a true metallurgical sense. The problems are two-fold, namely:

- Relative soft nature of the ore that generates fines and slimes on handling and wet processing
- High content of finely disseminated aluminous minerals associated with the hematite matrix

This makes beneficiation difficult and may involve a complicated processing circuit.

Role of Geological Mapping The nature of ore drawn from the mines changes with time as the excavation proceeds toward the deeper bench and/or in case the excavation is advanced to another mine site. In view of this, it is important to modify the existing plant circuit periodically in order to improve the productivity of ores and their quality.

The geological planning (for excavation of ores) has a crucial role in the exploitation of ore from different benches of the mines. This needs to be reviewed, supported by the data pertaining to the mineralogical characteristics of the ore, and updated periodically. It is essential to understand the nature of both the iron-bearing and gangue minerals that occur in iron ores and their association in the ores before the beneficiation option is defined and flow-sheet is developed.

The lean-grade and complex iron ores with high LOI content (goethite) are more difficult to beneficiate as the alumina present in the goethite matrix appear to be less amenable to physical separation in view of the unfavorable liberation characteristic: the gangue minerals and combined water are intimately associated with the iron-bearing minerals.

2.7.3 Processing of Hematite Ores

The processing of hematite-bearing iron ores depends on their nature.

a. **Flaky Ore and Blue Dust:** The flaky ore and blue dust in the run of material (ROM), in view of their finer size and higher grade, may not require processing through the wet circuit. The processing of these ores through the wet circuit results in a substantial amount of material loss in the form of slime. These materials are crushed and dry-screened to below 10-mm-sized fractions to generate product that is commonly termed "dry circuit materials" (DCM) in the mines.[22]

   The DCM is blended with classifier fines for the production of sinter-grade fines. Classifier fines are produced in the wet circuit from a mixture of hard ore, soft ore, and flaky ore by crushing, screening/sizing, and classification (**Figure 2.10**). The size of classifier fine below 10 and above 0.15 mm. The finer size fraction (below 0.15 mm) comes out as slimes after the screening and classification. This fraction is/can be processed further to recover additional iron value.

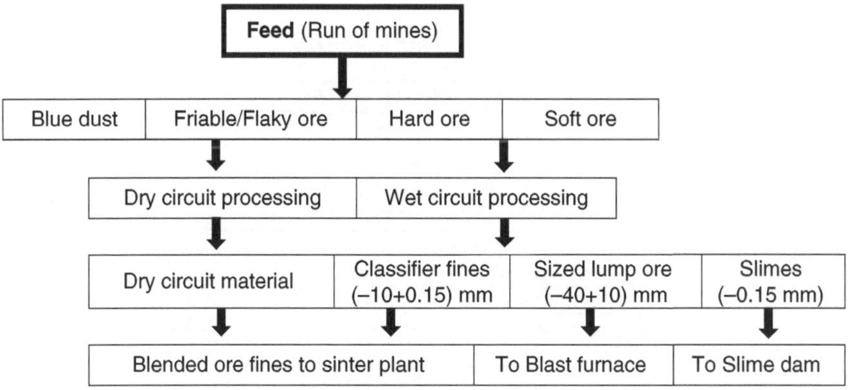

FIGURE 2.10 Schematic of dry- and wet-circuit operations for ores constituting flaky ore and blue dust (after Ref. [22])

b. **Simple Hematite Ore** A group of hematite ores does not possess much complexity. These ores are massive too hard and soft laminated and contain a lower percentage of LOI and alumina; the content of goethite, limonite, and laterite in the ores is low. These are processed entirely through the wet circuit; however, they may not require intensive beneficiation. The beneficiation process may include: (i) Crushing of ore to 40 mm in size, (ii) Scrubbing, washing, and size classification to produce lump ore (size in between below 40 and above 10 mm), sinter grade fines (size below 10 mm and above 0.15 mm), and fines (below 0.15 mm).
c. **High Alumina Hematite Ore** The beneficiation of high alumina iron ores, in many cases, is challenging in view of their poor liberation characteristics. The alumina-bearing minerals (kaolinite and gibbsite) are finely disseminated and associated with hematite in preference to siliceous minerals.[9a,9b] This makes the beneficiation difficult, in particular, for Indian iron ores having complex characteristics from the standpoint of alumina elimination. Simple techniques involving crushing, scrubbing, and washing have their own inherent limitations. Only a part of alumina, which is present in the ore as adhering clayey material, gets removed. Besides, the fine silica can be partially removed through washing. Therefore, there is preferential removal of silica over alumina, which naturally increases the Al_2O_3/SiO_2 ratio in the washed product.

In the case of low grade hematite ores, the beneficiation process may involve grinding of coarse sized fraction for adequate liberation of Fe constituent from gangue ones. A typical flow-sheet (**Figure 2.11**) may include:

- Multi stage crushing of ore.
- Scrubbing, washing, and size classification to generate coarse, medium and fine sized products. Gravity separation of the classified products to generate concentrates and tailings from the above three size fractions.
- Regrinding of these tailings products and magnetic separation to generate pellet grade concentrate, and finally tailings, which is dumped.

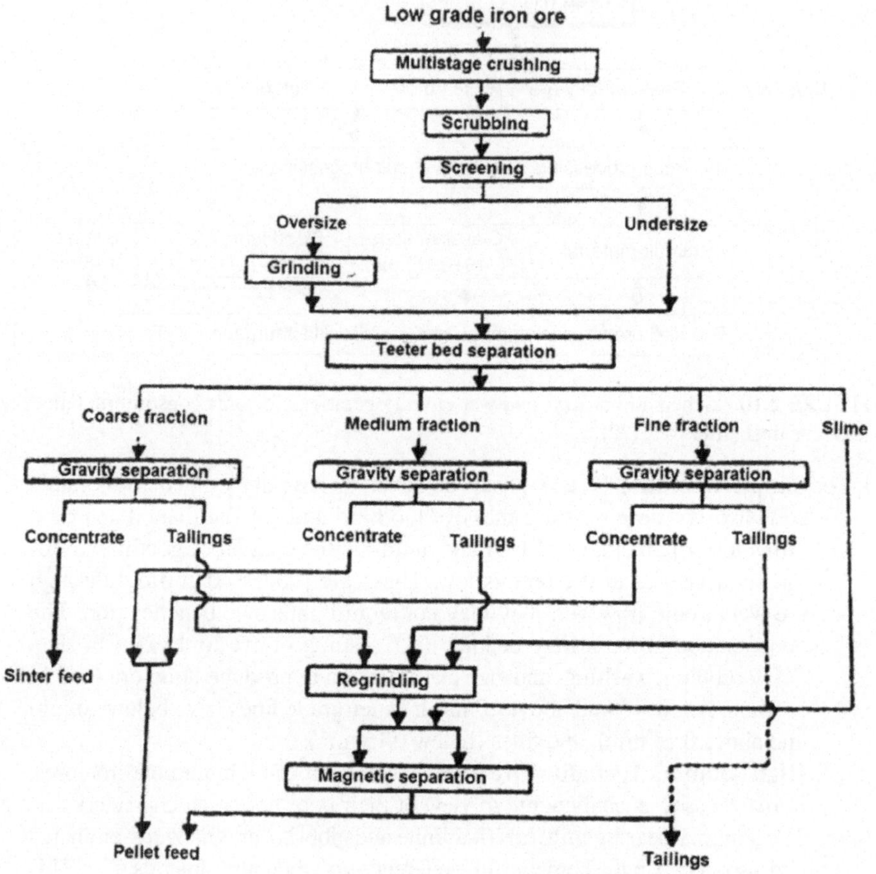

FIGURE 2.11 Process flow-sheet to beneficiate high alumina Indian hematite iron ore (after Ref. [23])

2.7.4 Processing of Goethite Ore

Problems in relation to the development of a suitable technique for the processing of goethite ore are:

a. High content of combined water associated with the goethitic ore. This makes the beneficiation process difficult to make suitable concentrates as feedstock for the blast furnace.
b. The alumina present in the goethite matrix appear to be less amenable to physical separation, which causes problems in achieving the grade concentrate.

The composition of goethite ore with respect to its alumina content varies as the iron is being replaced by alumina (**Sec. 2.4.3**); the alumina content in the case of

Brazil and Australia is lower than that of the Asia Pacific regions. Accordingly, the processing techniques should vary in these two cases. In the latter case, it requires specific beneficiation treatment[24, 25] to simultaneously handle both high LOI and high Al_2O_3 content in the ore bodies. The Australian Marra Mamba and pisolite ores are classified as low-grade ores and contain a higher amount of higher silica (compared to alumina) as well as combined water.[6,7]

The pre-concentration using reduction roasting is one of the possible approaches, widely adopted, that transforms goethite to hematite and eventually maghemite, which is paramagnetic/ferromagnetic at room temperature and can be recovered by magnetic separation. The reducing agents are usually C, CO, and H_2. The temperature maintained in reduction is usually between 700°C and 900°C. The hematite Fe_2O_3 is weakly magnetic and hard to separate; hence, it is essential to maintain the reducing atmosphere during the roasting process so that the reoxidation of maghemite phase back to hematite could be avoided. The surface area of the goethite ores is increased after reduction due to decomposition of combined water and its evolution. Murukami et al.[26] have observed that below 1200 K, the reduction of Marra Mamba and pisolite ores (carbon composite proceeded faster than that in hematite) ore-carbon composites. This can be attributed to the larger surface area of the ores after decomposition of the combined water. The usage of pine tree powders, the biomass as renewable solid fuel, has been attempted with reference to the reduction roasting operation. The goethite was reduced and magnetized into maghemite by biomass with dosage less than 20 wt. % and temperature less than 650°C.[27]

2.7.5 Processing of Banded Hematite Quartzite (BHQ)

These ores typically consist of bands of iron-bearing hematite or magnetite intergrown with quartz. The constituting minerals, hematite and quartz, are fine gained: about 80% of hematite grains carry extremely fine-grained inclusions of quartz. In view of this, the material is difficult to treat. Laboratory to pilot-scale investigations using medium- and high-intensity magnetic separators have resulted in partial success.[28] The process to recover iron values from these ores on a commercial scale has yet to be explored. Difficulty is also encountered as comminution at a fine size range is required.

2.7.6 Iron Ore Slimes

The ore washing plants produce slime, which is accumulated in tailing dams. Slimes to the tune of 10 to 15% of the processed ore are generated. They vary in their chemical constituents, depending on the source (the nature of ore) being treated as well as the complexity of the beneficiation process. The processing of soft ore generates a larger amount of slime; whereas the generation of slime is lower in cases where the hard to massive ores are treated. About 85 to 90 wt. % of the slimes happens to be below 150 microns in size. The slime is largely rich in Fe content, which varies between 55% and 60%, whereas alumina percent in the slimes varies between 2.8 and 6.

Benefits and Limitations The slimes generated from the processing of ore consist of iron value. This has drawn considerable attention in recent years for having merits in its processing.

a. Utilization of the mines' resources in an effective manner for sustainable development.
b. Higher mine output in terms of marketable products. Any additional recovery of Fe from the slimes leads to an enhanced economy of the mines.
c. Reduction in the environmental impact of iron ore mining as a consequence of less residue material (tailings) for storage and disposal.

Technically, as the slimes' material is in the sub-sieve range, it need not be further ground, so the cost of comminution is absent on one hand, whereas on the other hand, its processing has to be advanced. This is in view of:

a. The concentration criterion, defined in Sec. 2.6.2, is not favorable in the sub-sieve range. Hence, the process of conventional gravity separation will not work. The advanced gravity separation techniques based on multigravity, such as a multigravity separator, have been in use and can tackle the fines; however, they are on a pilot scale so far.
b. In case of flotation, the collection efficacy of particles by air bubbles also gets considerably reduced at their sizes in the sub-sieve range.

In either of these cases, the phenomenon of heterocoagulation prevails when the fines of sub-sieve range are treated affecting their separation efficacy and making the beneficiation difficult. The removal of heterocoagulation through stable dispersion Sec. 2.6.2 of mineral slurry is a prerequisite of several beneficiation techniques. A simplistic beneficiation process flow-sheet could be conceived by incorporating dispersion as a preseparation technique.

c. The beneficiated product is concentrate, which has to pass through the dewatering stages. The operational cost involved in the filtration of concentrate, in view of its finer particle size, is high.

Beneficiation Techniques: The beneficiation of slime can be carried out in a staged manner using a hydrocyclone and WHIMS. The slime is classified in a hydrocyclone. The cyclone underflow is treated in a wet, high-intensity, magnetic separator to achieve the grade. The capital and operational costs involved in the hydrocyclone are less than those involved in WHIMS. However, the treatment of slime in a hydrocyclone alone cannot result in the desired benefit. Therefore, combined beneficiation technology consisting of classification in a hydrocyclone followed by WHIMS can be incorporated for the treatment of slime.

Figure 2.12 compares the separation efficiency of different gravity concentration techniques vis-à-vis magnetic separation based on the laboratory studies.[29] The slime sample, assaying Fe: 56.5 wt.%, SiO_2 5.2 wt.%, and Al_2O_3 6.7 wt.%, could be enriched to 2.5 wt.% alumina with material yield of the concentrate around 40%

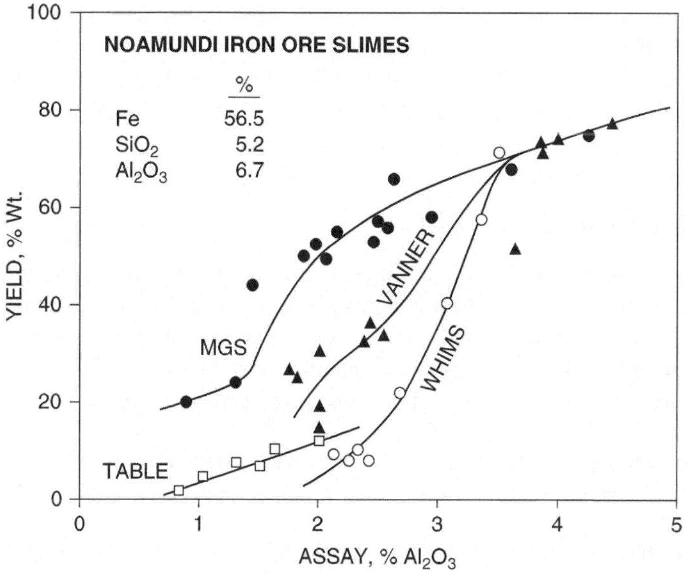

FIGURE 2.12 Efficiency of various beneficiation processes in relation to iron ore slimes (after Ref. [29])

using vanner compared to material yield more than 60% (at the same grade) when multi-gravity separator (MGS) was applied. The effect, using multiple "g" in the case of MGS, is evident in the case of ore slimes, which have a lower CC of around 2. Application of MGS in slime processing, however, has a drawback in view of its limited capacity. The equipment developed so far is of laboratory to pilot scale, which obviously cannot treat the slimes of high tonnage. Therefore, WHIMS remains the possible choice for the treatment of slimes.

The ultrafine particles (in slimes) have poor flotation behavior. In this case, the selectivity is reduced considerably as the extremely fine particles coat on the coarser ones. Besides, the flotation rate decreases due to entrapment of particles and increased froth stability. The consumption of reagents is excessive.

The aforementioned problem, encountered in flotation, can find possible solutions through (a) control of surface activity, (b) wettability, (c) modification of surface properties of ultra-fines, and (d) increase in the probability of adhesion between gas bubbles and mineral particles, which is facilitated when the particles are agglomerated or are attached to large hydrophobic particles that act as carrier particles.

Typical High Silica Ore The beneficiation of high silica ore slimes from typical mines in Karnataka, India, using the flotation technique has been attempted.[30] The sample size was 96.8 wt.% below 40 micron, and the chemical composition was as follows:

Fe(T)	SiO_2	Al_2O_3	CaO	MgO	LOI
56.34	16.13	1.50	0.14	0.08	1.62

Starch was used to depress iron oxide, and calcium chloride was used to activate silica. The pH of the suspension was maintained in the 10.5 to 11 range. The silica content was reduced to 1.23 wt. % and 1.07 wt.%, respectively, in the concentrate from around 16% in feed using cationic collectors, Liquid B-50, and n-dodecyl-amine hydrochloride, in the feed sample at their optimum dosage.

2.8 HANDLING OF IRON ORES

Handling of iron ore products is an important issue in the mines. The ore is fed to the crusher in the mines and, following its processing, the beneficiated products viz. lump ore, sinter-grade fines, and concentrate are conveyed to the stock piles at the mines through belt conveyors and subsequently loaded into the rakes for their transfer to the blast furnace and sinter plant sites. It is important to have:

a. Safe handling of the material without any/with minor spillage at the same time
b. Higher speed of handling for greater plant productivity

Angle of Repose and Angle of Surcharge Important material handling parameters are related to the characteristics: bulk density, angle of repose, and angle of surcharge. Table 2.4 lists bulk density and angle of repose of iron ore and coke. Angle of repose of a material is the angle that the surface of a normal freely formed pile makes to the horizontal. Angle of surcharge of a material is the angle to the horizontal that the surface of the material assumes while the material is at rest on a moving conveyor belt. This angle is usually 5 to 15 degrees less than the angle of repose (**Figure 2.13**).

Flow-Ability of Material The material flow-ability is an index that:

- Is measured in quantitative terms as the cross section of the material (weight) that can be safely carried on a belt per unit time.
- Decides the safe angle of the belt inclination.
- Is decided by the angle of repose and angle of surcharge. Iron ore fines are considered to be average flowing with an angle of surcharge between 20° and 25° and an angle of repose between 32° and 37°.

The flow-ability of material is an important issue in iron ore processing as it affects the plant productivity and quality of the products (lump and fines). The quality of ore fines with respect to size, chemistry, and consistency gets adversely affected if

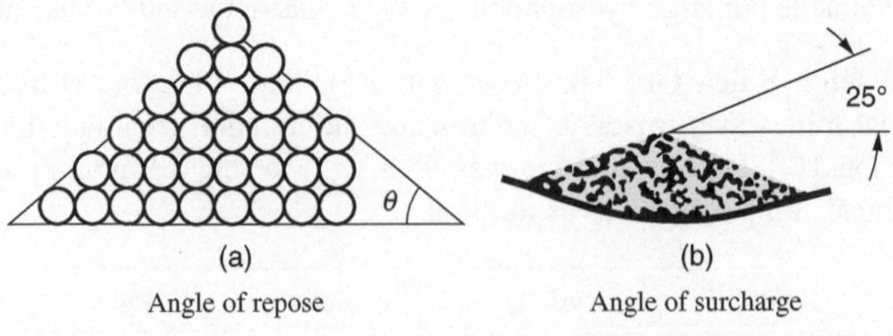

FIGURE 2.13 Angles of repose and surcharge (schematic) in relation to material handling

TABLE 2.4
Bulk Density (kg/m³) and Angle of Repose (Degree) of Important Raw Materials

Material	Lump ore	Fines ore (#)	Blue dust	Flaky ore	Coke
Bulk density, kg/m^3	1800-1900	~2300	~2540	~1910	~611
Angle of repose, deg.	33-35	31-38	~35	~36	35-38

the ore fines have poor flow-ability. The problem of ore flow-ability becomes more relevant due to: (a) fluctuation in weather conditions and (b) difficult genesis of ore in terms of its high LOI and alumina content (that of lateritic and goethitic).

The ore fines encounter the problem of flow-ability when the moisture content increases above 8%, especially in the rainy season. The consequences are:

a. Jamming of the material; moisture also is retained in the stockpile yard. It is difficult to unload the material into the rake.
b. Lower carrying capacity of the belt conveyor, which decreases the plant productivity.
c. A decrease in the screen effectiveness. Higher percentage of the undersized ore (fines) report to the lump-grade ore. In case the screen aperture is increased, a significant portion of lump ore report to the sinter-grade fine.

The flow-ability is determined by the material characteristics, such as the size and shape of the fine particles, lumps, roughness or smoothness of the surface of the material, proportion of fine and lumps present, and moisture content of the material.

Blending and Storage of Ores The ore fines (size below10/6.3 mm) generated after processing may have a heterogeneous chemical composition. This is blended on the ground at the mine site before it is dispatched to sinter plants. By base blending, the variation in chemistry and granulometry of ore fines can be minimized. The ore fines, along with other raw materials, are homogenized in a separate base blending plant called the raw material bedding and blending plant (RMBBP) also at the sinter plant.

The sampling of ores and sample preparation are important issues that need to be carried out at the mine site before the ores (lump, fine, and concentrate) are dispatched to the iron-making works. This can be done using the standard procedure, ASTM E877-13.[31]

REFERENCES

1a. R. K. Goel, "Sinter plant," (July 2016) Source: http://www.meconlimited.co.in/writereaddata/MIST_2016/sesn/tech_2/6.pdf
1b. Bureau of Indian Standard (BIS), New Delhi, Classification of hematite iron ore (IS: 5442–1992, reaffirmed 2003).
2. VT Corp **901** Mumbai :[http://www.vtcorpindia.com/iron ore-pelletization

3a. J. M. F. Clout and W. A. Manuel, "Mineralogical, chemical, and physical characteristics of iron ore," Chapter 2 in *Iron Ore: Mineralogy, Processing, and Environmental Sustainability*. Ed. Liming Lu. Elsevier and Woodhead Publishing, Cambridge, UK, 24 July, 2015.

3b. Graeme Churchard, Banded Iron Formation (Source: Dales George, Uploaded by: P. D. Tilman) https://en.m.wikipedia.org/wiki/File:Banded_iron_formation_Dales_Gorge.jpg (Uploaded 15 Aug. 2013)

3c. Andre Karwath aka Aka) (Licensee: Creative Commons Attribution–Share Alike 2.5 Generic) https://en.m.wikipedia.org/wiki/File: Black-band ironstone (aka).jpg (uploaded 26 Aug. 2005).

3d. IBM, *Iron & Steel–Vision 2020*, Indian Bureau of Mines, Nagpur, 2011.

3e. U.S. Environmental Protection Agency, Office of Solid Wastes, 1985 (a), December Report to Congress, "Wastes from the extraction of and beneficiation of metallic ores, phosphate rock, asbestos, overburden from uranium mining, and oil shale," EPA/530/SW-85-033. Washington, DC. Local cit. N. D. Rao and S. K. Rai, "Availability of iron ore for the expanded Indian steel industry," in *Awakening of the Indian Steel Industry: 2005-2020*, Ed. Amit Chatterjee. Jamshedpur Indian Institute of Metals, Tata Steel, 2006, 17–25.

3f. Geoscience Australia (Australian Government): Australian atlas of minerals, resources mines and processing centers www.australianminesatlas.gov.au (Iron Ore–AIMR 2011)

4. R. K. Upadhyaya, S. K. Sit, T. K. Chakraborty, M. V. S. Rao and S. Verma. "Role of geology in improving quality of iron ores at Joda: Six Sigma approach," *Tata Search*, 2008, 101.

5. C. Klein, "Some Precambrian banded iron-formations (BIFs) from around the world: Their age, geologic setting, mineralogy, metamorphism, geochemistry, and origins". *American Mineralogist*, October 1, 2005, 90(10), 1473–1499. Bibcode:2005AmMin.90.1473K. doi:10.2138/am.2005.1871. https://en.wikipedia.org/wiki/Magnetite

6. M. S. Magalhaes, P. R. G. Brandao, and R. P. Tavares, "Types of goethite from Quadrilatero Ferrifero's iron ores and their implications in the sintering process," *Mineral Processing & Extractive Metallurgy (Trans. Inst. Min. Metall. C)*, 2007, 116, C54–C64.

7. L. D. Santos and P. R. G. Brandao, "Morphological varieties of goethite in iron ore from Minas Gerais Brazil, *Minerals Engineering*, 2003, 16, 1285–1289.

8. S. Hembram and R. P. Bhagat, "Mineralogical characterization of iron ores having varying degrees of LOI from Gua and Kiriburu Mines," *Proc. XIII Inter. Seminar on Mineral Processing Technology* (MPT 2013), December 10-12, 2013, org. by CSIR-IIMT Bhubaneswar and IIME, Vol. 1, 45–50.

9a. B. Gururaj, N. Prasad, T. R. Ramachandran, and A. K. Biswas, "Studies on composition and beneficiation of a fine-grained alumina-rich Indian iron ore," Part a, *Proc. 13th Int. Miner. Process. Congr.*, Warsaw, Vol. 2, 1979, Amsterdam: Elsevier, 447–471.

9b. B. Gururaj, J. P. Sharma, A. Balwada, S. C. D. Arora, N. Prasad and A. K. Biswas, Dispersion-flocculation studies on hematite-clay systems, *Inter. J. Min. Process.*, 1983, 11, 285-317.

10. S. M. Rao, B. Roychoudhury, P. V. T. Rao, and A. D. Baijal, "Lowering alumina in Noamundi iron ores," *Tata Search*, 2001, 101.

11a. D. R. Rathburn and J. K. Mok, "Fine dry grinding of iron ore for pelletizing," *Trans. SME/AIME*, 1970, 247, 335–339.

11b. K. H. Peters, K. Kersting, and H. Beer, "Ways and means of reducing the energy consumption in iron ore sintering," *Proc. 4th Inter. Symp. on Agglomeration*, Toronto, Canada, Ed. C. E. Capes,: Iron & Steel Society, 1985, 157–167.

12. D. W. Furestenau, T. W. Healy, and P. Somasundaran, *Trans AIME*, 1964, 229, 34.

13. S. R. Balaji and I. Iwasaki, *Trans. AIME*, 1969, 244, 401.

14. S. V. Krishnan and I. Iwasaki, "Sodium silicate as a dispersant in selective flocculation of iron ores," *Trans. SME-AIME*, 1983, 272, 1984.

15. A. F. Colombo, Concentration of iron oxides by selective flocculation-flotation in *Advances in Mineral Processing*, Ed. P. Somssundaran, New York: AIME, 1986, 695-713.
16. S. Mathur, P. Singh, B. Moudgil, Advances in selective flocculation technology for solid-liquid separation, *Inter. J Mineral Processing*, 2000, 58, 201-222.
17. K. H. D. Humbolt, "Experience of KHD Humbolt Wedag in iron ore processing," *Iron Ore Workshop*, org. by Tata Steel and IIME, Jamshedpur Chapter, June 16–17, 2006.
18. N. D. Rao and S. K. Rai, "Availability of iron ore for the expanded Indian steel industry," in *Awakening of the Indian Steel Industry: 2005–2020*, Ed. Amit Chatterjee: Indian Institute of Metals, Tata Steel, 2006, 17-25.
19. Borah, S. C., M. Das, and R. Sengupta, "Beneficiation of graphite deposits of Arunachal Pradesh by gravity separation methods," *XXVI Inter. Mineral Processing Congress* (IMPC 2012), New Delhi, September 24–28, 2012, org. by IIME and IIM, Paper No. 0372, Session Poster.
20. N. N. Datta, and S. Mohiyddin, "On the beneficiation of Indian iron ores using chemical additives," *Proc. Inter. Conf. on Chemical Metallurgy*, (ICCM-91), Bombay: The Indian Institute of Metals, 1991, 729–761.
21. X. Yan and D. Witchard, "Sintering of blends containing magnetite concentrate and hematite or/and goethite ores," *ISIJ International*, 1998, 38, 1069–1076.
22. B. Roy, Utilization of India's iron ore resources–present situation and future possibilities, *Proc. XIth International Seminar on Mineral Processing Technology* (MPT 2010), 15–17 December, 2010, NML Jamshedpur, 534–541.
23. S. M. Rao, B. Roychoudhury, P. V. T. Rao, and A. D. Baijal, "Lowering alumina in Noamundi iron ores," *Tata Search*, 2001, 104.
24. A. Banerjee and T. K. Ghosh, "Mineralogy and mineral chemistry of fine iron ore generated from Joda captive mine of Tata Steel," *Tata Search*, 2011, 29–35.
25. C. E. Loo and D. J. Wong, Fundamental insights into the sintering behavior of goethtic ore blends, *ISIJ Inter.*, 2005, 45, 459–468.
26. T. Murukami, T. Nishimura, and E. Kasai, "Lowering reduction temperature of iron ore and carbon composite by using ores with high combined water content," *ISIJ Inter.*, 2009, 49, 1686–1693.
27. Y. Wu, X. Xang, Y. H. Li, P. Zhang, and Z. Y. Bao, "Biomass reduction roasting of goethite ore at a low temperature," *XXVI Inter. Mineral Processing Congress* (IMPC 2012), New Delhi, September 24–28, 2012, org. by IIME & IIM, Paper No. 0410, Session 12A-05.
28. K. K. B. Bhattacharya, R. Sarkhel, P. Saha, T. Alex, and N. Chkravorty, "Gravity separation as a method for lowering of alumina in iron ore slimes," abstract volume, *Inter. Symp. on Mineral Beneficiation—Recent Trends & 2000 AD*, Nagpur, October 3–5, 1994, 34.
29. P. K. Banejee, S. K. Rai, M. China, G. S. Dhar, and A. D. Baijal, "Beneficiation study of iron ore slimes at Tata Steel: A proposal," *Tata Search*, 2007, 83–86.
30. R. P. Bhagat, A. Vidyadhar, N. Kumari, K. V. Rao, and P. Sreedevi, "Study on flotation of iron ore from India," *Proc. International Mineral Processing Symposium* (IMPS 2012) Bodrum, Turkey, October 10-12, 2012, Paper No. 221.
31. ASTM E877-13, "Standard Practice for Sampling and Sample Preparation of Iron Ores and Related Materials for Determination of Chemical Composition and Physical Properties," *Annual Book of ASTM Standards*, Vol. 3.06 (Metals: Test Methods and Analytical Procedures), West Conshohocken, PA: ASTM International, 2013, 253–261.

3 Iron-Making Processes

ABBREVIATIONS

BF	*Blast furnace*
BOF	*Basic oxygen furnace*
CCS	*Cold crushing strength*
CSR	*Coke strength after reaction*
DRI	*Direct reduced iron*
ISP	*Integrated Steel Plant*
JSW	*Jindal Steel Works*
RDI	*Reduction-degradation index*
RI	*Reducibility index*
SAIL	*Steel Authority of India Limited*
SL/RN	*Stelco-Lurgi/Republic Steel-National Lead*
SR	*Smelting reduction*
TI	*Tumbler index*

3.1 PREAMBLE

Iron for steel making is produced through:

- Blast furnace route
- Alternate iron-making route

Blast Furnace Route The blast furnace (BF) has remained the major process of iron making for more than 200 years. The BF operates as a countercurrent chemical reactor where the ferrous-bearing burden materials in the form of lump iron ore, sinter, and pellets descend through the BF's height and are reduced by the ascending reducing gas (CO). Technological efforts have been made over the years to make its operation highly efficient and capable of providing stable operation with a wide range of feed materials. Hot metal produced in the BF is subsequently processed in a converter, which is basically a basic oxygen furnace (BOF) to produce steel. This route, however, involves sintering and/or pelletization installations as well as coke-making operations and is termed as "BF-BOF". The process is highly capital intensive. Besides, the BF requires raw materials of stringent specifications in terms of their physical-chemical characteristics. Coking coal is needed to make a coke strong enough to support the burden in the BF. Integrated steel plants (ISPs) of less than one million tonne (MT) annual capacity are generally not economically viable.

Alternate Iron-Making Routes The following processes have emerged in iron making as alternatives to the BF process.

a. Production of non-liquid iron—direct reduced iron
   Direct reduced iron (DRI) has emerged as an excellent substitute for scrap for electric furnaces. DR processes can be broadly classified into the following two groups. This is based on the type of reductant used.
   - Coal-based DR process, which uses solid reductant
   - Gas-based DR process, which uses gaseous reductant

   The DRI production throughout the world uses the following technologies, namely,
   - SL/RN process, which is coal-based rotary kiln process.
   - MIDREX direct reduction iron-making (referred to as the MIDREX process) which is gas based shaft reduction process.
   - ITmk3, (Nuggets) Technology etc.

   Out of these, the first two are the only major production technologies. Gas-based processes account for 78.8% of the world's DRI production; out of that, MIDREX alone contributed 63.2% in 2013.[1]
b. Alternative method for liquid iron (hot metal)
   Smelting reduction (SR) processes, namely, COREX smelting reduction process. This is VAI iron-making techmology.

   FINEX process, developed by Aiemens VAI and POSCO. This is essentially a combination of FINMET's Fluduzed Bed and COREX's Melter Gasifier.

   Hismelt, etc., are based on using non-coking coal and thus obviate the need for coke oven batteries and sinter plants as iron ore fines can be used directly, except COREX. The process of Corex is part of an integrated steel plant (JSW Steel) in the southern part of India.[2]

3.2 REDUCTION OF IRON OXIDE

3.2.1 Sequential Steps of Reduction

The reduction of hematite takes place in stages. The iron ore in its hematite state gets converted to a metallic state by losing its oxygen. The intermediary reduction products do have different crystallography and chemistry, as mentioned here (**Figure 3.1** and **Table 3.1**).

Wustite has a structure similar to NaCl (cubic closed packing of anions). It is usually non-stoichiometric (iron deficient) and denoted as Fe_XO where x is approximately

	+22 % vol.		+4 % vol		slight contraction	
Fe_2O_3 (hematite)	→	Fe_3O_4 (magnetite)	→	FeO (wustite)	→	Fe (metallic iron)
HCP		Cubic		Cubic		BCC or FCC

FIGURE 3.1 Different phases of hematite reduction

TABLE 3.1
Chemical Composition of Hematite and Intermediary Reduction Products

Constituents	Hematite (Fe_2O_3)	Magnetite (Fe_3O_4)	Wutite (FeO)	Metallic Iron
% Fe	69.9	72.4	77.7	100.0
% Oxygen	30.1	27.6	22.3	0.0

between 0.83 and 0.95. It is an important intermediate in the reduction of iron ores but is metastable below the temperature in between 560 and 570°C (833 and 843 K) and decomposes to α–iron. Wustite is paramagnetic at room temperature.

The reaction is kinetically controlled. Important steps (**Figure 3.2**) in this regard are:

a. Transport of CO gas from bulk gas phase to Fe-CO interface
b. Molecular diffusion of CO gas through Fe layer to the reaction interface
c. Adsorption of CO gas at the interface, reaction between iron oxide and CO gas
d. Desorption of the product gas CO_2 at the interface and diffusion of CO_2 to the surface
e. Desorption of CO_2 from the Fe-CO interface to the bulk of gas

3.2.2 Reduction Chemical Reactions

The iron oxides in different forms, namely, iron ore (lump), pellets, and sinter, are reduced through a series of indirect and direct chemical reactions as follows. The standard heat of reactions at 298 K, calculated from the thermodynamic data[3] are also shown.

Indirect Reduction Iron oxide (hematite) is reduced by the reductant, carbon monoxide, through a sequence of chemical reactions, mostly in solid state. These reactions are called indirect reduction as carbon does not take part in reduction directly. The CO gas that takes part in reduction is generated by the gasification of carbon.

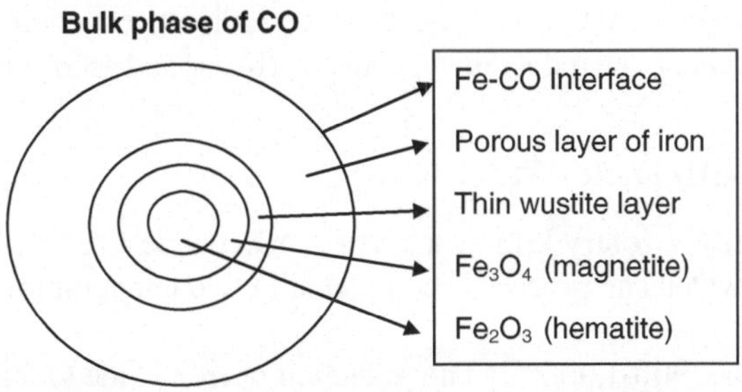

FIGURE 3.2 Steps in reduction reactions of Fe_2O_3 compound

Below 900°C

$$3Fe_2O_3(s) + CO(g) = 2Fe_3O_4(s) + CO_2(g);\ \Delta H^0_{298K} = -52.43\ \text{KJ/mol of CO} \quad (3.1)$$

The reaction is exothermic

$$Fe_3O_4(s) + CO(g) = 3FeO(s) + CO_2(g);\ \Delta H^0_{298K} = 40.46\ \text{KJ/mol of CO} \quad (3.2)$$

The reaction (reduction of magnetite to wustite) is highly endothermic and requires a considerable amount of heat. The wustite (FeO) is metastable and is reduced to metal iron as follows:

Begins at 850°C

$$FeO(s) + CO(g) = Fe(s) + CO_2(g);\ \Delta H^0_{298K} = -18.54\ \text{KJ/mol of CO} \quad (3.3)$$

This reaction is mildly exothermic but requires an extra mole of CO for the reaction to proceed: $FeO + 2CO + CO = Fe + 2CO + CO_2$. This is due to the fact that the free energy change in the reaction is positive and increases with increasing the temperature. The said reaction is thermodynamically feasible only when there is a decrease in free energy, which means that the reaction cannot proceed in the forward direction. But the reaction is made to move in the forward direction by keeping a large excess of CO over CO_2. Hence, the reaction is kinetically controlled rather than thermodynamically controlled.[4]

Direct Reduction In addition to the indirect reductions mentioned previously, iron oxide is directly reduced by C in coke according to the reaction, which is highly endothermic:

$$FeO(s) + C(s) = Fe(s) + CO(g);\ \Delta H^0_{298K} = 153.9\ \text{KJ/mol of C} \quad (3.4)$$

3.3 DIRECT REDUCTION PROCESSES

The objective of the process is to produce sponge iron, which is used as a feedstock in steel making. Temperature maintained in the process is usually greater than 1,000°C However, the production cycle is long. The process suffers as more reducing agent is needed, and heat consumption is more than the indirect reduction (BF) route. The process is dependent on the reductant (a) coal and (b) gas and accordingly categorized.

3.3.1 Coal-Based Direct Reduction

This uses, basically, a rotary kiln as a reactor and coal as a reductant and has been developed worldwide. The process is completed in two temperature zones:

a. Preheat zone (900-1,000°C) The reduction of iron oxide to ferrous oxide:

$$Fe_2O_3 + CO = 2FeO + CO_2 \quad (3.5)$$

b. Metallization zone (1,000-1,050°C): Final reduction to metallic iron occurs:

$$FeO + CO = Fe + CO_2 \tag{3.6}$$

The processes are highly sensitive to the types of raw materials used. Coals with higher reactivity are preferred as CO_2 is rapidly converted to CO, thereby maintaining reducing conditions in the metallization zone.

The main advantages of coal-based processes are as follows:

a. These do not require high-grade coal; rather, the process uses non-coking coal that is available in abundance.
b. These can be installed at a lower capacity and at places where small reserves of coal and iron ore are available. Modules of small-scale operation are available.

The rotary kiln process suffers from limitations, such as:

a. The rotary kiln operation has lower productivity (0.5-0.9 t/m^3/day). This is due to longer reduction time in the production of DRI.
b. These consume high energy, in the range of 16-21 GJ per T of DRI. For comparison, the energy consumed in production of hot metal through the blast furnace route is only 6-8 GJ/T. Energy requirement for DRI production depends mainly on the quality of coal and to a lesser degree to the reducibility of iron ore. The counter-current rotary kilns used for DRI production in India only allow about 60% of the heat to be used for reduction, while the remaining 40% is discharged from the kiln. This offers a significant potential for energy efficiency improvements.
c. Due to the presence of residual char and ash in the product, these cannot be fed to the steel-making furnace in a hot condition. Since there is no melting and formation of slag phase in DRI production, all the gangues of iron ore remain in the DRI. This needs to be separated via a slag in electric arc furnace (EAF). This increases the electric energy consumption of the EAF compared to steel scrap melting.
d. The process can only handle iron ore above 6 mm size, whereas below 6 mm size fraction needs to be discarded/agglomerated.

In view of these limitations, the rotary kiln operation has lower operational economy. However, the process is popularly/widely used in view of its low capital investment and due to the ease of diversification. Southeast Asia—including India, in particular, due to its large reserve of non-coking coal—has adopted these technologies often.

3.3.2 Gas-Based Direct Reduction

Iron ore lumps and/or pellets are reduced in the solid state using the gaseous reducing gases CO and H_2. The processes based on gaseous reduction are confined to the areas where natural gas is available in abundance at a reasonable cost.

Midrex is the most widely used process in DRI production in the world. In this process, the reductants are CO and H_2 gases that are generated by reforming natural gas at approximately 950°C in the presence of catalysts (Ni or calcined alumina). A steam oxygen mixture is used to reform the natural gas. The main components of the process are:

a. Shaft furnace
b. Gas reformer
c. The cooling system

The DRI furnace is a steel vessel with an internal refractory lining in the reducing zone. The charge solids flow continuously into the top of the furnace through seal legs. The reduction furnace is designed for uniform mass movement of the burden by gravity feed through preheat, reduction, and cooling zones of the furnace. The cooled DRI is continuously discharged through seal legs at the bottom of the furnace.

The gas-based processes have merits over the coal-based processes as follows:

a. Less capital cost: The capital cost per T of installed capacity in the case of gas-based plants is 1.8-2 times lower than that of coal-based plants. However, extra capital is necessary for the installation of catalytic reformer.
b. High productivity: The throughput rates for gas-based DRI plants are much higher than those of coal-based plants.
c. Better product quality: The quality of the DRI product from the gas-based plants in terms of metallization and carbon content is better than that of coal-based plants.
d. Energy efficiency: The gas-based processes are highly energy efficient. The energy consumption is lower—in the range of 10.5-14.5 GJ per T of DRI compared to 16.0-21.0 GJ per T for coal-based processes.
e. Better plant availability: In the case of the coal-based process, the ash in coal leads to the ring formation; therefore, the plant often encounters maintenance problems. On the other hand, gas is a clean source of energy and a reductant. Therefore, the maintenance problems in the case of gas-based plants are less and the availability of plant is of longer duration.
f. Environmental pollution: Since gas is a clean source of fuel, a gas-based plant is advantageous in DRI production compared to coal based. Natural gas-based DRI production also leads to lower CO_2 emissions. This amounts in the range 0.77-0.92 ton of CO_2 per ton of steel depending on the operational parameters.

3.4 BLAST FURNACE (BF) IRON-MAKING PROCESS

The integrated blast furnace and oxygen steel-making (BF-BOF) route is still a dominant process for worldwide steel production as a major portion of the world's total output of crude steel is contributed by this route.

The blast furnace has three distinct functions:

a. Essentially, a blast furnace (BF) is a countercurrent reactor whose main aim of operation is to remove oxygen from iron oxides (reduction of ferruginous burden materials—lump ore, sinter, and pellets)-.
b. To convert iron oxides into liquid iron (smelting).
c. To produce very hot reducing gas (combustion chamber).

3.4.1 Operation

A blast furnace (for iron making) is a circular shaft 30 m in height and with an internal diameter of 10 m; although blast furnaces with bigger sizes are now available. The furnace is a huge refractory-lined steel shell, and the refractory material is fireclay brick. Iron ore in lump form, agglomerates (sinter and pellets), and coke, which are often referred to as blast furnace burden, are charged into the top of the furnace.[5] The top of the furnace is sealed by means of a double bell and hopper that permits the solid material to be charged without allowing the furnace gases to escape to the atmosphere. The modern blast furnaces have now adopted a bell-less top.

Carbon in coke gets oxidized to CO by oxygen that is made available through air at high temperature. The air is preheated in large thermal regenerators called "stoves" and is injected at high pressure into the lower part of the furnace, called the bosh zone, through a series of evenly spaced nozzles called "tuyeres."

The hot air reacts with the coke in this zone to form CO, which ascends through the furnace and reacts with the descending ferrous burden to form Fe. The burden materials, as these descend toward the bottom of the furnace, undergo numerous chemical reactions with the ascending reducing gases. The reduction of hematite with CO takes place at a temperature of 900-1,000°C mostly in the upper part of the furnace, called stack zone, and hence in a solid state (**Sec. 3.2.1**). The heat resulting from the oxidation of coke causes the chemical reactions and also smelts the metal in the lower part of the furnace. Finally, the smelted liquid iron and slag are collected at the bottom of the furnace, which is called the hearth.

Figure 3.3 shows a schematic of the blast furnace along with the flow of materials. The ferrous burden—namely, sinter and pellets—is produced by agglomerating the fine ores and concentrate, respectively. Coke is obtained after carbonization of coal in the coke oven.

3.4.2 Material Handling and Charging

A larger portion of the total iron-making facility is required for material handling: the flow of materials to and from the furnace. The burden materials are received, blended, and stored in the ore yard. These are then delivered to a "stock house" where they are segregated in storage hoppers. The furnace charge is accurately weighed and delivered to the furnace top either by a "scale car/skip system," by a system of conveyors, or by a combination of both. At the furnace top, the materials are

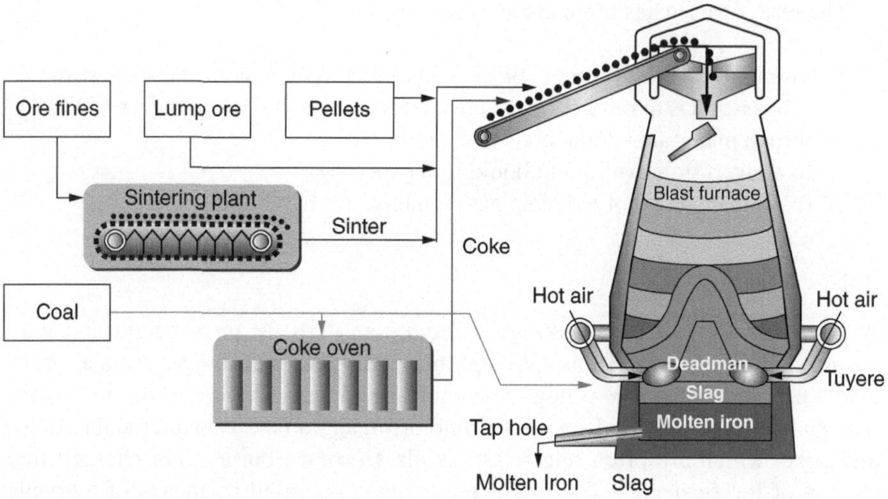

FIGURE 3.3 Schematic of material flow in blast furnace iron making

charged to the furnace in a prescribed sequence either through a system of "bells" or via a chute bell-less top. Both systems will maintain the furnace top under pressure during charging operations.

3.4.3 Raw Materials

The source of iron oxides, which is reduced in the furnace to produce iron, are iron ores in the form of lump (above 10/6.3 mm, below 40/35 mm size), pellets, and sinter. The iron ore, which is mined from the earth's crust is never pure as it contains a lower iron content than the theoretical values, with impurities (called "gangue") that are primarily silica and alumina.

Hence, this must be processed to increase its iron content through the technique called "beneficiation." The ore fines (below 10/6.3 mm size) cannot be used directly in the furnace and are usually beneficiated (ref. **Secs. 2.8 and 2.11**). The beneficiated products are agglomerated to produce, depending on their sizes and grade, sinter or pellets.

Pellets The pellets are an agglomeration product in the case of fine (below 100 micron) concentrate containing 63-65 wt.% iron. The concentrate is generated following the intensive beneficiation of iron ore, especially that of lean variety and slimes. The concentrate is balled in the form of spherical-shaped agglomerate and fired in a furnace to produce strong pellets.

Sinter The ore fines, which are coproduced in the mines along with the lump ore and are coarser in size, may not require elaborate beneficiation treatment in cases where the fines are moderately rich in iron content. The ore fines are used to make "sinter." The ore fines, along with small-sized coke (called coke breeze), sand-sized fluxes, and other steel plant waste materials that contain some iron materials, are proportioned in their ratios according to the targeted product (sinter) chemistry and then mixed together. The raw material mix is then placed on a Dwight-Lloyd sintering strand, where it is ignited by a gas-fired furnace and fused by the heat from the coke fines into larger-sized pieces that are from 10-40 mm.

Fluxes Any fluxes (limestone and dolomite) are not charged into the furnace directly; rather, these are incorporated through mostly sinter and, in minor amounts, through pellets.

Coke The coke, charged into the furnace, has the following functions:

- As heat source: Coke, on combustion, provides sufficient heat to raise the temperature to a level so that various (reduction) reactions take place in the furnace. Besides, the melting of the generated products, metal and slag, could occur.
- The coke, as reductant, reduces iron oxide to iron. The C present in the coke reacts with iron oxide directly (known as direct reduction) and on oxidation gives CO, which reduces the iron oxide (known as indirect reactions) to form iron.
- It is the only material that remains solid till it reaches the bosh zone of the furnace where it burns and gets oxidized. The coke supports the entire burden materials, which descend through the furnace and provide permeability.

In view of the aforementioned functions, chemical purity (% C) of coke and its strength are its prime important properties. The strength of coke under cold condition, known as micum index, and following reaction, known as coke strength after reaction (CSR), are measured following the standard procedure.

The coke is produced from a blend of a special variety of coals that have a caking property, called coking coal. The coal is crushed and ground into below 3-mm-sized fractions and then charged into an oven. As the oven is heated, the coal is cooked so most of the volatile matter such as oil and tar are removed. The cooked coal, called coke, is removed from the oven after 14-18 h of reaction time. The coke is cooled, crushed, and screened into pieces ranging from 25 mm to 80 mm.

3.4.4 Products

The reduced product, liquid iron, trickles through the coke bed and reports to the lower portion of the furnace, called the hearth. Slag is another product that, in liquid form, also reports to the bottom of the furnace. The volume of slag ranges from 150-300 kg per ton of hot metal produced and depends on the composition of burden. The slag, being lighter, is accumulated above the molten iron. The amount of slag and its chemical characteristics depend on the amount of impurities in the ferrous burden and coke. These liquid products are removed periodically from the furnace in the "cast house." The iron is directed to refractory-line hot metal cars for transfer to the steel-making facility. Slag is usually directed to slag pits where it is granulated to produce cement

Product Impurities: The molten iron consists of (besides Fe) carbon and impurities like silicon, sulfur, manganese phosphorous, and titanium. Typical hot metal chemistry is as follows:

Iron (Fe)	= 93.5–95.0%	Carbon (C)	= 4.1–4.4%
Sulfur (S)	= 0.025–0.050%	Titanium (Ti)	= 0.02–0.06%
Manganese (Mn)	= 0.55–0.75%	Phosphorus (P)	= 0.03–0.09%

The hot metal, iron, should be low in S and P for the production of quality steel. The phosphorous in hot metal burns during the steel-making process; hence, dephosphorization (removal of phosphorous) of hot metal may not be essential. The main problem that an integrated steel plant faces is with sulfur. It affects the product (steel) quality. Besides, the element S has a tendency to be built up within the furnace lining, which decreases the lining life of blast furnace. The major sources of S are coke and lump ore.

Sulfur from hot metal can be removed externally through the addition of $CaCl_2$

$$FeS + CaCl_2 = FeCl_2 + CaS \tag{3.7}$$

The process is called external desulfurization and is practiced in some plants.

In addition to that, desulfurization of hot metal is partially achieved in the furnace itself through the slag-metal interaction. The element sulfur is distributed between the slag and metal phases.

Typical chemical composition of slag is as follows:

CaO	35–40%	Al_2O_3	10–15%
SiO_2	30–35%	Fe	0.2–0.4%
MgO	4-10%	S	~1%

These constituents make up over 95% of the (slag) mass. The balance is made up of FeO, MnO, MnS, CaS, and alkali silicates. Another product of the iron-making process, in addition to molten iron and slag, is hot gases. These gases exit the top of the blast furnace and proceed through gas-cleaning equipment where particulate matter is removed from the gas and the gas is cooled. This gas has a considerable energy value and therefore is suitable as fuel. It is burned as a fuel in the "hot blast stoves" that are used to preheat the air entering the blast furnace to become "hot blast." The remaining gases are sent to different units of iron and steel plants to generate heat, including to the sinter plant for ignition.

Typical composition of BF gas is as follows: CO = 23–25%; CO_2 = 14–16%; H_2 = 3–5%; N_2 = 52–56%.

3.5 BLAST FURNACE (BF) IRON-MAKING REACTIONS

3.5.1 Blast Furnace Zones: Reactions and Burden Properties

Figure 3.4 shows a section of an iron blast furnace that is divided into different zones, each having different functions including the chemical reactions, which are described here. The physical and chemical characteristics of burden materials do affect the blast furnace performance. Accordingly, the quality requirement of the burden materials across the different sections (**Figure 3.4**) of the blast furnace varies.

Stack/Granular Zone: This is the top portion of the furnace. The burden materials, as they descend across this zone, get preheated by hot gases. The material gets reduced in solid state by the ascending reduced gas in a countercurrent gas solid reaction manner through the sequence of reactions (3.1–3.3). The CO gas that takes part in reduction is generated by the gasification of carbon in the raceway. CO gas is also generated (in this zone) through the oxidation of CO_2 (generated through the reactions 3.1–3.3) by the descending coke: $CO_2 + C = 2\ CO$ (ref: Eq. 3.10).

A porous surrounding iron layer with interconnected pores is formed on reduction of the ferrous burden material.

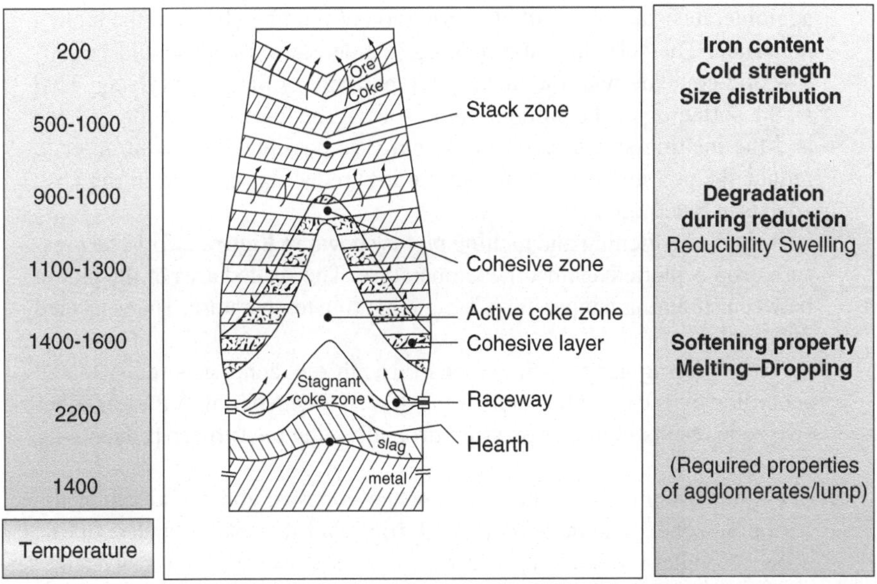

FIGURE 3.4 Various zones of blast furnace: approximate temperature (in degrees C) and deciding burden properties

Important properties of the agglomerates that affect the furnace operation in this zone are:

- ***Chemical Composition*:** Fe content of the agglomerates is significant.
- ***Size Distribution*:** The agglomerates should have less fine fraction. Appropriate particle size, close grain size distribution in the original condition, and favorable grain shape of the agglomerates are important in this respect.
- ***Strength in Cold Condition*:** The agglomerates should have adequate strength in cold condition so that these do not disintegrate during their (material) handling and also in their descent in the upper part of the furnace. A lower generation of fines when the agglomerates disintegrate less in the stack zone ensures high burden permeability.
- ***Degradation during Low Temperature Reduction (RDI):*** The agglomerates should have high resistance to low temperature breakdown during their reduction in the stack zone. This is to ensure lesser generation of fines during the reduction process.
- ***Reducibility*:** The agglomerates should have high reducibility (RI) so that a greater gas efficiency in the BF could be achieved.

Cohesive Zone: This is the lower part of the furnace where the temperature is high (greater than 1,200°C). The pre-reduced burden material softens and finally is reduced in this zone. At this temperature, the pellets swell as they get reduced. The flow of gas across this zone is affected by the size and shape of the cohesive zone, which is decided by the softening and melt-down characteristics of the burden materials. The temperature range in this zone starts from the softening of the burden materials (ore and agglomerates) and ends with the formation of liquid globules of the molten materials. The softening and melting (temperature) range should be narrow and coincide with the melting temperature of the primary slag. That is, the softening of the burden should start at the lower part of the furnace and the melting should conclude as early as possible. This is in order to reduce the pressure drop on the ascending (reducing) gas flow in the bosh zone (see **Sec.3.6.6** also).

A typical softening and melting plot is shown in **Figure 3.5**.[6] The pressure drop is plotted against the temperature. The shaded area of the plot in between softening temperature, Ts, and melting temperature, Tm, is termed the S-value.

The reducing gas, which ascends through this zone, gets redistributed according to its ease of flow across the descending material. A thicker cohesive zone results in higher pressure drop affecting the furnace productivity adversely.

Active Coke Zone: This zone is below the cohesive zone, also called the "dripping zone." The liquid hot metal drips over the coke particles. In this zone, carbothermic smelting reduction of oxide dissolved in the slag occurs.

The phenomena of melting, dripping, and final reduction of the burden material occur as the burden descends further. Slag is formed in the 1,200-1,450°C temperature range.

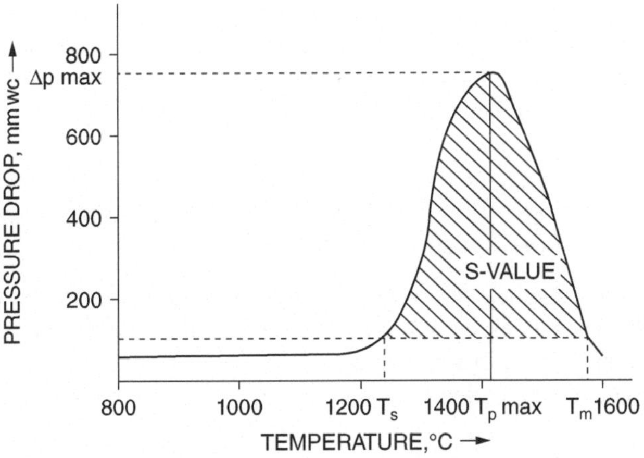

FIGURE 3.5 Softening-melting plot of a typical burden material (after Ref. [6])

The gangue constituting/nonferrous oxides also get reduced:

$$\begin{gathered} MnO + C = Mn + CO \\ SiO_2(S) + C = SiO(g) + CO;\ SiO + C(S) = Si + CO \\ P_2O_5 + 5C = 2P + 5CO \end{gathered} \tag{3.8}$$

The burden material should have:

- High softening and melt-down temperature so that these phenomena take place at the lower part of the furnace
- Small difference between the two temperatures in order to have a thinner cohesive zone
- Less amount of non-dripped material

Raceway: The reducing gas is generated in the raceway through the combustion of coke in the 1,250-1,550°C temperature range. As hot air blast (21-25% O_2) is blown into the hot coke through the tuyeres, carbon is oxidized to CO_2.

$$\text{(Combustion Reaction)}\, C + O_2 = CO_2 \tag{3.9}$$

The CO_2 gas is reduced to CO through an endothermic reaction in the presence of excess C at higher temperature:

$$\begin{aligned} \text{(Gasification reaction)}\, CO_2(g) + C(s) = 2\ CO(g)\ \Delta H^0_{298K} \\ = 172.42\,\text{KJ/mol of C} \end{aligned} \tag{3.10}$$

The product of this reaction, carbon monoxide, is necessary to reduce iron oxide (**Sec. 3.2.2**).

Hearth: Separation of metal and slag in liquid form takes place in the hearth. Above 1,600°C hot metal is desulfurized. The desulfurization process is influenced by the volume of slag and its characteristics: basicity and MgO content. The chemistry of burden materials is the deciding factor in this regard.

3.5.2 Solution Loss Reaction and Shaft Efficiency

In addition to the combustion (of coke to form CO_2) reaction in the raceway, the carbon in coke when it descends through the furnace also combines with the ascending carbon dioxide gas in the upper part of the furnace, according to the following endothermic chemical reaction:

$$CO_2(g) + C(s) = 2\ CO(g)\,(\Delta H^0_{298K} = 172.42\,KJ/mol\ of\ C) \tag{3.11}$$

The reaction is highly endothermic (toward the right) and proceeds toward the right at a temperature above 670° C and toward the left below 670-°C.

The reaction is called a *Boudouard* or *solution-loss reaction*. This is because part of the CO_2, which is generated in the lower part of the furnace (at higher temperature), combines with C to generate CO, which may not participate in the reduction and escape through the BF top therefore increasing the coke rate. The reaction has an important role in deciding the furnace efficiency since carbon is never utilized fully in the furnace.

The shaft efficiency (η_{CO}) is decided by the composition of the gases that come out from the top of the furnace and is given by the ratio: $CO_2/(CO + CO_2)$. Its value lies between 0.25 and 0.46.

The composition of the BF gas determines the shaft efficiency. A higher-than-normal amount of CO in the BF gas indicates ill-functioning of the furnace. The parameters, type of reaction, and the temperature at which the reduction takes place determine the shaft efficiency. The shaft efficiency is influenced by the following factors.

a. **Direct Reduction of Iron Oxide:** The direct reaction takes place at a higher temperature and in the lower part of the furnace:

$$2Fe_2O_3 + 3C = 4Fe + 3CO_2 \tag{3.12}$$

The CO_2 gas, generated through the reaction, is not stable at higher temperatures and reacts with C to produce CO gas:

$$CO_2(g) + C(s) = 2CO(g) \tag{3.13}$$

The CO gas, thus formed, may escape through the BF gas, hence reducing the shaft efficiency. Apparently, the direct reduction reaction increases the thermal load in the bosh/hearth zone (consuming C). It is estimated that for 1% decrease in direct reduction, around 6-7 kg C could be saved.

b. **Indirect Reduction of Iron Oxide:** The shaft efficiency gets affected by the way the indirect reduction takes place:
   - In case iron oxide is reduced at a lower temperature (relatively upper part of the furnace) by CO, the CO_2 formed will be stable and does not get altered before it leaves the furnace.
   - In case the reduction of iron oxide is delayed until the ore has come to the lower part of furnace where the temperature is high (above 670°C), any CO_2 produced will be re-carbonized to CO in accordance with the following solution-loss reaction:

$$CO_2(g) + C(s) = 2\ CO(g) \tag{3.13}$$

   The CO gas will escape through the BF gases.

c. **Dissociation of Carbonates:** The CO_2 gas does not necessarily have to come from the reduction of iron oxide; it may come from the calcination reactions also:

$$CaCO_3(s) = CaO(s) + CO_2(g),\ \Delta H = +179\ KJ/mol \tag{3.14}$$

$$MgCO_3(s) = MgO(s) + CO_2(g)\ \Delta H = +102\ KJ/mol \tag{3.15}$$

   The eliminated CO_2 reacts with C to form CO similar to that in the previous case and may escape through the BF gases. Since the fluxed sinter is incorporated in the furnace with desired chemistry to completely meet the lime and MgO requirement of slag, this aspect (calcination reaction) has remained theoretical, to stress the importance of using fluxed sinter.

d. **Burden Reducibility:** The importance of burden reducibility can be very well appreciated in view of the aforementioned. The agglomerates must reduce at the upper part of the furnace so that the CO_2 generated through the reaction are stable (which is only possible below 670°C).

3.6 BLAST FURNACE PERFORMANCE: FACTORS AFFECTING

The performance of blast furnace iron making is decided by (a) the furnace productivity, (b) coke rate, and (c) hot metal quality. The factors that influence these are interinfluencing. Important among those related to burden are described as follows.

3.6.1 Chemical Composition of Burden Materials

The chemical composition of burden materials affects the iron-making process, through blast furnace route in particular, in the following ways:

a. Its direct effect on the performance of iron-making process.
b. The chemical composition of agglomerates affects their properties, which in turn, influence the iron-making performance.

Fe Content: If the burden materials (lump ore and/or agglomerates) that are charged into the furnace are poorer in Fe content, more energy is required per unit of iron produced. This is because:

a. A higher acidic gangue constituent requires more flux to make the slag neutral (CaO/SiO_2 around 1); hence, the slag volume is increased.
b. A much higher proportion of heat is consumed in melting the gangue constituent and must be available at low temperatures (upper part of the furnace).

Conversely, if the burden is rich in Fe content, the energy requirement per unit of iron produced is low, and a high proportion of the heat is required at high temperatures so that the gases leaving the smelting zone are relatively cool and closer to temperature of the continuous descending solids.

Alumina Content: The ratio $Fe/(SiO_2 + Al_2O_3)$ together with basicity of agglomerates/ore determine the amount of slag produced when smelted in the furnace.

The furnace productivity deteriorates on account of increasing the slag volume because it requires extra hearth space when the agglomerates with higher alumina content are charged.

The **heat** requirement in the blast furnace increases when the agglomerates with higher alumina are charged into the furnace. The contributing factors are:

- Alumina component in iron ore/agglomerates is refractory by nature; hence, a higher temperature is required in order to get it melted.
- A higher alumina in the agglomerate increases the slag volume; hence, extra heat is required in making the additional gangue fluid.
- The input of alumina through the ore/agglomerates increases the slag viscosity and hence requires more heat to make it fluid.

A higher viscous (less fluid) slag, on account of its increased Al_2O_3, affects the hot metal quality as its desulfurization capacity deteriorates.

In view of this, it is essential to reduce the alumina content in the burden materials. As deposits of good iron ore (Fe above 62 wt.%) are depleting fast, the technologies of enriching the lean grade ores to remove alumina through intense beneficiation need to be developed (see **Secs. 2.8 and 2.11**).

Basicity The silica (SiO_2) and alumina (Al_2O_3) in BF slag are known as its acidic constituents. Their sources are: lump ore and coke (ash constituent). The source of these acidic constituents is also agglomerates since these are produced from ore fines/ore concentrate, coke breeze, fluxes, etc. All these acidic constituents report to and form part of the slag. The total input of acidic constituents through the burden needs to be neutralized by incorporating magnesium oxide (MgO) and calcium oxide (CaO), known as basic constituents, in fluxes through the burden so that the slag has a basicity ratio that remains near to one.

However, direct charging of limestone and dolomite in the furnace is not practiced now-a-days. This is because the calcination of these materials is highly endothermic (Eqs. 3.14 and 3.15), and If these material are calcined inside the furnace, they require heat from inside the furnace through the combustion of coke. Hence, it is imperative to incorporate the lime and MgO required for fluxing the acidic gangue, silica, and alumina of the burden, by regulating the chemical composition of agglomerate more so than that of sinter.[7,8]

MgO Content: The fluidity of slag and its desulfurization capacity are of prime importance for the smooth functioning of the furnace. These characteristics of slag are primarily affected by the MgO content and can be achieved through incorporating MgO in fluxed sinter,[9] also in fluxed pellets. This is more important for blast furnace performance where the slag is more viscous due to a higher alumina content in the iron ore.

Consistency in Quality Parameters: The performance of a blast furnace with respect to its productivity, quality, and cost depends not only on the (physical, chemical, and mineralogical) characteristics of the burden materials but also on their consistency. A higher consistency in quality parameters means their lesser fluctuation over the period. A blast furnace is fed with the numerous categories of burden materials, each having different physical and chemical characteristics. Any variation in these materials leads to unstable furnace operation and inconsistent hot metal quality. In view of this, the burden materials are required to be homogenized in order to get a hot metal product of consistent chemistry. This is one of the most important tasks before the furnace operators—to maintain the consistency in quality parameters (i.e., regularity of the physical, chemical, and mineralogical properties). In this respect, homogenizing of the burden materials in a separate base blending plant called a "raw material bedding and blending plant" (RMBBP) is carried out.

Elemental Impurities The burden materials are never pure. The impurities in the oxide form of various elements, depending on their nature, affect the iron-making process, and the hot metal quality in particular, and could be categorized in different groups as follows.

a. Those that are not reduced in the furnace, which also does not affect the composition of hot metal/pig iron. Alumina constitutes a large part of gangue and requires lime (CaO)/MgO to flux it, which forms slag in the furnace. The alkalis (NaO, KO, etc.) are driven off with flue gas/dust, also built-up within the furnace.
b. Those that are partially reduced in the furnace and yield elements that enter the hot metal/pig iron. Impurities such as SiO_2, S, Ti, Mn, Cr, Co, etc., come under this category. Silica constitutes a large part of gangue and is fluxed with lime and/or MgO to form slag. Silica is readily removed in the steel-making process; hence, it is not considered to be of much importance as far as its effect on the quality of the final steel product.
c. Sulfur compounds introduced in the furnace enter the hot metal as FeS and affect the hot metal quality. Major sources of S are coke and iron ore.

The element exists as compounds that are inorganic and organic. The organic S burns to SO_2 during sintering, while the inorganic one remains. S in hot metal deteriorates its quality. As it cannot be reduced/eliminated during the steel-making process, desulfurization is necessary.

d. Those that are always reduced in the BF and often alloy with iron. Compounds of As, P, Ni, Cu, Cd, Sn, etc., come under this category.
e. P is a major impurity in iron ore. Its reduction is important as a blast furnace is not capable of reducing the P, and entire P reports to the hot metal. However, P is readily removed from iron in the BOF steel making.

3.6.2 Blast Furnace Burden

The raw materials charged to the blast furnace constitute its burden. The quantity (in kg) of burden to produce 1 ton of hot metal is referred as "burden weight." The major contributions to the burden (other than that theoretically required) come from the impurity of raw materials including coke and the direct charging of limestone/dolomite, if any, into the furnace.

a. **Chemical Composition** The burden materials, which are fed into the blast furnace, are never pure. The impurities (gangue constituents) add to the burden and thus yield a greater formation of slag.
b. **Charging of Limestone** The carbonates, when charged into the furnace, add to the burden as these contain CO_2. As an illustration, the decrease in burden weight is 44% ([44/[56+44]) when lime (through fluxed sinter) is used instead of direct charging of limestone according to the reaction: $CaCO_3 = CaO + CO_2$ [40 + 12 + 48] = [40 + 16] + [12 + 32].

Both of these two components (CO_2 in fluxes and impurities constituting slag) in the furnace burden contribute to the increased coke rate (**Figure 3.6**). Since the calcination of carbonates is highly endothermic, the contribution to the coke rate with respect to the addition of CO_2, through direct charging of limestone, is higher than that when gangue constituents are added.

Charging fluxed sinter eliminates raw limestone from the furnace burden and, hence, results in better productivity and decreased coke rate due to decreased burden weight (ref. **Sec. 4.4**).

3.6.3 Size of Burden Materials and its Distribution

The size parameter of burden materials influences the BF performance in two ways acting in opposite directions.

Reduction Kinetics: Smaller particle size is desirable for rapid reaction

Burden Permeability: The particle size of the burden should not be so small that it hampers the flow of ascending gas through the burden, which is descending.

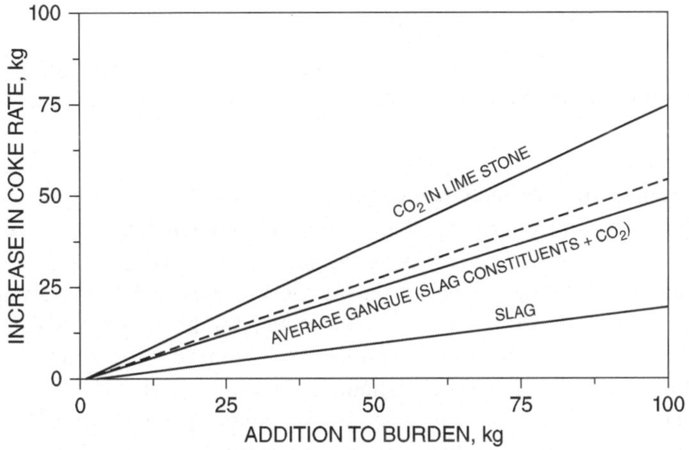

FIGURE 3.6 Contribution to the burden by slag and CO_2 sources (after Ref. [10]). Typical result from Indian BF is also shown (dotted line) (after Ref. [7])

Sources of fines in the burden material are:

- Those that are directly charged into the furnace
- Those that are generated in the furnace shaft by the reduction process

Hence, the quantity of fines that are carried along with the burden material and charged to the furnace must be restricted to as little as possible. This can be achieved by:

- Producing sinter/pellets of good cold strength (should have a higher tumbler index) so that these do not disintegrate during transport from the plant to the blast furnace site and do not generate fines (should have less abrasion index)
- Proper screening of the burden materials so that input of fines in the blast furnace is as low as possible

The control of fines by screening is possible to an extent before the burden material is charged into the furnace. Once the material is charged into the furnace, one does not have control and has to rely on the strength of the burden material. The burden material is subjected to subsequent degradation even in cold condition (without reduction) as it descends inside the furnace in the stack zone. This results in a higher generation of fines in the stack region, which adversely affects the permeability of this zone and sometimes results in hanging.[11]

The low temperature breakdown of the agglomerates during reduction is equally important in this respect as an extra amount of fines is generated during their reduction at low temperature. The agglomerates should have a lower reduction-degradation (RDI) value (**Sec. 4.3.2** and **9.5**).

3.6.4 Cold Strength and Reduction Parameters of Agglomerates

The strength of agglomerates (TI and CCS) under cold condition and their reduction properties are important parameters (Sec. 3.5.1) that affect the iron-making process. These are detailed in Chapter 4 (ref. **Sec. 4.3**).

3.6.5 Permeability of the Stack Zone

Higher permeability of the burden facilitates using a higher blast volume, which results in higher production. Important parameters that affect the burden permeability are:

a. **Content of fines** (usually below 5 mm) in the burden material. The burden materials, apart from the sized lump, may also contain fines that are carried with the burden (in view of material handling) and/or generated with the breakdown of burden materials during their reduction. A higher amount of fines in the burden will reduce its void fraction for the transport of the reduction gas and also tends to concentrate at the point of impact on the burden surface charging. The particles should not be so small that burden becomes impermeable to the furnace gas or an excessive amount of ore is blown out with top gas at high pressure. Ideally, the burden should flow down the stack zone of the furnace like a fluid in "plug flow," and the gases should pass through the burden with the same velocity all the way across the furnace.

   An aerodynamics analysis of the Bokaro and Durgapur blast furnaces shows the inverse correlation between the stack zone permeability and content of below 5-mm-sized fractions in the burden materials.[12]

b. **Size distribution** of burden materials. The burden should have appropriate particle size, close grain size distribution and favorable grain shape in its original condition.

   The proportion of fines to lump in the furnace burden as well as their size ratio do affect the burden permeability. **Figure 3.7** (top portion) schematically shows how the voids are filled when sized particles, large (l) and small (s), are packed at their different proportion viz.

   I and III: Uniform sized packing of large (I) and small (s) particles.

   II: Packing of mixed sized particles.

It is apparent from the figure that the voidage has the highest value in case of I and III, that is, with uniform-sized packing of spherical material: either large (l) or small (s). The voidage decreases when the large-sized particles (l) and small-sized particles (s) are mixed in different proportions, and it has the minimum value at the volume ratio Vs/Vl is around 0.3.that is, when the mixture contains about two-thirds of the large particles. This is schematically shown in **Figure 3.7** (bottom portion) and is represented by the dotted line. At the minimum porosity, the mixing ratio is such that the amount of small particles is just enough to fill the spaces formed between the large particles. It is evident that for a good and uniform permeability and low resistance to gas flow in a mixed bed, the size fractions of the particles should be as narrow as possible (**Figure 3.7**, see **Sec. 6.4.2** also).

Such representations in the case of sinter and ore are also schematically shown in **Figure 3.7**.

The charge to furnace consists of basically two components—namely, coke and (ferrous) burden (lump, sinter, and pellets)—of varying sizes in alternate layers. For

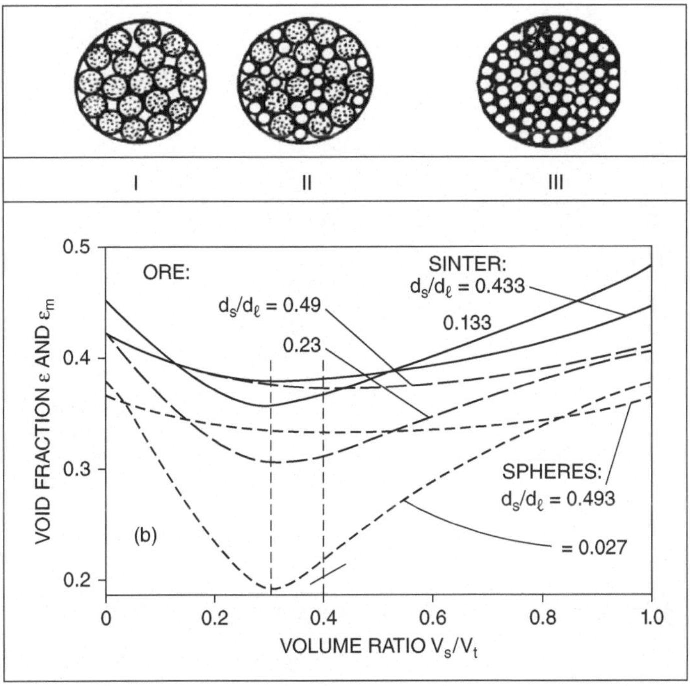

FIGURE 3.7 (Top) Packing of large (l) and small (s) sized particles at their different proportions (schematic) (Bottom) Schematic how the void fraction changes with volume ratio, V_s/V_l (after Ref. **[11]**)

each component, an equivalent or representative diameter can be calculated; hence, each layer could be considered as a bed of equivalent size from the furnace aerodynamic point: dB, and dC for mean size of burden and coke, respectively. It is reported that for improved aerodynamics of the stack zone, the ratio (dB/dC) should correspond to the optimum value,[11] which is around 0.25 for the Durgapur and Bokaro furnaces of Steel Authority of India Limited (SAIL), India.[12]

The resistance to flow of gases through such beds has been formulated by Carman, Ergun, and others (see **Sec. 6.3.3**). The Ergun's equation relating to the pressure drop per unit height of the bed with the flow of gas through the bed, neglecting the first term of (Eq. 6.3), can be represented by (and with modification):

$$\Delta P/h = \gamma[(1-\varepsilon)/\varepsilon^3][p/d]V_b^2 \tag{3.16}$$

where,

γ = friction factor,

ε represents the porosity or voidage and the fraction of bed volume occupied by a solid is (1-ε),

d, ρ = particle size and its density, and

V_b = gas flow rate.

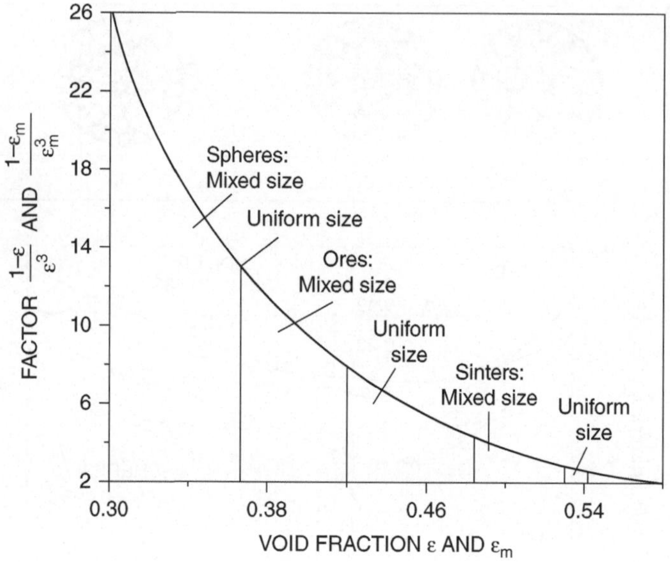

FIGURE 3.8 Void fraction of blast furnace charge material (after Ref. [11])

The factor $(1\text{-}\varepsilon)/\varepsilon^3$ when plotted against ε in relation to uniform and mixed beds (with subscript, m) of the furnace charge is represented by **Figure 3.8**.[11]

The BF productivity has been well correlated with the factor, ΔP*dB/(Vb*Vb)[11] where, ΔP: Pressure drop in the furnace, kg/cm^2

dB: Effective Size of the burden, mm, and

Vb: Blast volume, Nm^3/min.

Use of Agglomerates Sinter is subjected to abrasion following its transfer; also, during its charging into the blast furnace, fines are subsequently generated. Besides, the sinter consists of fines due to the poorer screen effectiveness. These fines adversely affect the furnace productivity and should be eliminated from the sinter before charging into the blast furnace. The empirical analysis of Bhilai (SAIL) blast furnace data substantiates that the elimination of fines content from the lump ore and sinter considerably influences the furnace productivity.[7,13]

The mean size of sinter produced in some of the sintering plants, say in India, however, is lower than that of lump ore. This incompatibility adversely affects the permeability of the stack zone and also makes a restriction for using a higher percentage of sinter in the burden (at the same coke size). Accordingly, improvement in the mean size of sinter is essential for improved productivity of the furnace[14] (**Sec. 7.9**).

Bed permeability of the stack zone is improved through the usage of pellets. This is because of their spherical shapes and open pores. The shape, size, and low angle of repose with respect to pellets provide conducing conditions for minimal segregation and an even charge distribution in the furnace, extending more toward the axis.

3.6.6 Permeability in Cohesive Zone

The pressure drop (ΔP) in cohesive zone is an important parameter and guided by the softening and meltdown characteristics of the burden materials. As soon as the burden material starts softening, its permeability is greatly reduced.

The shape, size, and location of the cohesive zone have a profound effect on

- *Furnace productivity*
- *Fuel consumption*
- *Hot metal quality*
- *Operational stability and lining life of the furnace*

A thinner cohesive zone is effective in lowering the pressure drop of the gas across it. Therefore, burden material should start melting at a relatively high temperature, and the interval between softening and melting should be as short as possible (so that they do not impede the gas flow). The configuration of cohesive zones is a function of the high temperature properties of the agglomerates vis-à-vis lump ore. Raw (lump) ores, depending upon their nature, soften and melt in a wide range of temperatures. The semi-viscous mass chokes the voids. A mismatch in the softening and meltdown temperature for lump iron ore and sinter is observed in some cases, especially in Indian blast furnaces. Lump ore starts softening at a lower temperature compared to sinter (**Sec. 3.5.1, Figure 3.5**). Sinter completes its melting at a higher temperature (than lump ore). In view of this, the width of cohesive zone increases. Hence, it is imperative to produce sinter with the softening–meltdown behavior matching with that of lump ore. Alternatively, the proportion of burden, lump ore, sinter, and pellet should be optimally selected so that, with their characteristics, a minimum of the resistance is faced by the ascending gases.

For a fluxed sinter as burden material:[6]

- The softening takes place in the temperature range 1,250-1,450°C.
- The melting takes place later: the typical temperature range is 1,400-1,500°C.

3.6.7 Formation of Slag and its Characteristics

All the gangue (non-Fe) constituents in the burden material constitute the slag following smelting in the blast furnace. The silica (SiO_2) and alumina (Al_2O_3) contents that enter with the iron ore, pellets, sinter, and coke are called "acidic" constituents of slag, whereas magnesium oxide (MgO) and calcium oxide (CaO) are termed as "basic." An important characteristic of slag is its basicity (B), which is determined as: $B = CaO/SiO_2$.

The formation of slag, its amount, and characteristics have a deciding effect on the furnace productivity, coke rate, and the quality of produced hot metal. The volume of slag should be kept minimum as enough heat (at the cost of BF coke) is required to melt it and maintain its fluidity. This calls for lower input of gangues through burden materials.

Important ***characteristics*** of the slag that affect the blast furnace performance are:

- **Chemical composition:** It is important for smooth functioning of the blast furnace that slag composition should be uniform in chemical composition.
- **Slag viscosity:** A highly viscous slag affects the permeability of the bosh zone. Hence, the formation of slag should be confined to a limited height (in the lower portion) of the furnace and should be stable.

  A higher percentage of alumina in slag makes it more viscous. This adversely affects the blast furnace working conditions by:

  a. Increasing the thermal load

  b. Decreasing the desulfurization capacity of slag.

  The slag viscosity can be lowered by raising the hearth temperature. However, this results in an increase of silicon in iron, and the steel-making process becomes difficult and uneconomical.
- **Slag fluidity:** The slag should be only fluid to provide good permeability in the zone of slag formation. The fluidity of slag can be increased by:
  - Diluting the slag through the addition of quartzite in burden, thereby adjusting the alumina-silica ratio in slag. This means a further increase in slag volume.
  - Higher percentage of MgO in slag and incorporation of MgO through sinter is preferable.[9]
- **Desulfurization capacity:** The slag in contact with hot metal removes its sulfur in the furnace itself through the reactions of its basic oxides' constituents, CaO and MgO:

  FeS + Ca (Mg)O = Ca (Mg) S + FeO, where the Ca (Mg) S becomes part of the slag.

  The correct combination of MgO and basicity ratio of slag provides it with the significant desulfurization capacity. The optimum slag volume is the minimum volume of the slag that is needed for desulfurization and removal of other impurities.

  The sulfur content of hot metal depends on the sulfur distribution ratio:

  [S] = 0.1 S/(1+0.001*L_s*V) %, where, S = sulfur load (sulfur input minus sulfur leaving the furnace with the gas), kg/THM; Ls = (S)/[S] the sulfur distribution ratio between slag and hot metal; V = slag volume, kG/THM.

Slag Basicity and MgO: The adverse effect of alumina content in slag on its viscosity can be reduced by choosing the correct composition of slag in terms of its basicity and MgO. It is also possible to modify the characteristics of slag by selecting the right combination of its MgO and basicity for enhanced desulfurization of hot metal. This is important and more preferable than external desulfurization for producing high quality hot metal. The desulfurization capacity increases with the increase in basicity of slag and the decrease in the slag viscosity.

A better slag characteristic can be achieved by incorporating higher MgO in slag without increasing slag basicity and Si in hot metal.[9] The MgO content increases the melting point of slag formed from gangue minerals.[15]

Requirement for Slag For smooth functioning of a blast furnace, the slag must have:

a. Minimum viscosity
b. Constant desulfurization capacity
c. Constant liquidus temperature

Usage of Fluxed Sinter: Fluxed sinter happens to be a convenient feedstock in the blast furnace iron making. The required amount of CaO and MgO can be incorporated through its charging in the furnace in order to modify the chemistry of slag for its desulfurization capacity and fluidity. Primary slag formed from fluxed sinter possesses lower viscosity and lower liquidus temperature and also a more uniform composition than in the case of raw ore and pellets or acid sinter. Hence, the use of fluxed sinter permits smoother operation of the furnace. This leads to lower thermal load in the bosh/hearth region. The slag characterization is partly attained in the sinter itself so that the transformation to the final slag is rapid but gradual and occurs at a relatively lower and narrower temperature.

Usage of Fluxed Pellets: Thickness of the cohesive zone decreases when fluxed pellets are charged into the furnace. This allows more gases to diffuse through the coke layers. The effects are (ref. **Sec. 4.2.2**): (a) decrease in the pressure drop across the bed, and (b) higher utilization of CO gas. This leads to increased productivity and decreased coke rate.

REFERENCES

1. World Direct Reduction statistics (2013), direct from Midrex, 2nd quarter, February 2014.
2. T. Umadevi, P. P. Kumar, M. G. Sampath Kumar, S. Kumar and M. Ranjan, Influence of carbon addition via Corex sludge on pellet quality at JSW Steel, *Iron Making & Steel Making*, 2008, 36, 421–429.
3. O. Kubaschewski and C. B. Alkock, *Metallurgical Thermochemistry*, 5th ed., Oxford: Permagon Press, 1979.
4. K. K. Prasad and H. S. Ray, *Advances in Rotary Kiln Sponge Iron Plant*, New Delhi: New Age International Publishers, 2009, Chapter 3.
5. Indian Bureau of Mines (IBM), Iron and Steel Vision–2020, Chapter 6 (Iron Making), Publication by IBM Nagpur, India, 115-119. (downloaded November 2018).
6. U. S. Yadav, D. N. Jha, and B. K. Das, "Influence of sinter proportion in burden on its softening-melting characteristics," *Tata Search*, 2006, 103.
7. R. P. Bhagat, H. S. Ray, and S. K. Gupta, Improvement in performance of the blast furnace at Bhilai through statistical analysis and improvement in burden preparation *ISIJ International*, 1991, 31, 669–676.
8. R. P. Bhagat, Statistical analysis of plant and experimental data—success & failures," *Proc. Short Term Course on Quantitative Approach in Process Metallurgy* (QAPM), Bhubaneswar, October 26–28, 1995, org. by RRL Bhubaneswar & IIM, A. K. Tripathi, P. Datta, and H. S. Ray, Eds., New Delhi: Allied Publ., 1995, 242–265.

9. R. N. Singh, U. K. Banerjee, S. Dharnipalan, and S. K. Gupta, "High MgO slag practice under Indian conditions," *Trans. IIM*, April 1979, 32(2), 90–100.
10. J. E. Astier, J. Michard, and P. Dancoishe, *JISI*, October 1961, 199, 143–148.
11. R. Jeschar, W. Potkee, V. Petersen, and K. Polthier, "Blast furnace aerodynamics," *Proc. Symp. Aus I.M.M.*, N. Standish, Ed., Wollongong, Australia, 1975, 136–147.
12. S. S. Gupta and S. K. Gupta *Steel India*, 1979, 2(2), 60–65.
13. R. P. Bhagat, H. S. Ray, and S. K. Gupta, "Performance analysis of a blast furnace at Bhilai Steel Plant" in *Progress in Metallurgical Research: Fundamental and Applied Aspects*, S. P. Mehrotra and T. R. Ramchandran, Eds., New Delhi: Tata-McGraw Hill Pub., 1986, 175–184.
14. R. P. Bhagat, "Effect of screening and recirculating fines from the blast furnace sinter on sinter productivity vis-à-vis improvement in mean size," *SEAISI Quarterly*, 1992, 21(4), 47–53.
15. Q.-J. Gao, F.-M. Shen, G. Wei, X. Jiang, and H.-Y. Zheng, "Effect of MgO containing additive on low temperature metallurgical properties of oxidized pellet," *J. Iron & Steel Research International*, 2013, 20(7), 25–28.

4 Agglomerates in Iron-Making Processes

ABBREVIATIONS

BF *Blast furnace*
DRI *Direct reduced iron*
PCI *Pulverized coal injection*
RDI *Reduction-degradation index*
RI *Reducibility index*
SAIL *Steel Authority of India Limited*
TI *Tumbler index*

4.1 PREAMBLE

The performance of iron making, through the blast furnace (BF) route in particular, is decided by:

a. Improvement in its productivity
b. Savings in coke rate
c. Improved quality of hot metal/direct reduced iron (DRI)

The iron-making process is highly capital intensive; and hence, any improvement in productivity increases its profitability. The iron-making process through the blast furnace route uses coke, which is produced through the carbonization of coking coal. This commodity is costly and is limited in resources. Hence, much attention is paid toward the savings in coke rate. Hot metal/DRI is feedstock of the steel making and its quality is equally important to produce steel at cheaper cost. The hot metal/DRI should have minimum sulfur so that the desulfurization step, which is otherwise required, is avoided.

The importance of agglomerates (sinter and pellets) in iron making lies in the fact that most of the commercially exploited processes use these as feed material (besides lump ore); hence, the quality parameters of the agglomerates decide the performance of iron making. The phenomenal growth of blast furnaces in dimension could be possible through the improved technology in production of agglomerates with superior qualities as regards:

- High mechanical strength and favorable aerodynamic characteristics
- High resistance to low-temperature breakdown during reduction
- High reducibility, also compatible kinetics of reduction with ascending reducing gases
- High and uniform softening temperature and narrow range of softening

4.2 AGGLOMERATES IN IRON-MAKING PROCESSES

4.2.1 Iron Ore Pellets in Direct Reduction (DR)

Iron ore pellets, besides lump ore, are used in the direct reduction processes, especially in the gas-based shaft furnace. The MIDREX direct reduction iron-making process (referred to as the MIDREX process) was developed and commercially introduced exclusively on the basis of the usage of pellets as feedstock. The pellets, when charged into the kiln/furnace, improve its performance. The contributing factors are:

a. Compared to lump ore, the usage of pellets in the process yields better metallization; Fe content in the DRI is as high as 95%.
b. Pellets have a limited size range and possess practically no fines. These have higher and more or less uniform mechanical strength and high abrasion strength. Besides, the pellets have lower gangue and higher iron content (Fe more than 63 wt. %) compared to sinter as it is made of beneficiated ore. The pellets usually have low phosphorous and sulfur contents.
c. The ore pellets undergo faster reduction reaction as these have higher porosity. The pellets, because of their spherical shape and high Fe content, undergo reduction reaction in the rotary kiln with less accretion. Besides, the formation of fused lump is minimum.
d. These factors, mentioned here, result in:
   - An increase in the rate of DRI (sponge iron) production by 25% to 30%, with the same amount of fuel, compared to lump ore.
   - A reduction in specific consumption of coal.
   - A longer campaign life of the furnace and reduced refractory cost are observed in case the pellets are used as feedstock in the DR process. This is on account of the previously mentioned factors.

Pellets have, however, the tendency to form clusters at higher temperatures, which affects the maximum bustle temperature that could be used for DRI production. Consequently, the shaft productivity is adversely affected. The sticking increases with the increase in Fe content and the decrease in gangue content of pellets. Therefore, high-quality DR pellets have a high sticking tendency.

The acidic oxides, SiO_2 and Al_2O_3, in the pellets have an adverse effect on the operational cost of DRI production. Every percent of acid gangue addition in the pellets is associated with an increased energy consumption of 30 kWh.[1]

4.2.2 Iron Ore Pellets in BF Iron Making

The usage of pellets in blast furnace iron making has merits and demerits as follows.

Merits The usage of pellets in comparison to lump ore or sinter results in improved furnace productivity and reduction in coke. This is on account of:

a. Slag volume and thermal load of the furnace decreases when pellets are charged. This is on account of their higher iron content (more than 63 wt. %) and uniform chemical composition.

b. Bed permeability of the stack zone is improved through the usage of pellets. This results in the improved furnace productivity. The contributing factors are:
   - The pellets have spherical shapes, are uniform, and have a close size range (9-15 mm in diameter). Compared to sinter, the low angle of repose (see **Table 16.1**) has been observed in the case of ore pellets. These provide conducing conditions for minimal segregation and an even charge distribution in the furnace, extending more toward the axis. However, when charged along with sinter, pellets have some drawbacks (see next paragraph).
   - Pellets have sufficient and high mechanical strength during transportation and are also able to withstand the impact on its charging. The effect is lesser fines content in pellets when these are charged into the furnace. Besides, the pellets can withstand more thermal stress under reducing conditions (lower RDI). The effect is less fines generation within the blast furnace stack.

c. **Figure 4.1** shows the JIS reducibility of different burden materials.[2a] The reducibility has been determined according JIS M8713.[2b] Pellets have greater reducibility compared to sinter and lump ore (**Figure 4.1**)[2a] due to:
   - Their uniform mineralogical composition in the form of an easily reducible hematite or hematite-bearing compounds. The basic pellets have higher reducibility (Figure 4.1) due to the formation of calcium ferrites and hematite phases through diffusion.
   - High and even micro-porosity 25–30% as compared to lump ore. The reducing gas, carbon monoxide, enters the pellets via the pores and affects the removal of oxygen.
   - Higher bulk density of pellets (3–3.5 t/m³) helps in improved heat exchange due to the longer duration of gas solid contact.

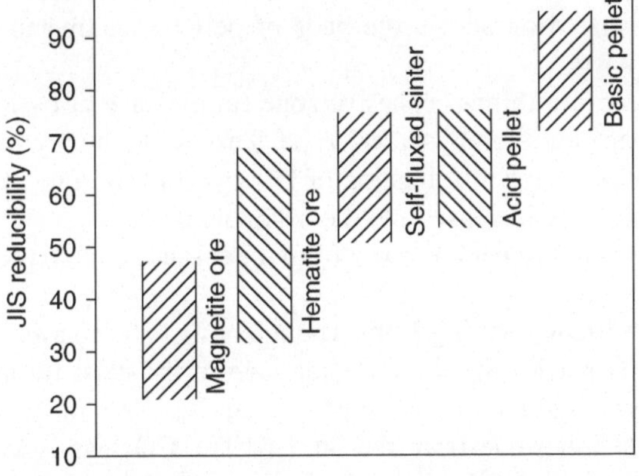

FIGURE 4.1 Comparison of reducibility values of ore, sinter, and pellets (JIS: after Ref. [2a])

- A better solid-gas contact is experienced in the case of pellets. This facilitates reduction, especially with pulverized coal injection (PCI), which is widely practiced in the Chinese mini blast furnaces.

These factors result in reduced specific consumption of coke and improvement in the furnace productivity.

Demerits The pellets have high swelling (over 20%) during reduction, particularly for acidic ones (see **Sec. 16.6**). Besides, charging pellets in conjunction with sinter in the blast furnace suffers from problems with respect to their segregation in the blast furnace top. The pellets have a lower angle of repose (see **Table 16.1**) than that of sinter; therefore, they have a tendency to flow toward the center of the furnace, while the sinter flows toward its periphery. This causes unstable operation. Additionally, the pellets have a tendency to sink into the coke layer, due to its higher bulk density, during burden descent.

Acidic versus Basic Pellets An increase in the silica content of the pellet charge to blast furnace leads to an increase in the unit cost of steel production. The acidic pellets have high swelling (over 20%) during reduction. On the other hand, the basic pellets demonstrate improved metallurgical properties in terms of reducibility and swelling and hence do have a positive effect in the blast furnace iron reduction. The addition of MgO helps in improving the high temperature properties due to the formation of high melting point slag and magnesiowustite phase during reduction.[3]

The effects of charging basic pellet burden vis-à-vis the acidic pellets are:[4]

a. The size and shape of cohesive zone changes when fluxed pellets are charged into the furnace compared to acidic pellets. The cohesive zone of the blast furnace becomes thinner when the fluxed pellet burden is used compared to acidic pellets. This allows more (reductant) gases to diffuse through the coke layers. The effect is also a decrease in the pressure drop across the beds of pellets, leading to an increased productivity.
b. The effect of a thinner cohesive zone (in the case of basic pellets) provides space for a lumpy (stack) zone. That is, the lumpy zone increases, which allows higher utilization of CO gas (in this zone). Consequently, shaft efficiency increases and the coke rate decreases.
c. Coke slits are thinner and away from the furnace wall in the case of basic pellets.

Restriction in Incorporating Lime Though basic pellets have improved metallurgical properties that favor their usage in the blast furnace, there are restrictions on the incorporation of the requisite amount of lime required to eliminate limestone from the burden. The CO_2 gas is evolved during induration of green pellets when it consists of calcite and magnesite components. This reduces the crushing strength of fired pellets. Therefore, though some CaO/MgO can be incorporated in the pellets that may not be sufficient to incorporate all that is required (through pellets) to raise their basicity sufficient in order to neutralize the slag, especially in the case of higher acidic gangue constituents in the furnace burden. Hence, to eliminate limestone/dolomite from the BF burden, charging it through sinter is essential in these cases.

4.2.3 Iron Ore Sinter in BF Iron Making

Sinter happens to be a predominant feedstock in the iron-making process through the blast furnace route. One important aspect of the usage of the sinter in iron making lies in the fact that while the lump ore, as burden material, is the product of nature with inherent variability in physical and chemical properties, the sinter quality can be regulated through the control of sintering condition and its requisite chemistry, which makes them important for their usage in a BF as burden.

Charging a higher percentage of iron-bearing materials through fluxed sinter results in better productivity and decreased coke rate because:

a. Burden weight decreases when direct charging of raw limestone and dolomite in the blast furnace is avoided (see **Sec. 3.6.2**) through the use of fluxed sinter. The carbonates, when (directly) charged into the furnace, add to the burden as these contains CO_2.
b. Lime and MgO in an amount, which is required in neutralizing the furnace slag, can be incorporated through sinter at high basicity (CaO/SiO_2). The direct benefit of charging the fluxed sinter in the blast furnace is based on the fact that the calcination reactions in the furnace, which are highly endothermic, are eliminated.
c. Besides, higher utilization of CO gas is expected through incorporating fluxed sinter/pellets as burden materials. This is because limestone/dolomite, in its original form, is eliminated through burden. The generation of decomposed CO_2 through the calcination ($CaCO_3 = CaO + CO_2$) reaction in the stack zone is absent.[5] In view of this, the shaft efficiency increases as the solution loss reaction ($CO_2 + C = 2CO$) is less. In addition to that, the temperature of stack gas is lower (see **Sec. 3.5.2**).

   Theoretical calculation shows that approximately 60–70 kg of C are saved per 100 kg of CO_2 removed or per 230 kg of $CaCO_3$ through incorporation of more flux in sinter.[5]
d. More uniform material distribution inside the furnace is achieved when sinter is used in the burden. This is because the angle of repose of sinter is less than that of lump (Table 2.5). This results in more uniform gas flow along the furnace axis. Sinter as burden has a higher void-age due to its irregular shape and size compared to lump.

   All these result in good bed permeability in the stack zone.
e. Fluxed sinter softens and melts at a higher temperature and in a narrower vertical zone. This results in improvement in high temperature gas permeability (in the bosh/hearth region). Dolomitized sinter has a higher softening

temperature than limey sinter. So far, the best high temperature gas permeability is achieved with fluxed sinter.

Fluxed sinter permits the use of high blast temperature and smoother furnace operation for the same reason.

f. The slag formation is partly attained in the sinter itself so that the transformation to the final slag is rapid but gradual and occurs at a relatively lower and narrower temperature. Primary slag formed from fluxed sinter possesses (a) lower viscosity, (b) lower liquidus temperature, and (c) more uniform composition than in the case of lump ore or/and pellets.

g. The desulfurization capacity of slag improves when fluxed sinter is charged in the furnace. Typically, the use of super-fluxed sinter results in active sulfur adsorption by the slag all over the path of its formation.[6]

4.3 PROPERTIES OF AGGLOMERATES

4.3.1 Cold Strength

A higher tumbler strength (TI) of burden materials is essential to ensure lesser generation of fines during their transportation and degradation inside the furnace on charging. Higher generation of fines in the stack region decreases the burden permeability and sometimes results in hanging on account of poorer gas permeability.

Pellets are hard to disintegrate during handling in cold condition, also during reduction. The sinter has less strength; hence, the amount of generated fines during its transportation from sinter plant to blast furnace is higher. This may require screening of the agglomerates at the skip point and recycling the fines to the sinter plant. Obviously, a higher recycling load proportionately decreases the input of fresh material to the sinter plant, thereby adversely affecting its productivity.

4.3.2 Reduction Degradation Index (RDI)

The degradation characteristic of the agglomerates while it is being reduced is important as it affects the permeability of the stack zone due to more generation of fines during reduction. This characteristic of agglomerates is referred as the reduction degradation index (RDI). Sinter has a much higher RDI (20–30%) than pellets (6–8%). Hence, the sinter RDI is of much concern with respect to blast furnace iron making.

The coke rate and furnace productivity are greatly influenced by the sinter RDI. This has been observed in the case of the Indian (SAIL- Bhilai, Tata Steel)[7] and Schwelgern and Ruhort (No. 6) blast furnaces. Nishio et al.[8] have reported a relationship between the RDI value of sinter and the permeability of Fukuyama BF No. 5. The permeability index decreases linearly with an increase in RDI value.

Understandably, the fines generated inside the furnace during reduction affect the permeability of the stack zone, increase the pressure drop, and disturb the gas distribution causing channeling of the furnace. All these factors result in a decrease in driving rate and adversely affect the CO utilization with the ultimate consequence of higher coke rate and lower productivity. A higher sinter RDI adversely also affects the size and shape of the furnace bosh region as higher generated fines in the burden impede the gas flow across this viscous zone. Besides, the desulfurization capacity of slag is affected.

It is important to produce the agglomerates that have lower RDI by modifying their chemistry and the sintering conditions.

4.3.3 Reducibility

The kinetics of reduction reaction are controlled by:

a. *Gas-Solid Contact.* The contact is between gas and the burden material in bulk phase as well as the diffusion of gas into the particles. The governing factors are: layer structure of the burden, porosity, and permeability of the layers.
b. *Reducibility of Agglomerates.* Reducibility of burden material is the ease with which the oxygen combined with it can be removed. The burden materials that descend through the furnace need to be quickly reduced with the ascending reducing gas, preferably at a lower temperature of the furnace (stack zone). In the case of poorer burden reducibility, the reduction is pushed toward the lower part of the furnace (in other words, this does not take place in the upper part) where CO_2 is not stable and converted to CO; thus it escapes with BF gas (see **Sec. 3.5.2**).

The reducibility index (RI) of burden materials is a major contributory factor to the gas efficiency in the BF. A higher reducibility of sinter/pellet provides the following in the blast furnace, that results in lower coke rate and higher productivity:

- A greater extent of indirect reduction
- Less solution loss reaction ($CO_2 + C = 2CO$)
- A decrease in residence time

Fluxed Sinter High gas efficiency is achieved in the BF with fluxed sinter. This is on account of higher CO utilization, which leads to a greater amount of CO_2 in the stack gas (see **Sec. 3.5.2**). The indirect reduction increases when the fluxed sinter is charged in the furnace. This results in lower thermal load in the bosh/hearth region and lessening of the stack gas temperature. It is estimated that approximately 6 to 7 kg C are saved from every 1% increase in indirect reduction.

4.3.4 Softening and Meltdown Characteristics

Softening and meltdown characteristics of burden materials affect the gas permeability in the lower region of the blast furnace (cohesive zone). A later start of softening shifts the softening phenomenon to much lower in this zone. Besides, a short interval between the softening and melting makes the cohesive zone thinner. These are important in improving the permeability in this zone. The burden material should start melting at a relatively high temperature, and the interval between softening and melting should be as short as possible, so that they do not impede the gas flow.

With respect to the softening and meltdown temperature of burden materials, a wide difference exists between the lump ore and sinter. In contrast to ore, the fluxed sinter softens and melts at a higher temperature and also in a narrower vertical zone (see Figure 3.5 also).

Typical figures (at 40% partial reduction) are[10]

Ore: Softening starts at 900°C and completes at 1,225°C.
Sinter: Softening starts at 1,075°C and completes at 1,260°C.

In order to reduce the difference between the softening and melting range, it is essential to:

- Change the proportion of ore vis-à-vis sinter
- Modify the softening characteristics of sinter

Effect of Alumina Content The reducibility index (RI) of sinter increases with the decrease in its alumina content. This is on account of:

- The formation of melt takes place at a higher temperature, exceeding 1,150°C during sintering.
- The amount of FeO containing melt decreases.
- The chance of pores being stepped up as the melt diminishes.

Yamaguchi et al.[9] have studied the influence of alumina on the reduction and meltdown behavior of sinter with reference to Kimitsu No. 3 and No. 4 blast furnaces. A decrease in Al_2O_3 content in the sinter to 1.50% compared to the base level of 1.85% increases the shaft efficiency $CO_2/(CO + CO_2)$ (see **Sec. 3.5.2**) due to an increase in the JIS-RI (reducibility index) by 5% in the stack (lumpy) zone of the blast furnace. The reducibility index was determined according to the Japanese Standards Association JIS M8713 (see Ref. [2b]).

4.4 QUANTITATIVE EFFECT ON BLAST FURNACE PERFORMANCE

An iron blast furnace can be considered a multivariate system, which is subjected to a large number of variables; many of them are inter influencing and may act in an uncontrolled manner. The criteria for assessing quality parameters of agglomerates in iron making on the basis of thermodynamic principles and mathematical modeling today are still not completely defined, and the operator has to rely on the empirical relationship derived from the analysis of operative data. One of the possible reasons for this is the fact that the furnace performance is material specific.

The empirical relationship for the response, coke rate, and productivity can be derived using statistical techniques to assess the quality parameters of sinter/pellets. The effect of variables on the response can be well represented to a certain degree of accuracy by a set of regression equations. Data that serve as the basis for the regression analysis can be obtained from operating log sheets and/or specifically designed experiments. However, experimentation on a complex system, such as an iron blast furnace, may not be permitted because of cost and time factors. Quite often, inferences need to be drawn from the operating log data physical model. The regression equations, established through the analysis of operating data, are of considerable value in a number of ways:[10,11]

- They provide a standard to appraise the current performance of a blast furnace by taking into account the external conditions. Though these are related to a particular furnace, the inferences drawn provide a firsthand guideline for the operators in the case of other blast furnaces' operation.
- They help in estimating the effect of changes in raw material supplies, burden preparation operation particles, and so on. These are perhaps not accountable on purely thermochemical treatments.
- Studies of variations in furnace economy and productivity by these techniques may assist in making comparisons of different furnaces.

Iron Content in Burden Empirical relations show that the coke rate decreases and the furnace productivity increases with the increase in Fe content of the burden. It has been estimated that with a 1% increase of iron concentration in the burden, the blast furnace productivity increases by 1.5–2.5% and the coke rate decreases by 0.8–1.2%.[12, 13]

The effect of increasing the Fe content of ferrous burden on the coke rate and furnace productivity has been represented by the following relationship.[14a] shown in **Table 4.1**.

Alumina Content in Burden Alumina in burden materials increases the slag viscosity. It is reported that a 1% increase in alumina increases the coke rate by ~2.25%, decreases the furnace productivity by ~4%, and increases the flux consumption by ~30 kg/THM.[14b] Similarly, in the COREX iron-making unit, 1% alumina increases the fuel rate by ~8 kg/THM, decreases the melting rate by 0.75 TPH, and increases the slag rate by ~8 kg/THM.[14c]

TABLE 4.1
Changes in Coke Rate and BF Productivity when Burden Fe Increases by 1%

Fe Content in BF Burden	Decrease in Coke Rate	Increases in Productivity
(50–55)%	(1.4–1.7)%	(2.2–2.6)%
(55–60)%	(1.1–1.5)%	(1.8–2.2)%

TABLE 4.2
Changes in Coke Rate and BF Productivity when Sinter in Burden Increases by 10%

Sinter in BF Burden	Decrease in Coke Rate	Increases in Productivity
(41–60)%	2.2%	2.3%
(61–80)%	2.0%	1.9%

Use of Self-Fluxing Sinter The benefits in terms of a decrease in coke rate and an increase in productivity that are achieved on switching over from raw lean ores to self-fluxing sinter have been realized since the early days of production.

The regression equation suggests that with an increase in sinter percentage in blast furnace burden, the coke rate decreases.[15]

$$K = 470 + 0.164B + 0.249L - 1.006S - 1.142P - 0.093T \quad (4.1)$$

where,

K = coke consumption rate (kg/thm),
B = net burden weight (kg/thm),

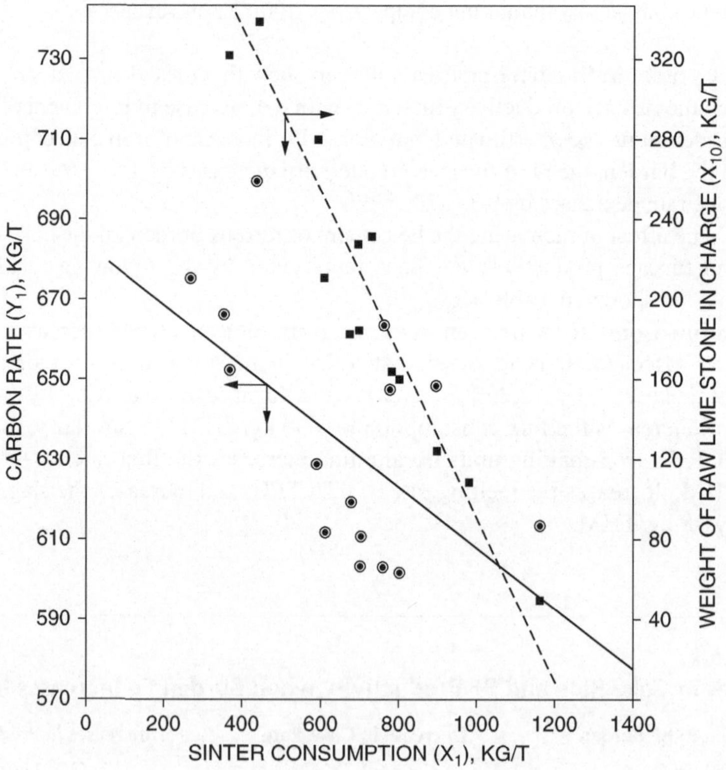

FIGURE 4.2 Effect of sinter consumption on weight of raw limestone and carbon rate (after Ref. [13])

L = weight of lime stone charged (kg/thm),
S = sinter percentage of gross burden weight,
P = pellet percentage of gross burden weight, and
T = blast temperature (c).

Another relationship showing the effect of increasing the proportion of sinter on coke rate and furnace productivity with reference to Russian furnaces[14a] is given in **Table 4.2**.

In relation to the Bhilai (SAIL) furnace, an empirical relationship has been established on the basis of 16 years of monthly average working data; this shows that around 20 kg of carbon/THM could be saved by eliminating 100 kg of raw limestone from the burden. **Figure 4.2** shows that the parameters of specific consumption of sinter (X1, kg/t) and the weight of limestone in the burden (X10, kg/t) were closely correlated. It also shows that the carbon rate (Y1, kg/t) is reduced with increasing sinter consumption in the blast furnace burden. Following relationships have been obtained (R = correlation coefficient):

$$X10 = 489 - 0.39 * X1 (R = 93\%), \text{ and } Y1 = 684 - 0.073 * X1 (R = 47\%)$$

REFERENCES

1. R. H. Heerema, H. Kortmann, T. Kater, and V. C. ven den Boogaard, "Improvements of acid, olivine and dolomite fluxed iron ore pellets using an organic binder," in *Proceedings of Fifth International Symposium on Agglomeration*, Brighton, 1989, P.A. Yong, Ed., The Inst. of Chem. Engineers, Publ., Rugby, U.K., 227.

2a. G. Suzuki, T. Satoch, and N. Sakamoto, "Use of iron ore pellets in Japan," National Seminar on Role of Iron Ore Pellets in Iron & Steel Industry. Under the aegis of Iron & Steel Davison of Indian Institute of Metals. Ranchi, April 15–16 1988. Org. by Steel Authority of India Limited, Kudremukh Iron Ore Company Limited, Mandovi Pellets Limited.

2b. Japanese Standards Association (JSA), Iron ores – Determination of the reducibility by the final degree reduction index (JIS M8713), 2009.

3. O. Sivrikaya, A. I. Arol, T. Eisele, and S. K. Kawatra, "The effect of calcined colemanite addition on the mechanical strength of magnetite pellets produced with organic binders," *Mineral Processing Extractive Metall. Rev.*, 2013, 34, 210–222.

4. J. J. Poveromo, "Blast furnace iron making and burden requirements," *COREM Pelletising Symposium*, Quebec City, September 2003.

5. A. K. Biswas, *Principles of Blast Furnace Iron Making: Theory and Practice*, chapter 10, CITY: Cootha Publication House, Brisbane, Australia, 1981.

6. H. Bonnekamp, H. D. Hass, W. Kleppe, and B. Pfotschner, "Plant trials on factors determining sinter quality and resulting blast furnace performance," *Iron and Steel Maker*, June 1980, 17–24.

7. A. Kumar, S. N. Sinha, and U. S. Yadav, *Proc. Trends in Beneficiation and Agglomeration of Iron Ore* (CORAS 2001), September 25–26, 2001, SAIL-Ranchi, 149–155.

8. G. Tiwari, Y. Venkateshwarlu, R. Mohanty, U. N. Behra: *Proc. Trends in Beneficiation and Agglomeration of Iron Ore* (CORAS 2001), 25–26 September, 2001, SAIL-Ranchi.

9. Y. Kajuyoshi, H. Ueno, T. Kawaguchi, S. Matsunaga, H. Oda, and S. Amano, "Influence of Al_2O_3 on reduction-meltdown behavior of sinter in blast furnace," *ISIJ International*, 1994, 34(12), 664.
10. R. V. Flint, "Effect of burden materials and practices on blast furnace coke rare," *Blast Furnace and Steel Plant*, 1962, 50(47), 47–58, 74–76.
11. J. M. Ridgion and A. M. Whitehouse, *J Iron & Steel Inst.*, 1956, 249–257.
12. E. F. Wegman, *A Reference Book for Blast Furnace Operators*, Moscow: Mir. Publ., 1984, 202.
13. R. P. Bhagat, H. S. Ray, and S. K. Gupta, "Improvement in performance of the blast furnace at Bhilai through statistical analysis and improvement in burden preparation," *ISIJ Inter.*, 1991, 31(7), 669–676.

14a. Y. F. Nikoulin, Private communication, R. K. Verma et al., Steel India, October 1987.

14b. Pradip, "Beneficiation of alumina rich Indian iron ore slimes," *Metals, Materials and Processes*, 1995, 6(3), 179–194.

14c. JSW Steel, Internal communication (Ref. S. Dwarapudi, S. Mohan Rao, and N.D. Rao, "Pelletization of high alumina iron ore fines: problems and prospects," *Tata Search, 2007*, 87–95) (ref. 11 of the paper).

15. L. V. Bogdandy, G. Lange, and P. Heinrich, "Prospects of development and limits of the blast furnace," *Stahl Eisen*, 1968, 88, 1177–1188.

5 Process of Sintering

ABBREVIATIONS

BF	*Blast furnace*
BOF	*Basic oxygen furnace*
BTP	*Burn-through point*
BTT	*Burn-through temperature*
CAP	*Composite agglomeration process*
CCS	*Cold crushing strength*
DPR	*Detailed project report*
DRI	*Direct reduced iron*
EOS	*Energy optimization sintering*
HPS	*Hybrid-pelletization-sintering*
ISP	*Integrated steel plant*
LD	*Linz-Donawitz (basic oxygen steel-making process)*
LOI	*Loss on ignition*
MEBIOS	*Mesoic-embedding iron ore sintering*
NKK	*Nippon Kokan Kabushiki Kaisha*
PCDD	*Polychlorinated dibenzo-paradioxins*
PCDF	*Polychlorinated dibenzo-furans*
RDI	*Reduction-degradation index*
RF	*Return fines*
RI	*Reducibility index*
ROM	*Run of mines*
VSS	*Vertical speed of sintering*

5.1 PREAMBLE

In the process of sintering, a shallow bed of the particles constituting the sinter mix is agglomerated by heat exchange and partial fusion of the quiescent mass. The sinter mix, besides the ore fines, usually contains materials like flue dust, return fines, limestone and/or dolomite, etc., in order to control the properties of the product as well as to utilize industrial waste. Heat is generated by combustion of a solid fuel admixed with the iron ore fines being agglomerated. The combustion is initiated by igniting the fuel exposed at the surface of the bed after which a narrow, high-temperature zone is caused to move through the bed by an induced draft applied at the bottom of bed.

The bonding is affected by a combination of fusion, grain growth, and slag liquidation. The generation of volatiles from the fuel and flux stone creates a frothy condition, and the incoming air quenches and solidifies the rear edge of the advancing fusion zone. The product "sinter" consists of a cellular mass of ore bonded in a slag matrix.

The sinter (cake) is subsequently crushed and sized. Normally, above 6.3/10 mm-sized fraction of sinter is used in iron making, while below 6.3/10 mm-sized fraction is recycled to the sintering plant as "return sinter fines" and constitutes the part of sinter mix material. The return fines have a lower fusion temperature than the rest of the raw mix and serve as seeds for initiating the sintering process.

The sintering process consists of the passage of a heat and reaction front through a packed bed of solids. It is an extremely efficient heat transfer process in which air entering the bed is heated by the finished sinter until it attains its peak temperature in the combustion zone. The gases leaving this zone yield their heat to the feed below. The physicochemical reactions in the sintering process occur in the solid stage. Hence, the constituents of the sinter product do not transform to the liquid stage. This necessitates the complete homogenization of the material before it is transferred to the sinter machine.

5.2 THE SINTERING PROCESS

5.2.1 Description of a Sinter Plant

The continuous sintering process is carried out on a traveling grate of the sinter machine. The machine mainly used for sintering nowadays is a Dwight-Lloyd sinter strand, which is a moving conveyor of perforated pallets. **Figure 5.1** shows a simplified diagram of a sinter plant.[1] The machine conveys a bed of prepared sinter mix over the traveling grate. Metal bars or rods are inserted longitudinally along the grate to enhance bed permeability and also to support the mix.

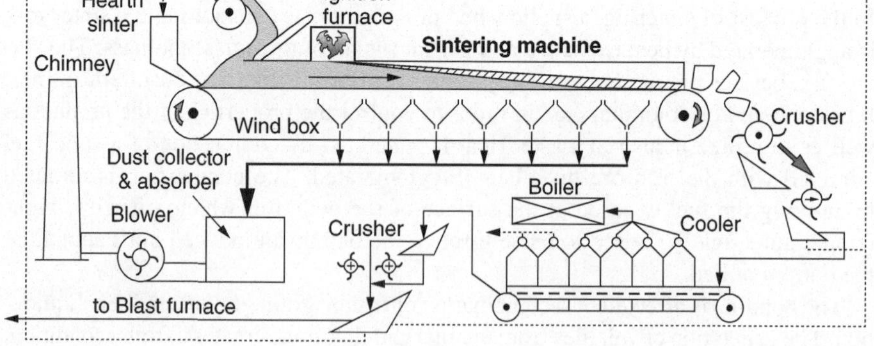

FIGURE 5.1 Simplified diagram of a sinter plant (after Ref. [1])

5.2.2 Raw Mix Preparation

A sinter plant receives iron ore fines having size below 6.3/10 mm. The ore fines, after processing in the mine (see **Chapter 2**), are stored in a premixing system, the bedding and blending yard of the sinter plant. The layer-wise stacking of ores and reclaiming is carried out in the yard to keep the composition variations at a minimum. Other raw materials, namely, flux and coke (each ground to below 3 mm in size), as well as return fines, constitute the sinter raw mix.

The return fines are undersized sinter (less than 6.3/10 mm in size), which is produced in the sintering plant after screening the sinter (mass) and recirculated to the plant. They have a lower fusion temperature than the raw mix and serve as seeds for initiation of the sintering process.

Coke breeze is used as a solid fuel in sintering. The size fraction of coke below 10 mm, that is generated in the coke ovens, is known as coke breeze. It is crushed to less than 3.15 mm in size for its use in sinter making. The coke breeze of a finer size (below 0.25 mm) as well as of a bigger size (above 3.15 mm) is not good for sinter making as incorporation of these fractions in the sinter mix causes a mismatch of flame front and heat front (see **Sec. 6.6.2**). The 3.15–0.25 mm size range of coke breeze, due to its superior performance, is termed as "golden size" by the sinter maker. The coke breeze is crushed, preferably in multistage, such as that practiced in Schwelgern (see **Figure 2.8**), to get the maximum of this size range.

Burnt lime is slaked into hydrated lime when it reacts with water in the sinter mix in the mixing and granulation stage and forms hydrated lime, which acts as a binder. The lime should be finer for complete hydration during the mixing stage (**Sec. 7.4.5**).

Raw mix preparation is primary but is an important step involving the granulation operation. The preparation of sinter mix is exercised in stages as follows.

a. Proportioning of the raw materials
b. Mixing of raw sinter mix and its balling

In the proportioning section, the raw materials are stored in separate bins. These bins are equipped with a variable speed weigh feeder discharge system. These are mixed in the correct proportions using weigh hoppers. The proportioning of sinter mix is carried out in sequence. Normally, the following sequence: iron ore mix, return fines, flux mix, coke breeze, and revert plant materials is followed.

The design of feed hoppers and feeders for distributing the prepared sinter mix into these hoppers is equally important because if the sinter mix gets compacted or segregated during handling and loading onto the grate (pellet), all of the advantages gained through good feed preparation may be lost. After this, the sinter mix is leveled.

The weighed materials pass along a conveyor to the "mixing drum" where water is added as a calculated percentage of the weight of the material. The mixing of these materials takes place in a mixing drum. The main purpose of the mixing drum is to homogenize the sinter mix. The parameters—namely the diameter of the drum, RPM, and the space factor—play a major role in achieving a higher degree of mixing. The sinter mix after homogenization is granulated in a "balling drum". The sole purpose of mixing and granulation is to add the required quantity of moisture in the mix and the size enlargement of fine particles of the sinter mix by interparticle adhesion via water bridging. In some old plants, mixing drums and granulation drums were separate. The plants that have come up in recent years have the mixing and balling operation in one unit. The balling drum ensures that fines are coated on the nuclei particles—thus higher sized balls. This facilitates in improving the mean size of the sinter mix. Water is sprayed to moisten the mix in the balling drum (also see **Figures 7.5** and **7.6**).

Customarily, this first stage of the procedure—namely, spraying of water into ore fines and tumbling—is called "granulation". Very little water is added in the mixing drum and major quantity is added in the balling drum.

In case the product from the sinter plant goes beyond its specification, the raw mix is required to be trimmed. Some of the ingredients are added to the mix before the mixing. These include lime fines, flux, and coke; 80–90% of the required flux and coke are added in the base mix, while the remaining portions are added along with return fines at the mixing and rolling stage.

The mixing and granulation of the sinter mix are affected by their pattern of movement in the drum mixture. In the low degree of filling, the frictional force between the drum wall and the sinter mix is large enough to overcome the granulating force of the sinter mix.

The raw sinter mix is subsequently loaded onto a permeable grate at the ignition end of the sinter machine. In transferring the sinter mix from the balling mixer to the grate of the sintering machine, it is essential to feed the material carefully so as to provide a uniform, homogeneous bed and to prevent compacting of the bed. A hearth layer of about 25 mm to 50 mm of coarse (already sintered) material is fed first onto the traveling grate (before the mix is charged). This is to avoid a direct drop of feed onto the grate and also to take adequate measure so that steel grate bars of the traveling grate do not get overheated. Feeding devices typically include a roll feeder in conjunction with chutes that act to avoid compacting of the sinter mix. For high bed permeability, any compression caused by gravitational forces or external densification of sinter mix should be avoided.

5.2.3 Sintering Process

Ignition: The surface of the sinter mix is ignited near the head or feed end of the traveling grate under the ignition hood, and the flame is forced through the mix under the influence of forced air and fans that remove the waste gases. The hood is fueled by a mixture of coke oven gas, blast furnace (BF) gas, and sometimes natural gas (see **Sec. 6.13**). The ignition hood temperature easily can be kept constant if the calorific value of the fuel is controlled to within predefined limits—about 4,000 to 6,000 kJ/m^3. In the case of only blast furnace gas, it is usually preheated.

Proper ignition of the sinter mix is important. Poor ignition results in spotty burning and may leave unsintered material over the surface of the bed. Conversely, too intense an ignition flame can result in slagging over the bed and reduced sintering rates. The radiant hood ignition furnace provides good ignition.

Sintering: After a short ignition period, heating of the bed top is discontinued. The sinter mix moves continuously over the wind boxes along on the traveling grate toward the discharge end of the strand. Air is sucked down by a draught fan through the ignited sinter mix layer through a battery of wind boxes to burn the admixed solid fuel by downdraft combustion. A narrow high temperature (combustion) zone is caused to move downward through the bed by an induced draft applied at the bottom of bed (See **Figure 5.1**). Air is infiltrated through the top of the bed and pulled from the sinter by a strand draught fan through a battery of wind boxes. The process of sintering thus continues as each layer of the combustion zone in turn is heated to about 1,300–1,450°C (1,573–1,723 K) while the material in the traveling grate passes through wind boxes beneath the sintering machine. The draught on the strand is maintained at a preset value by controlling the main fan louvers from pressure measurements in the wind main. This governs the point at which burn-through occurs.

In the combustion zone, bonding takes place among the grains, and a strong agglomerate is formed. The suction of air makes the combustion zone, developed initially at the top, travel through the bed, raising its temperature layer by layer to the sintering temperature. The cold air sucked through the bed cools the already sintered layer, gets itself heated, and preheats the layer below the combustion zone. The process completes when combustion zone reaches the lowest layer of the bed. The top layer of sinter is cooled on the strand itself during its travel to the discharge position as the air is being sucked through the bed.

Burn-Through Point (BTP): An important part of the sintering process is burn-through. This is where the sinter layer has completely burned through its section and is detected by temperature probes under the sinter bed.[2] The temperatures of exhaust gases that come through the different wind boxes beneath the sinter bed is measured to arrive at the BTP (see Sec. **6.6.6, Figure 6.16**). The maximum of the waste gas temperature from the beginning to the end of the strand is the BTP. The time to achieve the burn-through point is called burn-through time (BTT). Burn-through should be achieved one/two wind box(es) ahead of the discharge point in the traveling grate. This is to ensure the end of sintering process coincides with the discharge of sinter from the strand. If sintering is completed before, then the strand area is underutilized (though sinter is overcooled for a higher duration). If burn-through position is extended and the sintering is allowed to continue beyond or till the last wind box, the cooling of the sinter is not proper, which will cause weakening of the sinter and burning of the downstream conveyor belt. The BTP is controlled by adjustment of the pallet speed. The BTP can be advanced or lagged, which is caused by a change of bed permeability. With higher and more favorable bed permeability, the vertical speed of sintering (VSS) should be faster. The sintering time is shorter, the BTP will be advanced, and vice versa.

5.2.4 Cooling of Sinter and Downstream Treatment

At the end of the machine, the sintered material in the form of a cake is discharged into the hot sinter crusher. Hot sinter cake is crushed to a predetermined maximum particle size, which is finally discharged onto the sinter cooler where the sinter is cooled under an induced air draft to below 150°C, so that it can be handled on conveyor belts. A sinter cooler can be either a straight grate or rotary type; the latter is more commonly used. It is desirable to avoid a water quench as the quench adversely affects sinter properties. The exhaust air from these coolers is used for recovery of heat in some of the sinter plants.

During cooling, euhedral and skeletal crystals of magnetite and hematite continue to exsolve from the slag, with hematite predominating. After cooling, the sinter is passed into the discharge bunker. The sinter then passes to the cold screening area, where it is passed through crushers and screens to produce particles in a specific size range. Sinter below the required size, called return fines, is recirculated in the sintering process along with the hot fines, which are removed by the hot screens after the burn-through.

The sinter is transported to the blast furnace stock house by conveyors.

5.2.5 Power Consumption

The power consumed in the main waste suction fan depends on the suction applied through the bed and accounts for a major portion of the total energy consumed in a sintering plant operation. For a required sinter output, it is only possible to a limited extent to alter the draught and the volume of air drawn through the sinter bed. In many of the sinter plants, the speed of the fan is not variable (to control the draught). Hence, the waste gas fans are throttled by means of vanes until a reduction in the specific power consumption is brought about. In this way, a savings in power consumption up to 3 kwh per T of sinter produced has been achieved at the **Schwelgern** Sinter Plant. The plant's operational data show that the fan power consumption was reduced with the increase in sintering rate through the usage of a gas permeable mix[3] (**Figure 5.2A**).

The addition of lime in the sinter mix also decreases the fan power consumption. **Figure 5.2B** represents the dependence of fan power consumption on the addition of burnt lime in the sinter mix constituting fine ore and concentrate.[3]

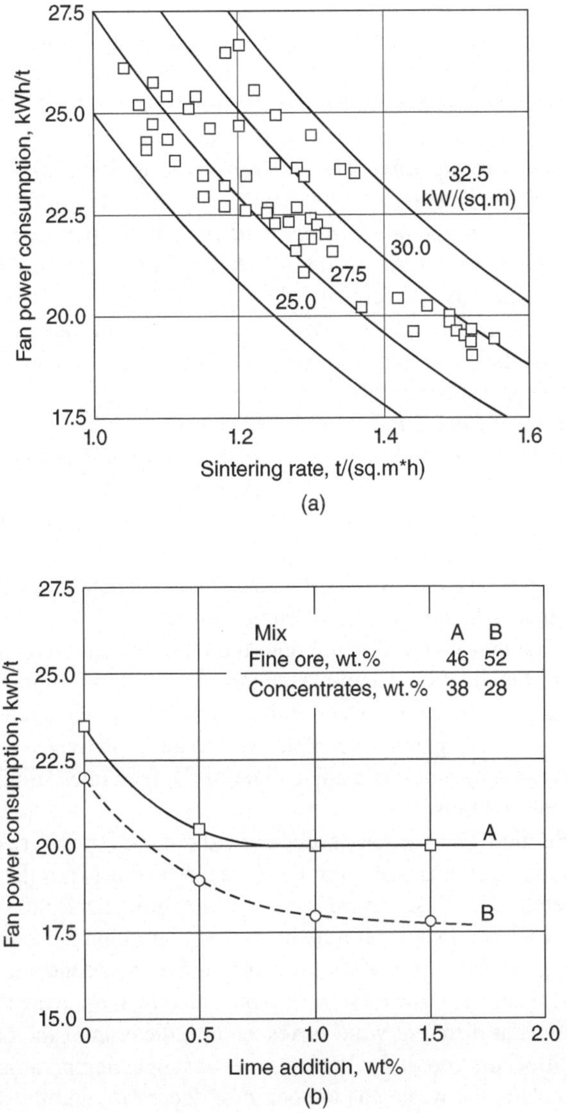

FIGURE 5.2 Fan power consumption as a function of (a) sintering rate, (b) burnt lime addition (after Ref. [3]).

5.3 CONTROL OF SINTER PLANT OPERATION

There are several key variables in the sinter plant operation that need to be controlled for the process optimization. These are described as follows.

Moisture Content in Mixing Drum The amount of primary water added should be proportional to the weight of the raw mix entering the mixing drum. This can be easily achieved using a proportional controller. The moisture content of the sinter mix is measured on the strand roll feed hopper and used to trim the secondary water flow rate. The mix permeability is also measured and used to modify the amount of water required. For greater accuracy, the moisture meter reading is used to trim the material/water ratio. This corrects the water flow rate according to the measured moisture content of the raw mix.

Ignition Hood Temperature The control of the ignition hood temperature can be carried out using:

a. *Fixed ratio air lead temperature control.* The temperature demand provides a set point for the air flow. The fuel follows the air flow in a fixed ratio making this control method fuel efficient.

b. *Variable ratio air lead temperature control.* The air flow is fixed and the output of the hood temperature controller feeds the ratio set point trim input of the gas controller.

Calorific Value of Fuel Gas The mix of blast furnace gas and coke oven gas is used to fuel the ignition hood burners (see **Sec. 6.13**). The calorific value of the fuel is controlled to a consistent value by a separate control loop.

Ignition Hood Pressure This is normally achieved by varying the setting of dampers in the wind boxes under the ignition hood. A single loop proportional, integral and derivative (PID) controller is generally used to automatically maintain the pressure at a desired value.

Bed Depth The mix material is fed onto the strand from the hopper by a roll feeder. The bed depth is set and kept constant by adjusting the cut-off plate, which is fitted with probes to sense the depth of material and automatically vary the roll feeder speed. The quantity of material in the feed hopper itself is held constant by automatic adjustment of the feed rates from the individual raw material bins.

Main Fan Suction The suction produced by the main fan is varied by louvers near the fan inlet, which are controlled by a fan suction controller. If the waste gas temperature increases above a safe working limit, a selector switch allows the waste gas over the temperature controller to position the louvers.

Burn-Through Point Control The burn-through point is controlled by altering the strand speed. The strand speed is controlled by measuring the waste gas temperatures at different wind boxes as an indication of the burn-through point. If it occurs too early, the average waste gas temperature rises. If it occurs too late, the waste gas temperature decreases, and the strand speed is slowed to compensate.

Cooler Speed After leaving the strand, the hot sinter is cooled on a rotary cooler. The speed of the rotary cooler is controlled to match the strand demand defined by the bed depth and the strand speed.

Discharge Bunker Level The level of the discharge bunker, where sinter is conveyed after cooling, is controlled by varying the outlet feed rate (usually vibros). The sinter, after the cold screening, passes over a belt weigher and returns with the hot fines to the return fines bin. The difference between the weight of the cold fines and the weight of the total fines produced gives a measure of the hot fines. Any abnormal variation in the rate of production of hot or cold return fines indicates a possible plant fault.

5.4 POLLUTION CONTROL AND WASTE HEAT RECOVERY

The sinter plants generate a substantial amount of harmful waste gases and solid particulates. In order to meet the increasingly stringent environmental standard, these need to be treated and/or recycled in operation.

The measures to control the pollutants are:

a. Modification in the existing processes and practices
b. Recycling of the gaseous pollutants and diversification of energy resources

The energy consumed in the sintering process constitutes about 10% of the total energy consumed in iron and steel works. The fuel requirement in the sintering process is in the range of 1,528–2,640 MJ per T of sinter produced and is much higher than the pelletizing process (which requires less than 50%). A higher fuel requirement in the case of sintering is due to insufficient fuel utilization and nonrecovery of sensible heat of the product. Therefore, the sinter plant provides a considerable scope for heat recovery. This will reduce the operational cost, besides, the efficiency of sinter is considerably improved.

5.4.1 Emission of Pollutants

A sintering plant is one of the most polluting units in total iron and steel works. The pollution is from the particulate as well as gaseous emissions. Between the two agglomeration processes, the sintering process emits a larger amount of off-gases and solid particulates than pelletization (**Table 5.1**).

Emissions from the sintering process arise from:

a. The material handling operation, which results in airborne dust.
b. Combustion reaction on the strand. The combustion gas contains sulfur dioxide, NOx, CO, CO_2, etc. The concentration of these substances varies with the quality of fuel, raw materials used, and combustion conditions.
c. Atmospheric emissions also include volatile organic compounds formed from the volatile material in the coke breeze, oily mill scale. Polychlorinated dibenzo-paradioxins/furans (PCDD/PCDF), commonly known as "dioxins" and "furans", are also formed from organic material.

PCDD/PCDF The processes by which PCDD/PCDF are formed in the sintering are complex. The PCDD/PCDF formation mechanism appears to start

TABLE 5.1
Environmental Emissions in Agglomeration Processes (per T of Agglomerate)

Process	SOx (g)	NOx (g)	CO (kg)	CO_2 (kg)	Particulate (g)
Sintering	1,500–,700	600–700	32–40	200–250	220–280
Pelletizing					
Hematite ore	150–250	450–550	0.8–1.2	25–35	120–160
Magnetite ore	120–140	180–220	~1.2	20–30	115–130

in the upper regions of the sinter bed shortly after ignition, and then the dioxin/furans and other compounds condense on cooler burden beneath as the sinter layer advances along the sinter toward the burn-through point.[4] The process of volatilization and condensation continues until the temperature of the cooler burden beneath rises sufficiently to prevent condensation and the PCDD/PCDF exit with the flue gas.

5.4.2 Measures to Reduce Pollutants

The pollution control measures include:

a. Reducing the constituents in the feedstock that emit pollutants after the sintering process
   - Coke constitutes N_2 besides carbon. In view of this, a decrease in solid fuel (say by 5 wt.%, equivalent to 3.5 kg/T of sinter) reduces the NO_X and CO generation.
   - Lowering of S in the feedstock—coke, in particular—reduces the SO_2 emissions.
   - The source of organic pollutants, PCDD/PCDF, is raw materials. Their formation increases with increasing carbon and chlorine content. Therefore, reduction of these components in raw materials, in general, decreases the formation of PCDD/PCDF.
   - Oil is a major source of the formation of PCDD/PCDF in sintering process. The mill scale, if used in the sinter mix, should have minimum oil content. This is an important measure in this respect.

b. Modification in process and operation parameters

   The dioxins/furans are formed in the sinter bed itself, ahead of the flame front as the hot gases are drawn through the bed. Any disruption to the flame front increases the formation of these organic pollutants.

   Addition of burnt lime helps to reduce the formation of PCDD/PCDF.

   The replacement of coke by anthracite coal reduces the PCDD/PCDF concentration in off-gas.

c. External cleaning

Water Scrubbers: These are used in sinter plants and have a lower collection efficiency for particulate (than an electrostatic precipitator, or ESP) but have a higher collection efficiency for gaseous emission.

Dioxins are reduced through wet scrubbing fabric filtration with the addition of lignite coke powder.

Heavy metals in the wastewater of scrubbers is taken care of through the methods of precipitation, neutralization, and filtration.

ESP and Bag House: Combustion gases are most often cleaned in ESPs and/or in multicyclones. These units significantly reduce the dust emission but have a minimal effect on the gaseous emissions.

Chain conveyors are installed below the process gas ESP hopper to pick up the discharged fine dust. The dust discharged is transferred to the ESP dust bin for dust treatment, and treated dust is recycled.

d. Recycling of off-gases. The off-gas from the sinter strand contains harmful organic and inorganic pollutants. The influx of these pollutants, mainly PCDD/PCDF, NOx, and SO_2, to the environment is significantly reduced when part of the off-gas is recycled, along with ambient air, to the sintering strand. The particulate matters that are contained in the off-gas also get collected through their recirculation to the sintering strand. The Emissions Optimized Process® (EOS) technology has been developed where part of the gases coming out through the wind boxes is recycled. (see **Sec. 5.4.4**).

5.4.3 Heat Recovery from Sinter Cooler

The heat energy that goes to the atmosphere through the exhaust is contributed mainly by two sources: about 24.3% from the waste gases and 30.6% from sinter cooler exhaust air. The recovery of heat from cooling of the sinter mass can be initiated by distributing the hot air from the cooler at approximately 300–330°C to the preheating hood and post-heating hood. Depending on the requirement, the distribution can be in the ratio 50:50 or more air in the preheating hood (**Figure 5.3**).

The heat recovered from the sensible heat of cooler exhaust is utilized to generate superheated steam with a temperature of approximately 280°C and a pressure of 9 kg/cm^2 and is applied in the generation of electricity. As much as 110 kg of superheated steam can be generated recovering 25% of output heat at the sintering plant.

5.4.4 Emissions Optimized Sintering (EOS®) Process

Outokumpu Technology developed a process in the 1990s based on the recycling technology called the "EOS Process."[5a,5b] The entire sinter strand is housed within an EOS hood. A part of the gas is recycled so that the off-gas volume is reduced by 40% to 50%. The recycling lessens the amount off-gas that is available for treatment through a smaller secondary gas treatment system. The EOS process takes advantage of the fact that only a part of the oxygen in the air is consumed for coke combustion. In spite of higher oxygen potential in the gases through the sintering bed, the combustion is never complete. The off-gas contains CO (see **Sec. 6.14.9**). Therefore, a partial stream of the off-gas is recycled via the hood and enriched with ambient

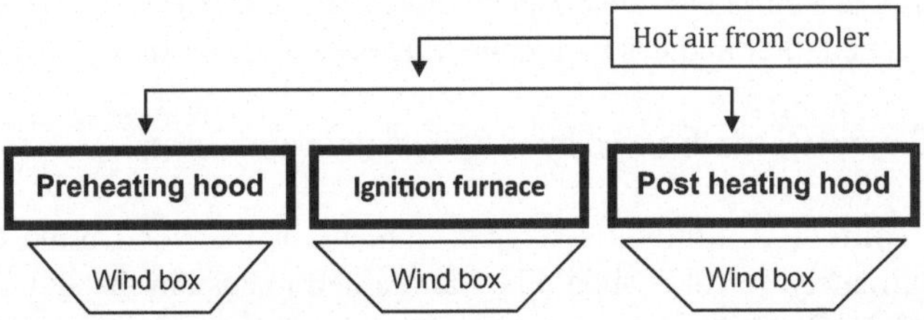

FIGURE 5.3 Recovery of sensible heat from cooler in pre-, and post heating hoods

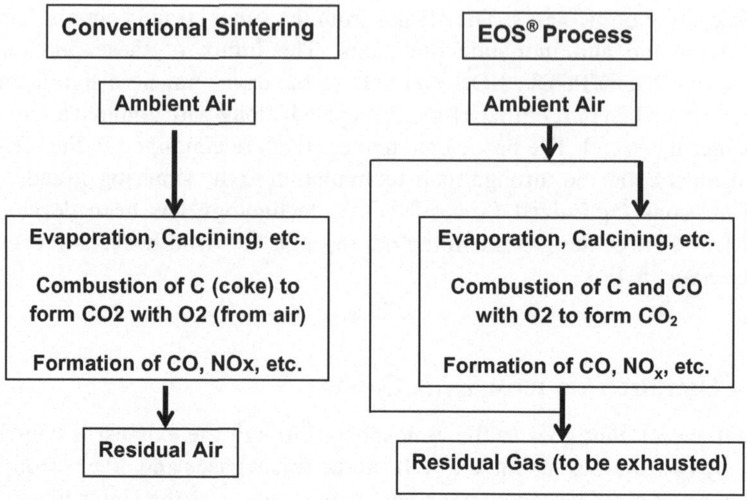

FIGURE 5.4 Schematic (comparison) of conventional sintering process and EOS® (after Ref. [5c])

air to an oxygen content of 13–14% and used as intake process air. **Figure 5.4** schematically compares the EOS® process with the conventional sintering process. In contrast to the conventional sintering, in the EOS® process a part of the flue gas is recirculated back to the sintering.

The EOS process offers distinct advantages over conventional sintering with regard to environmental and process performance. A higher air flow is required in conventional sintering as atmospheric air is used to transport heat within the sinter bed. The gaseous emissions are reduced through the EOS process as follows.

- **SO_2.** Sulfur is introduced into the sintering process through the solid fuel, which happens to be its major source. On combustion, it forms SO_2, which emits to the environment through the gases. With lower consumption of coke, the S emission will be considerably reduced.
- The dioxins, furans, and NO_X all are recycled through the sinter bed. The content of dioxins, furans, and NOX are all decomposed due to high temperature in the sinter bed.

  Solid particulates, when they go down through the sinter bed (along with the gaseous emissions), are collected inside the bed. The sinter bed acts like a gravel bed filter and has high collecting efficiency for the removal of dust.

The EOS process is influenced by the following parameters:

Oxygen content A minimum oxygen content in the hood atmosphere of approximately 11 vol.% should be maintained to sustain the sintering process. If the O_2 content is higher, a higher recycle gas volume would have to be handled; thus, the economic advantage would be partly lost.

CO The sinter off-gas contains up to 1% CO, which is recycled to the sinter machine. The gas is directly used as fuel in the process. This results in a replacement of solid fuel rate by 15–20 wt.%.

Specific Heat Content of Recycled Gas The air requirement in the case of conventional sintering is determined by the ratio of the specific heat of air and sinter mix in order to ensure satisfactory convective heat transfer. The specific air requirement is approximately 745 Nm^3/T of sinter mix. In the EOS process, the recycled gas has a higher specific heat due to its higher water vapor and CO emission.

5.5 RECYCLING OF STEEL PLANT SOLID WASTE

5.5.1 Scope for Recycling

Integrated steel plants (ISPs) generate several by-products that are rich in iron, flux, and fuel values, which are categorized as metallurgical wastes. Depending on the chemical characteristics of the input material, the generation of metallurgical waste fines varies. In the 1970s, the steel plants needed an average of 1.4 T of raw materials to produce 1 T of steel. The figure now has reduced to 1.15 T or less with the usage of rich ores and the adaptation of advanced technologies.[6]

The sintering process is an excellent way of recycling these wastes in various quantities. Recycling of the metallurgical wastes to the sinter plant decreases the input of fresh material, hence decreasing the cost of sinter production. Besides, this facilitates the sinter plant to meet the statutory regulations on the environment and to make an effective utilization of natural resources.

At the same time, these wastes contain hazardous constituents that cause damage to the environment if they are incorporated in the sintering process. In view of this, their recycling needs to be balanced. That is, these should be incorporated up to a certain limit. The quantities of different waste products to be recycled through the sintering route should be such that the performance of sinter plants and product (sinter) quality are not impaired.

5.5.2 Characteristics of Metallurgical Wastes

The metallurgical wastes are characterized in terms of their physical, chemical, and high-temperature properties, with each having important functions in the sintering process.

- **Physical:** The important characteristics of waste materials are: size, flow-ability, and grind-ability. These describe the ease in their pretreatment and recycling through the sintering process.
- **Chemical:** The chemical composition of waste materials decides their suitability for recycling through the sintering process. The chemical composition depends on local parameters like composition of input raw materials, technological status, etc. **Table 5.2** shows the typical chemical composition of steel plant wastes.

- **High-Temperature Properties:** The physical and chemical properties of waste materials at high temperature explain their behavior during the sintering operation and influence the sintering process and quality of sinter product.

5.5.3 Mines and Metallurgical Wastes: Categories and Functions

The metallurgical wastes contain various chemical constituents that may be categorized as (**Table 5.2**):

a. Desirable: The addition of these constituents add to the values in the sinter product's iron and sinter basicity (CaO/MgO) and reduce the requirement of additives that are added to the sinter mix.
b. Undesirable: These do not add to the values; at the same time, they do not harm the product quality.
c. Harmful: These constituents deteriorate the quality of the sinter product, which affect the performance of the blast furnace.

LD Slag: The basic oxygen furnace (BOF) process commonly termed as LD (Linz-Donawitz) process generates slag containing: CaO = 15–45 wt.%; MgO = 15–20 wt.%; Fe = 10–12 wt.%. Recycling it through the sintering process results in a savings of fluxes and iron but at the cost of plant productivity, fuel rate, and product quality. The consequences of which are as follows:

- Incorporation of LD slag in the sinter mix as substitute of CaO demands higher heat input (solid fuel). The CaO in LD slag does not exist in free form but as silicates and ferrites. The mineralogical composition of LD slag is: di-calcium silicates, calcium ferrites, and calcium wustite. The calcium silicates are high melting compounds (**Table 5.3**) and so are the solid solutions rich in silicates. LD slag has a higher melting point (than lime).[7] The temperature of the liquid at the iron oxide–slag interface is higher than that between iron oxide and lime.

TABLE 5.2
Categorization of Steel Plant Wastes and Recycling of Wastes Through Sinter

Material	Extent of Use, Kg/t of Sinter	Desirable Constituents	Undesirable Constituents	Harmful Constituents
LD Slag	10–20	CaO, Fe	SiO_2, Al_2O_3	P
LD Sludge	8–16	CaO, Fe	–	Zn
Mill Scale	~15	Fe	–	Oil
Lime Fines	~10	CaO	LOI	
Flue Dust	~2	Fe, C	SiO_2, Al_2O_3	P, S

Note: LD = Linz-Donawitz (basic oxygen)

TABLE 5.3
Melting Point of Calcium Silicates

Compounds	Melting Point K
$CaO.SiO_2$	1,817
2 $CaO.SiO_2$	2,403
3 $CaO.SiO_2$	2,343

- The sintering reaction gets delayed with the usage of LD slag.[8] This results in a decrease in sinter productivity and inferior sinter quality.
- The LD slag contains P_2O_5. This has a harmful effect on the sintering process and quality of sinter.

  The component, P_2O_5, stabilizes di-calcium silicate minerals present in the slag, which prevents the disruptive phase inversion.

  The sinter with incorporated LD slag contains higher P_2O_5, which deteriorates the hot metal quality when this is charged.

  The presence of P_2O_5 in LD slag has become a restricting factor in its recycling. A replacement of CaO to the extent of 10% through using the LD slag seems to be optimum and can be used without the problem of phosphorous buildup.[9]

LD Sludge: LD sludge is a high iron and high CaO material. It is a profitable material for recycling. Undesirable/harmful constituents are: higher percentage of alkali, Zn, Pb, etc., which restrict its usage in the sintering process. The loss in bed permeability during sintering is also observed due to its finer size. However, these are not so critical so as to avoid this waste material in its recycling.

Mill Scale: Mill scale is a major waste generated during rolling of steel. It constitutes 1–2 wt.% of the steel rolled. It is very high iron and low in impurity product. About 90 wt.% of the mill scale generated is of a coarser category, whereas the remaining 10 wt.% is oily mill sludge. The coarser (and non-oily) variety of mill scale is porous and flat in shape. It mainly consists of wustite and magnetite.

The high and unstable oil content of mill scale is undesirable. If added into the sinter mix, the oil condenses in the ESP, which reduces its efficiency. Besides, its presence in the sintering waste gas possesses a threat to catch fire in the precipitator and pipeline.[10] Hence, de-oiling (removal of oil) of the mill scale (sludge) is essential.

Incorporation of mill scale in the sintering process has the following merits/demerits:

a. (Merits): Mill scale behave exothermally in the sintering process. They are effective in reducing SOx emission and sinter reduction–degradation index (RDI).[11] A savings in energy consumption is observed when mill scale is incorporated in the sinter mix. This is due to its high Fe content.

b. (Demerits): Mill scale has a lower mean size compared to iron ore fines. The sintering time gets prolonged as a result of inadequate assimilation.

BF Flue Dust: The flue dust is associated with the BF top (flue) gas and is collected at the blast furnace yard in the cyclone/electrostatic precipitator. The flue dust generated in the blast furnace is transported to the bedding and blending yard where it is bedded on piles for its recycling.

Flue dust is a mixture of oxides whose major components are iron oxides and coke fines. High iron and carbon content in flue dust makes it very attractive for recycling. On the other hand, flue dust has a finer particle size, which impairs the bed permeability. This makes a restriction in its recycling through the sintering process. Flue dust has high alkali content. Therefore, the operation of a blast furnace is affected due to alkali buildup in the furnace (see **Sec. 3.6.1**) when sinter with high alkali content (through incorporation of flue dust) is charged.

Manganese Ore Fines During mining and processing of manganese ore, a substantial part of the total run-of-mines (ROM) is generated as fines and are dumped, especially during the earlier days of mining. Although manganese ore lumps are used and in great demand in the iron and steel industries, fines have had little application till now. One of the promising methods to conserve the manganese ore reserve is to utilize the fines through sinter.

The chemical composition of the manganese ore samples is not a significant parameter to affect the sintering indices, provided the fines are limited to 3.0% of the sinter mix.

a. The specific productivity of the sintering strand increases by 10.0% with the addition of the manganese ore fines up to 2.0%. However, there happens to be loss in shatter strength, which is compensated by optimizing the content of moisture and fuel in the mix.[12]
b. The specific productivity decreases with an increase in below 1 mm-sized fractions in manganese ore fines; however, the strength of sinter improves.

Lime Fines The process of BOF steel making requires high-purity soft burnt lime in greater amount in case the hot metal produced in the furnace is high in silicon and phosphorus contents. Around 20–30% of the total lime produced degrades to generate below 10 mm-sized fines that cannot be fed to the BOF converter. Sintering is an excellent way of utilizing of lime dust on account of two factors:

- It provides available lime (CaO-SiO_2) to the sinter mix and therefore replaces part of the limestone that is fed to the blast furnace.
- The mix temperature is increased by exothermic reaction of hydration and granulation improved by the binding effect of $Ca(OH)_2$, which can ease moisture condensation in the bottom layer of the sinter bed and improve granule strength (see **Sec. 6.5.6**).

- The sintering efficiency gets improved with the addition of lime; however, the effect is up to a certain limiting value that needs to be studied from case to case.

5.5.4 Economics of Recycling

The assessment of waste material recycling through the sintering operation should be viewed in terms of tangible impact on its techno-economics vis-à-vis its intangible impact on the environment.

a. The tangible part includes savings in disposal cost and the cost of materials it replaces after compensating for extra processing costs involved in waste usage. The requirement for pretreatment is decided by the type of material. The costs involved in transportation of waste material need to be evaluated based on the distance covered. The waste materials, which are in bulk, need to be transported far away from the factory premises, whereas minor waste materials can be dumped inside the works without any treatment.

   The cost involved and therefore the savings in cost due to the recycling of waste materials in the sintering process comprise the following:

   i. Cost of material it replaces through recycling
   ii. Processing cost to make it recyclable
   iii. Cost of disposal

b. The intangible part is on account of its impact on the environment. This is in view of the fact that any disposal of the waste, whether in iron and steel works' premises or away from them, causes environmental (soil) pollution. There are legal requirements for a thorough utilization.

   As per regulations in India and several countries, 100% waste generated in BF and BOF is to be recycled. The detailed project report (DPR) should envisage that provision for land has not been made available for solid waste disposal. Existing landfill capacities as well as the scarcity of primary resources and, in the long term, the rising price of metals make the recycling of these wastes attractive.

The techno-economics of waste material recycling depend on the mineral matter of the waste, replacement ratio, and costs involved in their processing vis-à-vis cost of disposal.

The technologies to maximize recycling within the steel works and downstream linkage for use of BF/BOF slag need to be established for the growth of the steel industry.

5.6 NONCONVENTIONAL/OTHER PROCESSES

The concentrates generated through the intense beneficiation of ores and slimes contain a higher amount of ultra-fines whose size is below 0.125 mm. Obviously, the sintering of ore blend containing a higher amount of ultra-fines is not a preferred

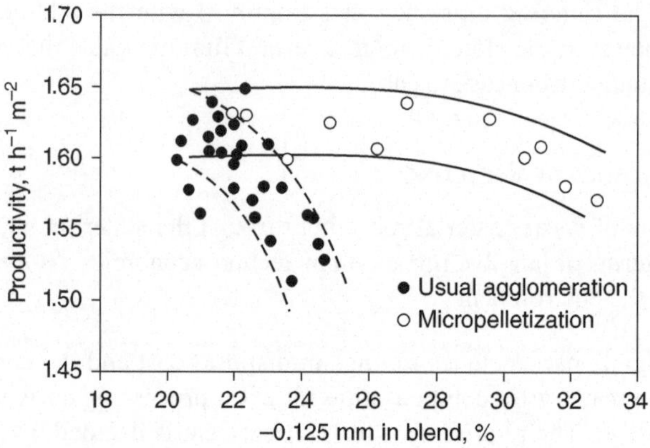

FIGURE 5.5 Effect of micropelletization on sinter productivity (after Ref. [13])

process of agglomeration. This is due to the fact that the bed permeability decreases significantly when a higher amount of ultra-fines is incorporated in the sinter blend which reduces strand productivity.

When these fines are pelletized and the micropellets, thus produced, are incorporated in sinter blend, the bed permeability (green, and also during sintering) improves significantly because of favored granulation. This increases the sinter productivity. A higher amount of ultra-fines can be incorporated in the sinter blend while maintaining the sinter productivity (**Figure 5.5**).[13] Several processes have been developed and/or are in use to utilize the finer fraction of ore/concentrate while avoiding the conventional pelletization. These include:

a. Hybrid pelletized sinter process (HPS) developed by NKK (now JFE Steel Corporation) in Japan
b. Pelletization-sintering
c. Composite agglomeration process (CAP)

These processes are based on the micropelletization of the finer-size fractions of ore (concentrate), while they differ on the manner the finer fractions and coarser fractions are treated to go for their rational usage optimally.

5.6.1 Hybrid Pelletized Sinter Process (HPS)

The hybrid pelletized sinter process (HPS), involving the agglomeration of fine-grained iron ores, was developed by NKK (now JFE Steel Corporation) in Japan in the 1980s. This was developed in order to make use of finer iron ores, including pellet concentrate (from lean ore and slime). The raw materials (ore fines/concentrate, limestone/dolomite, and burnt lime) are blended and made into 5–10-mm green pellets in a disc pelletizer. Fuel is then coated on the pellets in a secondary disc and a predrying zone is installed ahead of the ignition zone on the sintering machine. The products are irregularly shaped pellet aggregates.[14] **Figure 5.6** schematically shows the material flow in a commercial hybrid pelletization sinter

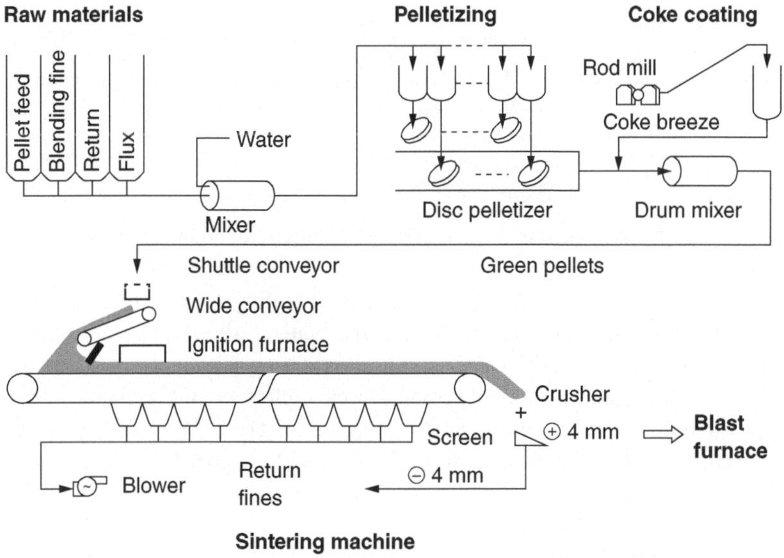

FIGURE 5.6 Schematic of commercial 6-Mt hybrid pelletizing sinter (HPS) plant (after Ref. [15])

plant (6 Mt capacity) at the NKK Fukuyama Steel Works.[15] From the trial, it was confirmed that it was possible to operate the plant at 4.7% silica content and 60% fine ore at higher sinter productivity, 1.65 t/(sq.m*h).[16]

5.6.2 Pellet-Sintering

The process involves combining the pelletizing and sintering processes. The concentrate generated through the beneficiation of iron ore is pelletized in a rotary disc using the binder and water to produce smaller sized (approximately 3–5 mm) green balls called "micropellets".

The micropellets thus produced are mixed with sinter-grade fines and subsequently sintered. The step of induration (heat hardening) of the green pellets, which is energy intensive, is therefore avoided through the process. The product of the pelletization-sintering process of satisfactory quality can be produced on a sinter-type strand at an acceptable production rate.

The process is important in cases of low-grade iron ores and slimes, which require intense beneficiation for an acceptable grade (Fe approximately 64 wt.%). The concentrate, thus generated, has unfavorable granulometry for sintering. At the same time, the concentrate may be coarser for its acceptance to the direct pelletization unless it is further ground.

Process: The process is comprised of (**Figure 5.7**):

a. **Pelletization** The high grade (Fe more than 65 wt.%) ore concentrate (grain size below 0.150 mm), subsequent to the beneficiation, is pelletized in a disc to produce green pellets of 3–5 mm in size. Crushed burnt lime (1-3 wt.%)

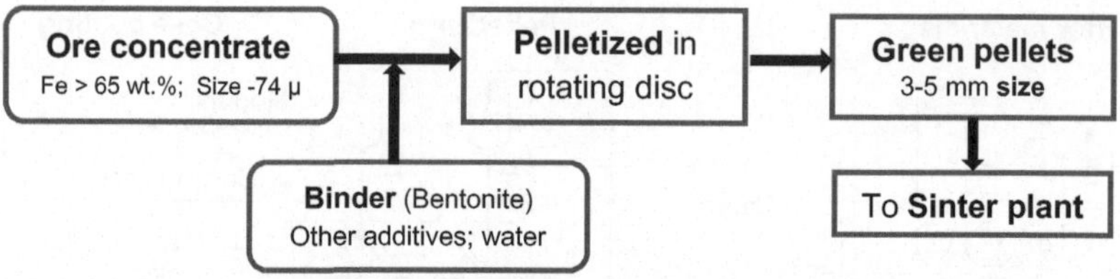

FIGURE 5.7 Schematic of pelletization of ore concentrate for producing pellet-sinter

and/or bentonite (0.5-1 wt.%) is added as a binding agent. The concentrate is intensively mixed with the binder and water (7-9 wt.%) in an agitating mixer and subsequently transferred to the rotating disc (similar to that used in a normal pelletization process). Green pellets of 3–5 mm in size are produced. A high agitating mixer can also be used. This reduces carbon variation in the mix from 2.96 to 0.43% due to higher mixing efficiency. FeO consistency of sinter increased from 80 to 100%.

Micropellets can also be produced using lime in the base mixture. Hydrated lime, formed when lime comes in contact with water, provides bonding in the green pellets. The green pellets are air-dried, when voids are formed within the green pellets as the water gets evaporated. This facilitates the penetration of CO_2 into the pellet during the treatment. The pellets are subsequently treated by the stream of flowing carbon dioxide; the result is a hard $CaCO_3$ phase is formed in the pellet.[17] The formation of calcium carbonate is primarily responsible for imparting good strength to the pellet.

$$Ca(OH)_2(s) + CO_2(g) = CaCO_3(s) + H_2O(g/l) \tag{5.1}$$

The transformation of hydrated lime to calcite is associated with volume expansion by 1.52% (molar volume of $Ca[OH]_2$ and $CaCO_3$ are: 33.59 and 34.10 cm^3/mol.). The volume expansion is likely to be accommodated by the internal pores of the pellets leading to a decrease in porosity and an increase in compactness of the pellets. The transformation of $Ca(OH)_2$ to $CaCO_3$, therefore, plays a predominant role in the strength development in pellets. The treated micropellets have a cold crushing strength (CCS) value of 4.1 kg/p; the abrasion loss is around 10% and drop number is approximately 50.[17]

It is important to note that the micropellets produced in the pelletizing plant should not disintegrate during their handling from the pelletization disc to the sintering plant, also during its granulation (along with sinter grade ore fines) in the rotary drum. In view of this, it is required to produce micropellets with adequate strength in their green form and to take part in sintering reactions in the continuous strand to have strong bonding with ferruginous and slag-forming ingredients.

The green pellets thus produced are transported to the proportioning section of the sinter plant and kept in a separate bunker.

b. **Proportioning** The green pellets of sinter mix are directed to the proportioning section in the sinter plant. The proportioning of sinter mix materials constituting sinter grade ore fines with size below 6.3/10 mm and Fe in 63–65 wt.% range, along with limestone, dolomite, lime, coke breeze (5–6 wt.% of mix), metallurgical wastes (1–2% wt. of mix), and return sinter fines (below 6.3/10 mm size), is carried out as per the usual practice adopted in the sinter plant. All the basic constituents, lime (CaO), and magnesia (MgO) required to maintain the basicity of pellet-sinter are incorporated through the addition of fluxes in the sinter mix. The proportioned material constitutes sinter mix.

c. **Granulation.** The green pellets and sinter mix materials are separately drawn from bunkers of the proportioning section. These are mixed in the mixing drum of the sinter plant and balled in the balling drum (water 5–6 wt.% of mix is sprayed throughout the length). The granulated sinter mix is discharged from the balling drum to the sinter machine (600–700 mm in bed height).
d. **Sintering** The sintering is, thereafter, continued according to the usual practice. The top portion of the charge is ignited. A vacuum (15–19 kPa) under the traveling grate is applied and the sintering process continued (layer-wise) till near the discharge end (wind box) of the traveling grate, and the sinter is broken and cooled in a separate cooler. The sinter (mass) is screened at a 6.3 mm-sized aperture. The undersize is recycled back to the proportioning section. The oversize (sound sinter) is sent to the blast furnace(s).

Study on Pellet-Sinter Making A laboratory study was undertaken in order to develop a suitable beneficiation and agglomeration scheme so that additional material value could be recovered from the lean grade ore fines (Fe: 55–58 wt.%). An intensive beneficiation process was adopted to generate the concentrate to meet the desired chemical specification (Fe: approximately 65 wt.%). However, the product of concentration was of an extremely finer size (51 wt.% below 44 micron). The pelletizing-sintering was adopted. The objective was to produce green pellets with the right strength that could sustain the impact during handling and mixing in the rotary drum for granulation in the sintering stage.

The concentrate, generated subsequent to the intense beneficiation, was pre-balled to approximately 3 mm in size (**see Figure 5.7**) in a pelletizing disc with the addition of 0.5–1.0 wt.% bentonite as a binder. The dried (green) pellets, thus produced, were mixed with (normal) sinter grade fines in various proportions to yield the ferruginous part of the mix. Thereafter, the pot sintering (bed height, 600 mm) was conducted in the 2–2.25 sinter basicity (CaO/SiO_2).[18a] The study showed that the fine-grained concentrate, when incorporated in the pre-balled pellet, could be partially utilized (up to 50 wt.%), with the rest being sinter-grade fines. The sinter productivity increased marginally compared to that in the case of 100% sinter fines. This was contributed by a reduction in sintering time and an increase in sinter yield. Other indices—namely, tumbler index and reducibility index—were more or less comparable.[18]

The pellet-sintering process has potential to be developed in commercial scale to extensively utilize iron ore microfines. The process could incorporate 60% of less than 100 micron iron ore fines in the pellet feed. The remaining 40% was the sinter feed, which typically contained 20% of less than 149 micron. In total, 68% of the feed to the process could constitute microfines compared to 20% in normal sinter feed.

Perceived Benefits Perceived benefits of the pellet-sintering process are as follows.

- The process envisages enhanced utilization of fines generated in mines (dry and washed) other than the normal process of pellet making.
- The process does not require grinding the concentrate further to achieve higher Blaine number (more fineness) for the conventional pelletization.
- The process facilitates shifting toward a cheaper means of fuel (coke breeze) in sintering the micropellets for the heat requirement from the gaseous/liquid fuel for the induration (of green pellets) in large amounts.
- The solid fuel rate in the sintering process can be reduced as coke can be coated over the green pellets.
- The required amount of basic constituents—namely, lime (CaO) and magnesia (MgO)—can be incorporated in the product of pelletizing-sintering to maintain the slag chemistry, whereas the process of palletization has limitations.
- The sinter produced through this process can be more reducible.
- An increase in strand productivity can be achieved compared to that when 100% sinter fines is used.
- Diversification of the process can be made as a smaller setup comprising pelletizing disc and a traveling/pot grate sintering unit would suffice to agglomerate.

Limitation The process has limitations and these need to be taken care of with respect to the following:

- The process requires granule formation of adequate strength so that the sinter bed does get collapsed and permeability can be maintained.
- The process requires the installation of disc pelletizers above the raw material surge bin. This increases the cost of erection and modification in the existing plants. If they are installed at the ground level, the process requires conveying of the green pellets to the raw material surge bin; in this case, the green pellets should have more strength.

5.6.3 Composite Agglomeration Process (CAP)

An innovative process, the "composite agglomeration process (CAP)"[14, 15] (**Figure 5.8**), has been developed. The CAP, similar to the pellet-sintering process, treats the fine-grained ore and coarse-grained ore separately: the latter portion is directed to the proportioning section of the sinter plant to follow the usual sintering process (along with the agglomerated fine-grained ore). However, unlike the pellet-sintering process, a part or all of fine-grained iron raw materials are balled into a primary mixture. The agglomerated (balled) fine ore, along with the sinter-grade coarser ore, are fed to the secondary proportioning and mixing unit (see **Figure 5.7**). Afterward, the

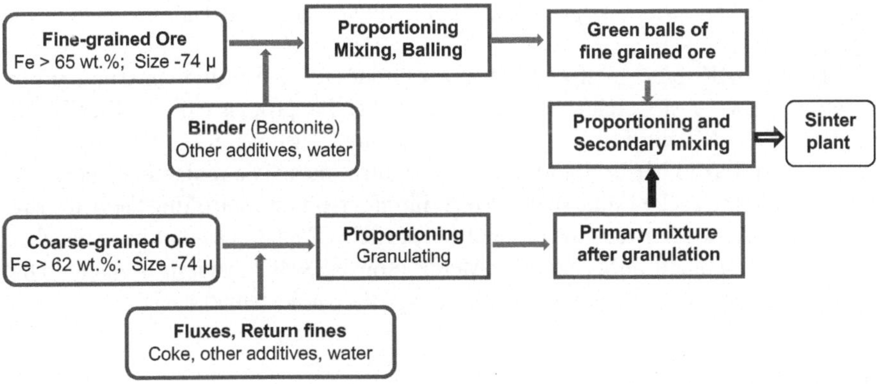

FIGURE 5.8 Schematic of composite agglomeration process (after Ref. [14])

secondary mixture is subsequently fed into the sintering machine. By ignition and down draft firing, the mixture is made into the composite agglomerate.

An important feature of this process: the composite is/can be composed of basic sinters and acid pellets in which acid pellets are embedded in basic sinter.

Advantages The new process has a series of advantages over traditional sintering or pelletizing such as the ability to use diverse iron-bearing materials, significant improvement in the material bed permeability and productivity of sintering machine, low energy consumption, and ability to prepare a low basicity charge with good metallurgical performance.[14]

5.6.4 Mebios Process

The Iron and Steel Institute of Japan (ISIJ) has recently proposed a research project called MEBIOS (Mosaic Em-Bedding Iron Ore Sintering) for production of porous meso-mosaic texture sinter. **Figure 5.9** is schematic of the process. The process arranges pregranulated dense pellets (*aging bed*) that do not easily deform during sintering in the ordinary sinter mixture (*induction bed*). Such an arrangement creates a ventilation route in the sinter bed. The coarser pisolitic ore was used for the

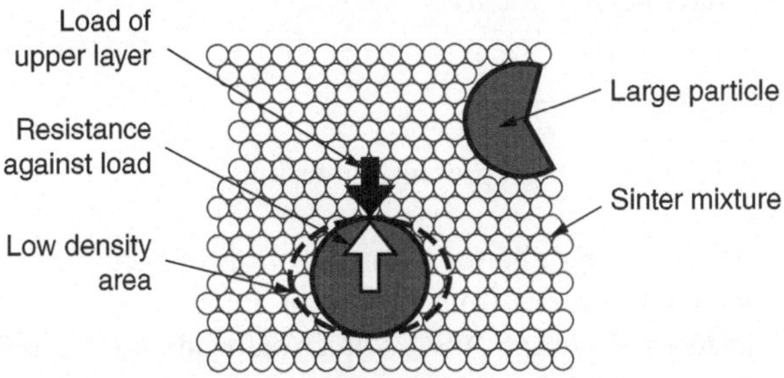

FIGURE 5.9 Conceptual illustration of the MEBIOS process (after Ref. [16])

induction bed and Mamba-type of ores consisting of fine particles were used as the aging bed.[19]

The MEBOIS process aims to form a ventilation route in the sintering bed by creating a low-density area around the large pellets. A process called RF-MEBOIS (Return Fines-Mosaic Em-Bedding Iron Ore Sintering) has also been developed, which is similar to MEBOIS, to utilize the return fines. The return fines particles, charged into the packed sinter bed, have a similar effect on controlling the bed structure due to friction between the dry and wet particles. The process has been installed in the sintering plants of Sumitomoto Metals (now NSSMC). An improvement in the sinter productivity has been confirmed through the implementation.[20]

5.6.5 New Charging Systems

The guiding factors in designing the improved charging system are as follows.

a. Coke breeze, in the sinter blend, is distributed uniformly in the normal sintering process. The sintering temperature in the upper bed region remains lower compared to the bottom portion of bed. Consequently, a higher amount of return fines are generated from the upper bed region compared to the lower bed region. In view of this, an appropriate segregation of coke is desirable to enhance the sintering temperature of the upper part (see **Sec. 10.7.3, 10.7.5**). That is, a higher amount of coke is required in the top layer of the bed (to compensate for the chilling effect), while the bottom portion requires a lesser amount of coke (to avoid overheating).
b. The bed permeability of the lower region is lesser compared to the upper region when the size distribution of feed material is uniform. This is due to the condensation of moisture at the bottom part and compression under the load of the bed in the upper portion (see **Sec. 7.5.2**). A coarser size distribution of feed material in the lower bed region is therefore desirable.

The segregation of feed material that can be achieved with the conventional drum charging system is limited. Charging of feed material in multiple layers has been attempted to achieve segregation in some of the sintering machines. However, such a system requires more than one charging unit for an individual (one) sintering machine. Therefore, systems like slit bars (**Figure 5.10a**)[21] and slit wires (**Figure 5.10b**)[22] have been introduced.

5.6.6 High-Pressure Sintering

Following are the features of high-pressure sintering.

- High-pressure sintering allows the use of positive pressure from the top. The applied pressure varies from 0.05 to 0.1 atm.
- Vertical speed of sintering (VSS), which depends on the infiltration of air through the bed, is approximately three to four times that of normal sintering.

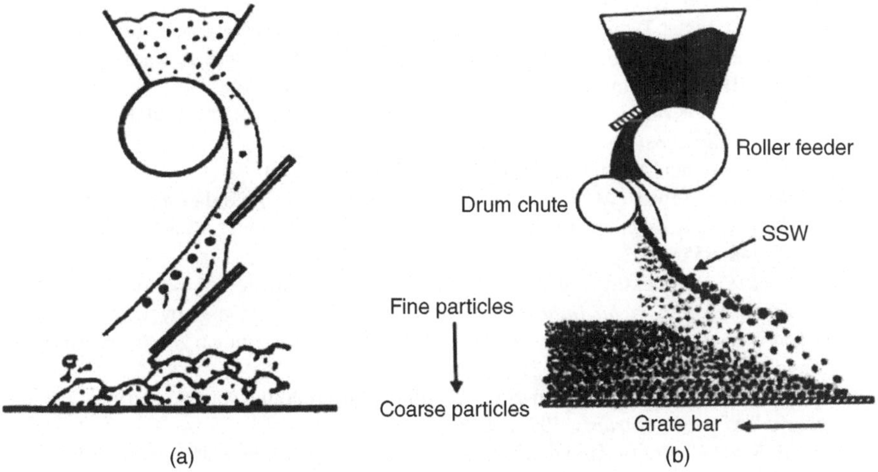

FIGURE 5.10 Schematic diagram depicting size segregation achieved by (a) slit bars (after Ref. [21]) and (b) slit wires (after Ref. [22]) in charging systems

- A high (positive) pressure differential gives better strength compared to normal sintering. Higher strength is due to increased compression of the sinter bed coupled with moderate VSS, 30–35 mm/min.
- Productivity increases by 1.5 to 3 times.

REFERENCES

1. S. V. Komarov and E. Kasai, "Simulation of sintering of iron ore bed with variable porosity," *Institute of Multidisciplinary Research for Advanced Materials*, Tohoku University, Phoenics User Conference, Melbourne, 2004.
2. X. L. Chen, X. H. Fan, Y. Wang, H. M. Long, T. Jiang, J. Shi. Q. Y. Song, and X. D. Yang, "Control guidance system for sintering burn through point," *Ironmaking & Steelmaking*, 2009, 36, 209–211.
3. K. H. Peters, K. Kersting, and H. Beer, "Ways and means of reducing the energy consumption in iron ore sintering," *Proc. 4th Inter. Symp. On Agglomeration*, Toronto, Canada, 1985, C. E. Capes, Ed., Ontario, Iron & Steel Society of AMIE CIM, 1985, 157–167.
4. L. C. Wang, W. J. Lee, P. J. Tsai, W. S. Lee, and G. P. Chang-Chen, "Emission of PCDD/F from stack flue gases," *Chemosphere*, 2003, 50, 1125–1129.

5a. T. Schwalm, W. Gerlach, and Lurgie Mettlurgie GmbH, Oberusel (part of Finnish Outokumpu Group), "New Developments in Agglomeration Technologies," *Conf. Proc. Asia Steel International Conference*, April 9–12, 2003, Jamshedpur India, Vol. 1, T. Mukherjee and A.S. Dhillon, Eds., Indian Institute of Metals and Zhongguo-Jinshu-Xuehui, Organisers, Indian Institute of Metals, Kolkata.

5b. Anonymous, 2014, "Traveling grate sintering" (cited September 10, 2014). Available from http://www.outotec.com/en/About-us/ourtechnologies/Agglomeration/Traveling-grate-sintering/#tabid-1.

5c. L. Lu and O. Ishiyama, "Recent advances in iron ore sintering", *Mineral Processing and Extractive Metallurgy (Trans. Inst. Min. Metall. C)*, 2016, 125, 132–139.

6. U. S. Yadav, B. K. Das, A. Kumar, and H. S. Sandhu, "Solid wastes recycling through sinter—Status at Tata Steel," *Proc. Conf. Environment & Waste Management* (EWM 2002), January 24–25, 2002, A. Badopadhyay, R. Kumar, and P. Ramachandrarao, Eds., CSIR-NML Jamshedpur: Jamshedpur, 81–94.
7. H. Noda, S. Krishimoto, A. Maki, O. Komatsu, N. Sakamato, and K. Takemoto, *McMaster Symposium No. 21*, 1993, 30–54.
8. K. Takemoto and A. Kumasaka, *CAMP-ISIJ*, 5, 1992, 1101.
9. B. Roy, R. S. Tewari, M. Shrivastava, D. K. S. Bhdoria, S. A. Balaji, *Proc. Inter. Conf. on Raw Materials and Sintering, RDCIS-SAIL*, RDCIS-SAIL, Ranchi: Ranchi Eds., 2001, 348–355.
10. L. Juckes, "Processing, utilization, and disposal of waste in the steel industry," (*Conference Report on Industrial Wastes*, June 3–6, 1996, org. by United Nations Economic Commission for Europe, Balatonszeplak, Hungary), *Ironmaking & Steelmaking*, 1996, 23, 384–386.
11. Kim C. Ho and S. M. Jung, "In-plant recycling of ironmaking waste material at Pohang works," *SEAISI Quarterly*, 1998, 29–34.
12. SAIL R&D Centre for Iron and Steel, *Study of the Utilization of Manganese Ore Fines in Iron Ore Sinter*, Report No. 12.02. 441.84, Ranchi, SAIL R&D Centre for Iron and Steel, 1984.
13. S. Kurosawa, H. Fukuyo, O. Komatsu, N. Sakamoto, and Y. Yamaoka, "Recent advances in sintering technology by NKK," *Proc. Ironmaking Conf. ISS-AIME*, 48, April 2–5, 1989, Chicago, 451–458.
14. T. Jiang, G. H. Li, H. T. Wang, K. C. Zhang, and Y. B. Zhang, "Composite agglomeration process (CAP) for preparing blast furnace burden," *Ironmaking & Steelmaking*, 2010, 37(1), 1–7.
15. Y. Niwa, N. Sakamoto, O. Komatsu, H. Noda, and A. Kumasaka, "Commercial production of iron ore agglomerates using sinter feeds containing a large amount of fine ores," *ISIJ Inter.*, 1993, 33, 454–461.
16. L. Lu, O. Ishiyama, "Recent advances in iron ore sintering," *Mineral Processing and Extractive Metallurgy (Trans. Inst. Min. Metall. C)*, 2016, 125, 132–139.
17. J. Pal, S. Ghorai, M. C. Goswami, S. Ghosh, D. Ghosh, and D. Bandyopadhyay, "Development of fluxed iron oxide pellets strengthened by CO2 treatment for use in basic oxygen steel making," *ISIJ Inter.*, 2009, 49, 210–219.
18. CSIR-NML, "Improvement in sinter productivity through deep beneficiation and agglomeration technologies for rational utilization of low grade iron ore and fines (GAP 0228)," Project sponsored by Ministry of Steel, Govt. of India, 2010–2013.
19. T. Otomo, Y. Takasi, and H. Sato, "Granule design for the sintering with less amount of liquid phase formation," *ISIJ Inter.*, 2009, 49(5), 659–666.
20. M. Matasumura, Y. Yamaguchi, M. Hara, C. Kamijo, T. Kawaguchi, and Y. Nakagawa, "Improvement in sinter productivity by adding return fine on raw materials after granulation stage," *ISIJ Inter.*, 2013, 53, 34–40.
21. T. Fukuda, M. Himeda, H. Yoshida, N. Sasaki, M. Sato, T. Nasun, S. Aso, and R. Haga, "The result of higher bed test operation in Hirohata no.1 sinter plant," *Tetsu-to-Hagane*, 1984, 70, 30.
22. K. Takeda and M. Oogami, "Recent development of iron making technology in JFE Steel," *JFE-Giho*, 2008, 22, 1–5.

6 Sintering Fundamentals

ABBREVIATIONS

BF	*Blast furnace*
BTP	*Burn-through point*
BTT	*Burn-through temperature*
CO (gas)	*Coke oven gas*
CV	*Calorific value*
FFS	*Flame front speed*
GI	*Granulation index*
SFCA	*Silico-ferrite of calcium and aluminium*
VSS	*Vertical speed of sintering*

6.1 THE PROCESS OF SINTERING

6.1.1 Preamble

The process of iron ore sintering is material specific. That is, the production rate of sinter and its quality parameters are dependent on the ore characteristics. Nevertheless, the process is guided by the basics of heat and mass transfer, which need to be studied. This is especially important while considering the present requirements of sinter qualities, which are becoming more stringent day by day. In view of this, one needs to look into the sintering mechanisms as well as the mineralogical and morphological properties of sinter when the characteristics of raw material changes, in particular with a higher proportion of ultrafines in the mix. Besides, the concerns of the environment have to be addressed, especially with reference to the generation of dust particles and SO_x, NO_x, and CO_X effluent gases. In this regard, it is of paramount importance to increase the efficiency of sintering. The research in these directions on the sintering fundamentals has gone through the evolutionary phases, and there is scope for further study to gather theoretical data that can be unified into a set of relationships that would permit the accurate prediction of sinter productivity and quality under a specified set of operating parameters applied to a variety of materials to be processed.

6.1.2 The Process

In the process of sintering, a shallow bed of the mix materials constituting iron ores and other ingredients—namely, fluxes (CaO, MgO bearing limestone and dolomite), metallurgical wastes, and so on—is agglomerated by heat exchange and partial fusion of the quiescent mass. Coke is added into the mix to provide heat for the propagation of a heat front through the sinter bed after the ignition zone.

Raw sinter mix is charged at the ignition end of the sinter machine. A hearth layer of about 25 mm to 50 mm of coarse (already sintered) material is fed first onto the traveling grate (before the mix is charged). This is to avoid a direct drop of feed onto the grate and also to take adequate measure so that steel grate bars of the traveling grate do not get overheated. Air is infiltrated through the top of the bed. Heat is generated by combustion of a solid fuel admixed with the bed of fines being agglomerated. The combustion is initiated by igniting the fuel exposed at the surface of the bed after which a narrow high-temperature zone is caused to move through the bed by an induced draft applied at the bottom of the bed. The process of sintering thus continues as the material in the traveling grate passes through wind boxes beneath the sintering machine. The product, sinter cake, is discharged at the (other) end of the machine. The hot sinter cake is crushed and sized. The material is finally discharged onto sinter cooler, which is again screened and sized. The return fines (below 10/6.3 mm), thus generated from the crusher in the hot stage and after cooler in the cold stage, are recirculated back to the machine, while the sized (above 10/6.3 mm fraction) sinter reports to the blast furnace.

Sintering is considered to be an extremely efficient heat transfer process in which air entering the bed is heated by the finished sinter until it attains its peak temperature in the combustion zone. The gases leaving this zone give their heat to the feed below. Important factors that influence the sintering process simultaneously are:

a. Thermal characteristics of the process
b. Flow of the gases in packed beds

Figure 6.1 represents the schematic of the sintering process along with the various zones during the process of sintering. High temperature generated in the bed results in the formation of a partially molten mixture comprising:

- Melt: The characteristic parameters are its volume and properties.
- Solid: The characteristic parameters are: (a) volume, (b) size, (c) voids volume, and (d) its distribution.

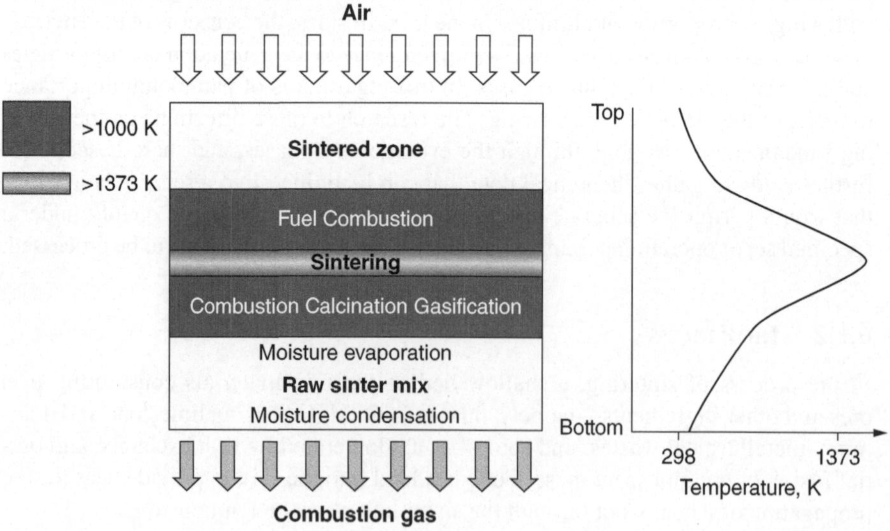

FIGURE 6.1 Schematic of the sintering process showing various zones (after Ref. [1])

The bonding is affected by a combination of fusion, grain growth, and slag liquidation. The generation of volatiles from the fuel and flux-stone (limestone and dolomite) creates a frothy condition and the incoming air quenches and solidifies the rear edge of the advancing fusion zone. The product "sinter" consists of a cellular mass of ore bonded in a slag matrix. Its physical properties will depend on how the variables related to melt and solid influence material flow and coalescence.

Properties of the partially molten mixture and solidification process during the sintering process will determine the following properties:

- Bonding phase: volume and its type, residual stress
- Unreacted solids: volume, size, and their distribution; pores, size, and their distribution; closeness/openness of pores; particle size distribution

6.1.3 Characteristics of Sintering Process

The process of sintering consists of the passage of a heat and reaction front through a packed bed of solids. However, the process is in contrast to a chemical reaction on the basis of following parameters:

- Heat is transferred through the convection and conduction in a narrow zone of the sintering bed.
- The gas mixture inside the packed bed behaves like fluid.
- Water of the sinter mix in the heating zone gets vaporized and subsequently (the vapor is) condensed. This makes the high temperature zone concentrated in a narrow zone. The vapor when condensed at the bottom of the sintering bed may affect the bed voidage (bed permeability) and, in worst case, cause the collapse of the bed.
- The carbonates get decomposed through endothermic chemical reactions.
- The coke gets combusted with the incoming air; the process does affect the bed voidage. Complete combustion does not take place, which results in CO gas in the exhaust. This exothermic reaction provides a major portion of the heat for melting and rendering mobility for solid-state diffusion.
- The gas components get converted homogeneously, and also are heterogeneously catalyzed.
- The solid material is melted partially and gets solidified in the mixture. The properties of melt, before it gets solidified, depend on the chemical composition of the fines layer adhered to the granules and also on the assimilation of nucleus particles.
- The bed is subjected to shrinkage after the sintering is accomplished.

6.2 SINTERING ZONES

A series of physically and chemically separate zones is formed across the sintering bed, which is contributed by the temperature distribution through the bed and, consequently, on account of the various reactions that take place.

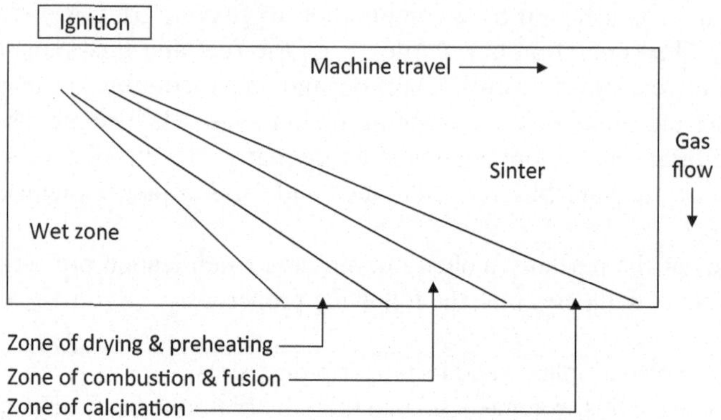

FIGURE 6.2 Schematic of various sintering zones in a traveling grate

The sintering zones that prevail following the temperature distribution across the material bed being sintered and the physical-chemical changes in these zones are shown in **Figure 6.2** and are described as follows.[2]

The standard heat ($\Delta H_{298}{}^{\circ}$) and entropy ($\Delta S_{298}{}^{\circ}$) of chemical reactions are also mentioned in the following tables. The free energy of chemical reactions ($\Delta G_{298}{}^{\circ}$) can be calculated by using the relation: $\Delta G = \Delta H - T\,\Delta S$.

a. **Zone of Sinter:** This represents the product of sintering process. A temperature gradient exists in this zone that ranges from near ambient (temperature) at top of the bed surface to the temperature of the fusion zone at the bottom of this zone. The sinter mass is stabilized upon the solidification of the fluid slag matrix. The physical and chemical changes, such as the oxidation of magnetite to hematite and grain growth of crystals of iron oxide, occur just after solidification. Magnetite is oxidized to hematite (Fe_2O_3) at greater than 400°C (673 K) in air. The mechanism of magnetite oxidation is the outward migration of iron.[3]

Reaction	**$\Delta H_{298}{}^{\circ}$ (KJ/mol)**	**$\Delta S_{298}{}^{\circ}$ (J/mol K)**
$2Fe_3O_4 + ½\ O_2 = 3\ Fe_2O_3$	–232.2	161.3

The hematite minerals sinter much more rapidly than magnetite. These changes determine the gross physical character of this zone.

b. **Zone of Combustion and Fusion:** Carbon in coke breeze gets oxidized to carbon monoxide and carbon dioxide in this zone. Heat is also evolved with the formation of slag through the exothermic chemical reactions. Besides, a large fraction of the sulfur (in sulfide form) is burned to sulfur dioxide within this zone.

The permeability of the combustion-melting zone in the bed affects the efficiency of sintering process with regards to productivity and yield of sinter as well as its quality control. As the flame front passes through this layer in the sintering bed, the slag constituting calcium-alumino-silicate minerals form at around 1,200°C (1,473 K). The slags, thus formed, dissolve ore particles along grain boundaries and take the finer iron oxides into solution. This slag also reacts larger ore particles to develop crystalline margins.

As various sintering reactions proceed, ferrites and silicate minerals are formed, also bonding is affected by the diffusion and slag. These are detailed in **Secs. 6.7** and **6.9**. During the precombustion stage, which is below the ignition temperature of the coke (around 1,100 K), the central position of the larger ore particles (approximately 5 mm in diameter) may not be heated sufficiently and thus (these) particles of sinter mix may enter the combustion zone with a heat deficit. The combustion of carbon in coke is also initiated.

Reaction	$\Delta H_{298}°$ (KJ/mol)	$\Delta S_{298}°$ (J/mol K)
$C\ (s) + O_2\ (g) = CO_2\ (g)$	−393.5	2.90
$C\ (s) + ½\ O_2\ (g) = CO\ (g)$	−110.5	89.24
$CO\ (g) + ½\ O_2\ (g) = CO_2\ (g)$	−283.0	−86.38
$C\ (s) + CO_2\ (g) = 2CO\ (g)$	172.5	175.66
$C\ (s) + H_2O\ (g) = CO\ (g) + H_2\ (g)$	131.3	133.56

c. **Zone of Calcinations:** The carbonates (limestone and dolomite) are calcined to yield lime (CaO) and MgO through endothermic chemical reactions with the sufficiently hot temperature of the gas stream.

Reaction	$\Delta H_{298}°$ (KJ/mol)	$\Delta S_{298}°$ (J/mol K)
$CaCO_3\ (s) = CaO\ (s) + CO_2\ (g)$	179.3	160.5
$MgCO_3\ (s) = MgO\ (s) + CO_2\ (g)$	101.5	174.9

In the case of calcium carbonate ($CaCO_3$), ΔGr is 0 at T = 1117 K or 844°C;

$$\Delta Gr = \Delta H_r - T\,\Delta S_r = 179300 - T * 160.5.$$

The reaction begins when the temperature is above the dissociation temperature of the carbonates in the limestone. The decomposition pressure P_{CO_2} has been found to be 1.00 atmospheric (1,01,325 Pa) at 1,117 K and 0.0333 atmospheric (3,374 Pa) at 973 K.

In the case of magnesium carbonate ($MgCO_3$), ΔGr is 0 at T = 580 K or 307°C;

$$\Delta G = \Delta H_r - T\,\Delta S_r = 101500 - T * 174.9.$$

The reaction begins when the temperature is above the dissociation temperature of the carbonates in the $MgCO_3$.

The reduction of hematite to magnetite is also initiated with the presence of CO. The reductant, CO, is generated through partial combustion of carbon and generation of water gas.

Reaction	$\Delta H_{298}°$ (KJ/mol)	$\Delta S_{298}°$ (J/mol K)
$CO + 3Fe_2O_3 = 2Fe_3O_4 + CO_2$	–50.6	46.3
$CO + Fe_3O_4 = 3FeO + CO_2$	–389.1	52.3

d. **Dry and Preheat Zone:** Within this zone, the charge has been heated sufficiently to a volatilize-free moisture. The hydrated iron oxides, such as goethite, undergo thermal dissociation. Besides, a part of sulfur is volatilized from sulfide minerals.

Reaction	$\Delta H_{298}°$ (KJ/mol)	$\Delta S_{298}°$ (J/mol K)
H_2O (l) = H_2O (g)	44.02	118.68
2FeOOH (s) = Fe_2O_3 (s) + H_2O (g)	52.2	155.4

e. **Wet Zone:** This is the lowest portion of the bed, which has essentially the same characteristics as the original mix. The gas stream during the (sintering) process transfers essentially all its sensible heat to the upper part of the zone, while the lower part of the zone is at the temperature of the original mix.

6.3 AIR FLOW AND PERMEABILITY

6.3.1 Permeability and Sintering Speed

The vertical rate of sintering (VSS) is an important index that affects the productivity of the sinter plant directly. It is defined as: shifting of sinter formation along the height of the layer of sinter charge over the grates per unit time per unit bed area.

It is a function of:

- The permeability of the bed and its height
- The suction pressure

The sintering rate will depend on how fast the air is blown in across the bed. This is dependent on the permeability of the bed, bed area, bed height, and suction. A sinter feed with good permeability from ignition until burn-through is required for efficient operation of a sinter plant.

The following simplified relationship holds well:[4]

$$P = Q/A * (H/\Delta P)^N \tag{6.1}$$

where:

P = bed permeability index,

Q = air flow across the bed (m^3/h at NTP),

A = bed area (m^2),

H = bed height (m),
ΔP = pressure drop (Pa), and
N = an empirical factor related to flow state and depends on material granulometry (value lies in between 0.55 and 0.65).

The velocity of gas infiltration varies in a complex manner with time as the sintering advances. This is in view of the fact that the bed is composed of various sintering zones with each having different bed permeability, which moves downward as the time progresses. The bed permeability includes pre-ignition permeability and (sintering) process permeability:

Pre-ignition permeability is the permeability of the raw mix before ignition.
Process permeability refers to the average bed permeability during the entire process of sintering after ignition.
Minimum permeability refers to the bed permeability, when it value is minimum.

The parameter, pre-ignition permeability, has significance in affecting the pressure drop during and after sintering. A close correlation between the rate of post-ignition air flow and flame front speed (FFS) has been observed. **Figure 6.3** shows the relationship between the rate of post-ignition air flow and flame front speed in the case of various ore mixes of low (L), medium (M), and high (H) bulk densities.[5] A fourth sinter mix (B) was a blend of these three ores and had bulk density similar to that for M.

The sintering process also experiences the bed permeability at its minimum level, which is called minimum permeability and is achieved a few minutes after ignition when the velocity of the air flow through the bed gets reduced. In the case of certain types of Indian ores having characteristics that vary from hard laminated to soft laminated and lateritic, a close relationship has been found among sintering time, the

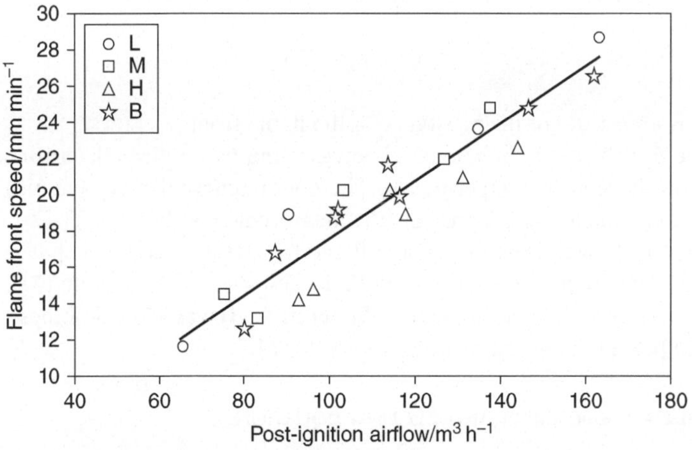

FIGURE 6.3 Dependence of flame front speed on post-ignition air flow (after Ref. [5])

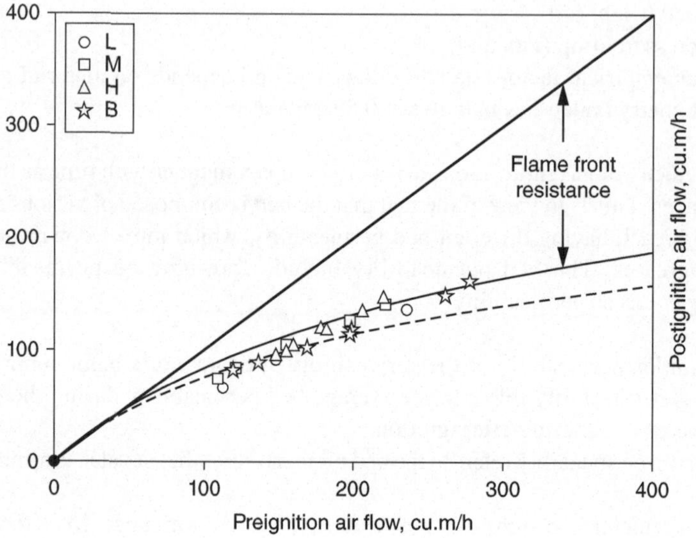

FIGURE 6.4 Relationship between the post-ignition air flow and pre-ignition air flow (after Ref. [7])

permeability of the green mix (Vg), and the minimum permeability (Vmin).[6] The relation between Vg and Vmin at 2.5 and 3.1 sinter basicity (CaO/SiO_2), respectively, as observed, is as follows (r = correlation coefficient):

$$\text{Vmin.} = 0.64\ \text{Vg} - 5.24\ (r = 0.84) \tag{6.2a}$$

$$\text{Vmin.} = 0.71\ \text{Vg} - 5.64\ (r = 0.79) \tag{6.2b}$$

The post-ignition air flow (V_s) has been correlated with the pre-ignition air flow (V_g), both in cubic meter/h, according to the following equation:[7]

$$V_s = Vg - K\,V_s^3 \tag{6.2c}$$

where K is a measure of the resistance of the flame front.

Figure 6.4 shows the relationship between the two indices determined experimentally in the sintering experiments. The bottom curve denotes the measurement at 10 KPa, and the top one represents the measurement at 16 KPa. The straight line, Y = X is also shown in the figure, and flame front resistance is computed from the difference. The ore mixes of various bulk densities: low (L), medium (M), and high (H) were used in the sintering process. Sinter mix (B) was a blend of these three ore mixes and had a bulk density similar to that for M.

6.3.2 Bed Permeability and Ergun's Equation

Several investigators have derived equations relating the flow of gas through a packed bed to the pressure drop across the bed, including Ergun.[8] With regard to bed permeability, air flow through the packed bed, before a sinter bed is ignited, can be described by Ergun's equation.

$$\Delta P/h = A * \acute{\eta}\, Vo(1-\varepsilon)^2/\varphi^2 d^2{}_p \varepsilon^3 + B * pV^2{}_o(1-\varepsilon)/\varphi\, d_p \varepsilon^3 \quad (6.3)$$

where:

d_p = the equivalent spherical diameter of the packing (m),
h = the bed depth (m),
ΔP = the pressure drop across the bed (Pa) and its
Vo = the superficial gas velocity (m/s),
ε = the void fraction of the bed (-),
$\acute{\eta}$, p = the dynamic viscosity of the gas (Pa.s) and its density (kg/cu.m), and
φ = the shape factor.

According to Ergun's equation, key properties that determine the gas flow across a packed bed of a given height are bed void age (ε), effective or mean diameter of the particles (d_p), and particle shape (φ). The equation shows that for a given pressure drop, air flow is increased if there is an increase in φ, ε, or dp.

Factors A and B in Eq. 6.3 have, according to Ergun, the values 150 and 1.75, respectively. The initial void-age may be estimated from the measurements of the bulk density and the real density of the mix at the loading stage on the machine. The value of d_p, the harmonic mean diameter of the green balls that compose the bed packing, is obtainable from a size analysis of the raw mix using the equation

$$1/d_p = f_1/d_1 + f_2/d_2 + \text{-------} + f_n/d_n \quad (6.4)$$

where f_n is the fraction of the balls between two sieve sizes with a mean diameter d_n. The balls are not perfectly spherical, and the effective diameter is related to the harmonic mean diameter by a shape factor φ.

6.3.3 Application of Ergun's Equation to Sinter Mix Bed

Ergun's equation pertains to the isothermal flow of an incompressible fluid across a packed bed. The equation may be applied for computation of the air flow to the bed of sinter mix before ignition. This is because changes in gas volume caused by the pressure gradient across the bed, in view of low suction pressure, have a negligible effect on gas flow.

Several investigators, including Rankin and Roller,[9] showed that the Ergun equation accurately predicted bed permeability over a wide range of moisture additions in the case of green sinter mix. They made calculations for the pressure drop against the gas flow across a packed bed of wet sinter-mix material according to Ergun. The calculated values of pressure drop, while assigning the values of A and B (i.e., 150 and 1.75, respectively), according to Ergun's equation, differ from the measured ones.

As another way to approach the Ergun's equation in order to predict the volume flow correctly, the values for A and B can be adapted from the method of least square fit instead of assigning them fixed values as proposed by Rankin and Roller.[9]

Ergun's equation can be rewritten as:

$$\Delta P = C\,V_o + D\,V_o^2, \text{or } \Delta P/V_o = C + D\,V_o \qquad (6.5)$$

where:

$$C = A\,\acute{\eta}\,(1-\varepsilon)^2 * h/\varphi^2 d_p^2 \varepsilon^3, \text{ and}$$
$$D = B\,p\,(1-\varepsilon) * h/\varphi\, d_p \varepsilon^3 \qquad (6.6)$$

Therefore, $\Delta P/V_o$ when plotted against V_o yields a straight line. The values of C and D can be obtained from the intercept and slope of the straight line.

Calculation of Mean Size and Voidage These may be calculated using Ergun's equation. From Eq. 6.6,

The parameter

$$(0.07)(C/D) = 0.07 * [150\,\acute{\eta}\,(1-\varepsilon)^2 * h/(\varphi^2 d_p^2 \varepsilon^3)]/[(1.75\,p\,(1-\varepsilon) * h/\varphi\, d_p \varepsilon^3)]$$
$$= [6\,(1-\varepsilon)/\varphi * d_p] * [\acute{\eta}/p] \qquad (6.7)$$

$$(0.07)(C/D) = SSA * [\acute{\eta}/p] \qquad (6.8)$$

where, SSA, the specific surface area of the granules and can be computed from the ratio, [C/D] using Eq. 6.8.

The value of void fraction (ε) can be found using the following relationship:

$$SSA/D = [6\,(1-\varepsilon)/\varphi * d_p]/[1.75\,p\,(1-\varepsilon) * h/\varphi\, d_p \varepsilon^3]$$
$$\text{or } \varepsilon^3 = [1.75 * p * h/6] * [SSA/D] \qquad (6.9)$$

The value of equivalent spherical diameter, d_p, can be computed as follows:

$$dp = 6\,(1-\varepsilon)/\varphi * (1/SSA) \qquad (6.10)$$

The shape factor, φ, needs to be assigned a certain value. In the case of a sphere, the value is 1.

Annexure VI.1 lists typical calculation for spherical equivalent diameter and void fraction of granules.

Validation of Ergun's Equation Mitterlehner et al.,[10] while adopting this exercise, have shown that the calculated values of pressure drop match very closely with the observed ones against various gas flow (**Figure 6.5**).

Similar results have been observed in batch sintering experiments using the iron ores from Bhilai Steel Plant (SAIL), India.[11] The experiments were carried out over a wide range of sinter mix characteristics: ore types, moisture, and basicity ratio. The suction pressure and velocity of air infiltration across the bed before it is ignited (pre-ignition permeability) were measured. Three sets of data were generated at different suction pressures

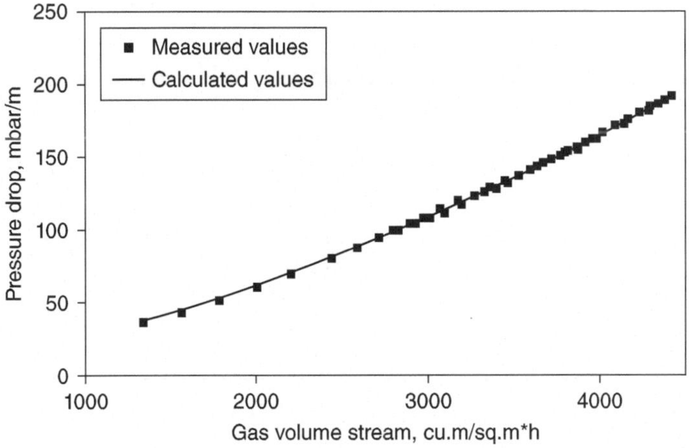

FIGURE 6.5 Relationship between gas flow and pressure drop (after Ref. [10])

in each experiment to validate Ergun's equation.[11] The values of C and D were computed from the two sets of data using Eq. 6.5. The pressure drop, ΔP, was computed from the third set of data by fitting the aforementioned parameters (C and D) in the equation (6.5) and comparing it with the observed one. The value ΔP, thus calculated (from the third set), matched closely with the observed one.[11] This shows that the pre-ignition permeability can be accurately predicted by the Ergun equation over a wide range of moisture addition, sinter basicity and ore type.

Typical Study from SAIL Mines The size of the granules (d_p) and void fraction (ε) in the case of Indian ores[11] have been calculated from the relationship of pressure drop and air flow across the bed before ignition (green permeability) at 2.5 sinter basicity for different types of ores over a wide range of moisture. The results at 0.9 shape factor are shown in **Table 6.1A**. This refers the calculated equivalent diameter (d_p) and bed porosity (ε)

TABLE 6.1A
Calculated Equivalent Diameter (d_p) and Bed Porosity (ε) against Moisture Content of Sinter Mix at Sinter Basicity (CaO/SiO_2 = 2.5) for Rajhara and Dalli Ores

Moisture in sinter	Rajhara ore		Dalli ore	
mix, %	d_p, mm	ε, (–)	d_p, mm	ε, (–)
6.5	6.3	0.095	5.2	0.118
7.0	6.4	0.096	5.0	0.122
7.5	6.5	0.098	5.6	0.108

TABLE 6.1B
Sieve Analysis of Rajhara and Dalli Ores

SIZE (mm)	Weight %	
	Rajhara ore	Dalli Ore
+10 mm	6.5	10.0
−0 mm + 5 mm	13.4	19.6
−5 mm + 3 mm	10.8	20.4
−3 mm + 1 mm	13.8	19.7
−1 mm	55.5	30.3

against moisture content of sinter mix at sinter basicity ($CaO/SiO_2 = 2.5$) for Rajhara and Dalli ores. **Table 6.1B** refers the sieve analysis of Rajhara and Dalli ores.

Rajhara ore is hard laminated to massive hematite ore; whereas Dalli ore is soft laminated, having high laterite (alumina) content. The dependence of granule size on sinter feed moisture in case of Rajhara ore is not as sensitive as in the case of Dalli ore. In the case of Dalli ore, d_p increases significantly with an increase in moisture in the range 7.0–7.5. Mean diameter of granules with Rajhara ore, in spite of its lower (mass average) size, is higher than that with Dalli ore.

Limitation of the Applicability The applicability of Ergun's equation has limitations in the sense that the basis on which the granule characteristics are determined refers to the bed (green) permeability in the wet condition. In practice, the bed structure changes as the sintering proceeds, which affect the velocity of gas infiltration and the sintering speed.

6.4 STRUCTURE AND POROSITY OF BED

6.4.1 Bed Structure

The structure of the bed during sintering is important as it determines the bed permeability. The bed shrinks as the sintering proceeds. The zone of combustion-melting is the most sensitive one to affect the sintering bed structure on account of the sinter cake load upon this zone. The structural changes of the bed that occur during sintering are affected on account of following:

a. The water vapor, initially generated at the top of the bed, gets condensed at the lower portion of the bed once the sinter mix is ignited resulting in bed permeability to reach a minimum value. The minimum permeability is arrived few minutes after ignition when velocity of the air flow through the bed gets reduced.

b. The structure of sintering bed is influenced by the manner the sintering proceeds at its various stages (**Secs. 6.9** and **6.10.1**) namely,
    - Dissolution of the minerals from the mix ingredients
    - Formation of pores within the sinter matrix
    - Cohesion
    - Formation of minerals in sinter
c. The increasing temperature in the sintering bed affects the strength of granule, its size distribution, and its shape. These changes affect the bed packing and void age.

6.4.2 Bed Porosity

While the diameter and shape factor mentioned in Ergun's equation can be measured according to the characteristics of granules, (but) it is difficult to measure the bed porosity. This is because the fine particles in the sinter mix fill the spaces between the granules, which affects the bed porosity and gas permeability. The influence of below 1 mm-sized fractions in granules on the bed porosity has been worked out and the effect is shown in **Figure 6.6a.**[12]

The porosity has been calculated from the measurement of bulk density of sinter mix and true density of granules. Porosity is expressed by the following equation:

$$P = 1 - (\rho_a/\rho_r)$$

P = Porosity (—).

ρ_a, ρ_r = Apparent and real density

The spaces between the granules are easily filled with fine grains; hence, the content of fines in the sinter mix granules affects the porosity and the bed permeability before ignition and also during the sintering.

Size Distribution The void-age of the sinter bed depends on the proportion of small-sized fine particles (S) to coarser ones (L) as well as their size ratio. The latter may constitute coarser-sized ores as well as sinter mix granules. As schematically shown in **Figure 6.6b,** when large (L) and small (S) particles are mixed in different proportions, a porosity minimum occurs. This is when the mixture contains about two-thirds of the large particles. Apart from that, the size difference between the large and small particles affects the porosity; the minimum in porosity becomes more distinct with the larger the (size) difference. At the minimum porosity, the mixing ratio is such that the amount of small particles is just enough to fill the spaces formed between the large particles. (See **Sec. 3.6.5, Figures 3.7** also)

The sinter mix is composed of ore and (return fines) components having coarser sizes (up to 8 mm) and nonferruginous ones (up to 3 mm) consisting of flux and coke breeze. Therefore, there is a gap between these two constituents, and the particles with a size less than 1 mm can easily fill the voids according to **Figure 6.6b** (see **Sec. 3.6.5 also**).

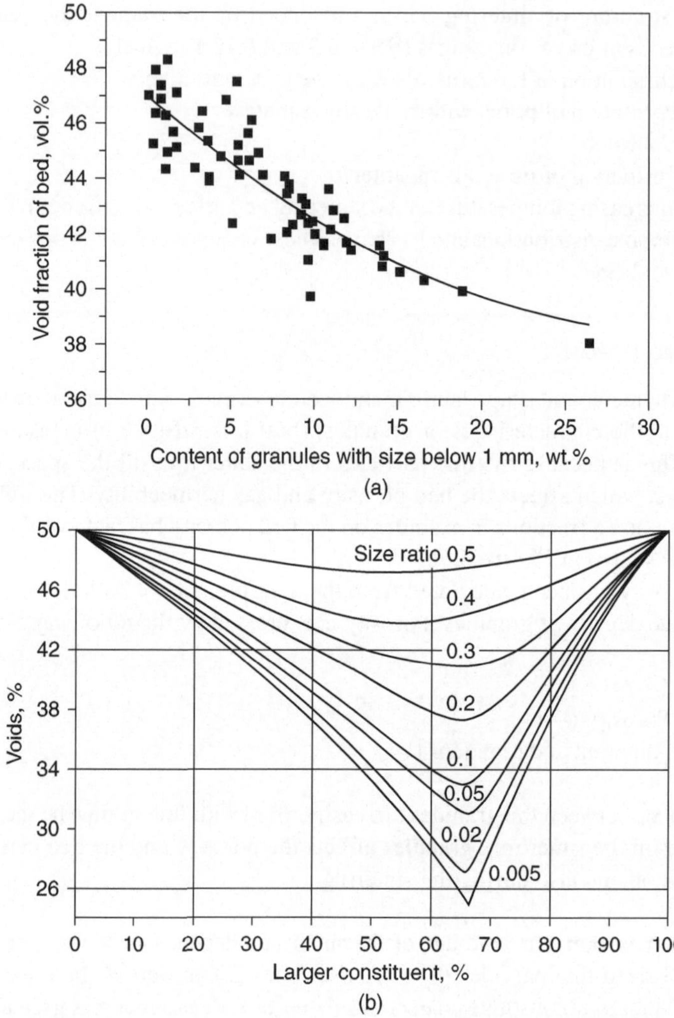

FIGURE 6.6 (a) Effect of –1 mm-sized fraction in granules on its void fraction (Ref. [12]) (b) Effect of inclusion of finer fractions on the void fraction in a two-component sized material of different size ratio. Packing porosity in a two-component system with different mixing ratios and with varying size ratios between the large and small particles is also shown (after Ref. [13])

6.5 GRANULES AND GRANULATION

6.5.1 Effect of Granule Characteristics on Productivity

A sinter feed with good permeability from ignition until burn-through is required for efficient operation of a sinter plant. In view of this, the sinter mix is required to have proper granulation. Kasai et al.[14] confirmed that pre-ignition permeability was related to the mean size of granules, which was, in turn, related to moisture addition.

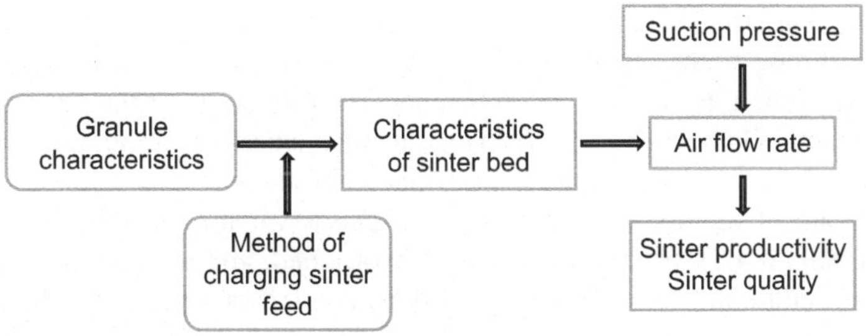

FIGURE 6.7 Schematic showing how the sinter productivity is influenced by granule characteristics

This is more important for the iron ores with high ultrafines content that have low pre-ignition permeability.

The granulation process, in addition to influencing coke location and access to oxygen during combustion, will determine the structural properties of the bed formed for sintering.[15]

The granulation process is required to be optimized to achieve the following objectives:

- To increase the sinter productivity. The size of granules affecting the bed permeability is important in this regard.
- To improve the sinter quality: mean size of sinter, cold strength, and reduction parameters namely, reduction-degradation index (RDI) and reducibility index (RI).

Figure 6.7 schematically represents the effect of granule characteristics on sinter productivity. The shape of granules and their size (distribution) along with the method of their charging on the strand determine the structure and permeability of the bed that undergoes sintering. Consequently, the sinter productivity is affected vis-à-vis the suction pressure of wind boxes.

With the improvement in granulometry of the sinter mix, the sintering efficiency gets enhanced—the phenomenon has a positive effect on the sinter quality also.[5] In this regard, the granulation step is critical to the overall sintering process.

6.5.2 Characteristics of Sinter Mix Granules

The granules of sinter blends are characterized by:

a. Their shape and size distribution parameters. These are:
   - Equivalent spherical diameter (d_p) and size distribution
   - Shape factor (φ)
   - Void age (ε)

   The shape of granules and their size distribution, including their mean size,[13] are key determinants of green bed permeability and the strength of the sintering bed. The speed of flame front in the sintering process is determined by the rate the air from the top of the surface is sucked into the sintering bed. Hence, the green bed permeability is one of the most important factors in deciding the sintering rate and hence the strand productivity.

b. Their strength: The granules must have sufficient strength to withstand the load of the sinter cake and thrust due to air flow (suction pressure) (see **Sec. 6.4.1**). For high permeability of a bed, any compression caused by gravitational forces or external densification of sinter mix should be avoided.

6.5.3 Granulation: Size Classification of Mix Particles

The (raw material) fed to the granulating unit has a widely dispersed size distribution that ranges from up to 5/10 mm in top size down to sub-sieve size ultrafines and moisture content in the range of 5–8 wt.% of the mix. The principal of granulation is to reduce the content of fines by the simple expedient of depositing or layering them onto coarse particles. The granulated product in turn is larger in size and narrower in size distribution than the original feed. In this process, the green permeability of the granulated product is greatly improved.

The prominent growth mechanism in the case of the granulation (of sinter feed) may be termed as "auto-layering" in which the coarse size components of the feed act as seeds or nuclei onto which the fine particles of the same feed can layer. The role of intermediate-sized particles in the auto-layering scheme is somewhat ambiguous.[16] Fouri et al.[17] have suggested that the fractions of ore fines having size in between 0.7 and 0.2 mm essentially plays no role in granulation. Lister and Walters[18] undertook a middle position; the intermediates act partly as nuclei and partly as layering material. The median partition size—the size at which a particle has equal probability of acting as a seed or depositing in the layer—increases with moisture content.

The sinter mix, which undergoes granulation, consists of various types of ores, besides fluxes and other ingredients, each having different size parameters with respect to their role in granulation. Very fine particles are classified as "adhesive ones" as these adhere to the quasiparticles which undergo granulation; whereas the coarser sized particles do not take part in granulation and termed as "nuclear ones." A definition of nuclear and adhering particles based on their size is, however, vague as it is material specific. Nonetheless, with respect to granulation, the mix particles are approximately classified, based on their size, as follows (see **Sec. 6.14.3**):

a. Adhering Particles: Below 0.25 mm-sized fractions in base mix are generally considered to behave as "adhering" particles.
b. Nuclear Particles: Nuclear particles are generally greater than 1 mm.[7,11]
c. Intermediate Particles: The iron ore particles with sizes ranging from 0.25 mm to 1 mm are defined as middle size particles (MSP), which are neither nucleus nor adhesive fines. Their proportion should be controlled (i.e., their proportion should not be high).

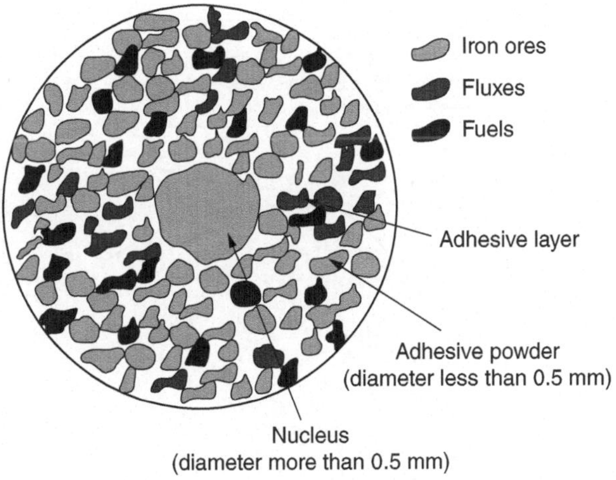

FIGURE 6.8 Model of a granule showing adhesive particles layering onto a nuclear particle (after Ref. [12])

The adhering fine (below 0.25 mm) particles will layer onto the surfaces of all nuclear and intermediate particles when sufficient water is present (**Figure 6.8**). Different adhering fines are expected to behave differently during granulation due to the differences in their moisture absorbing capacity and apparent density.[12]

If water addition is increased, the intermediate particles can be embedded into the adhering fines layer. However, because of their sizes, they are not incorporated into the layer securely. They detach easily when the granules roll over each other in the cascading mix. For a defined drum and water addition level, it is believed that a steady-state exists, where the rates of granule growth and degradation are balanced. If the residence time in the drum is increased further that does not result in an additional increase in the size distribution of the granulated mix. The particles between these two sizes (or intermediate size range) tend to pick up little adhering material and remain unattached (**Figure 6.9**). In view of this, their proportion in blending should be controlled. That is, the proportion of middle size particles (0.25 mm to 1 mm) should not be high.

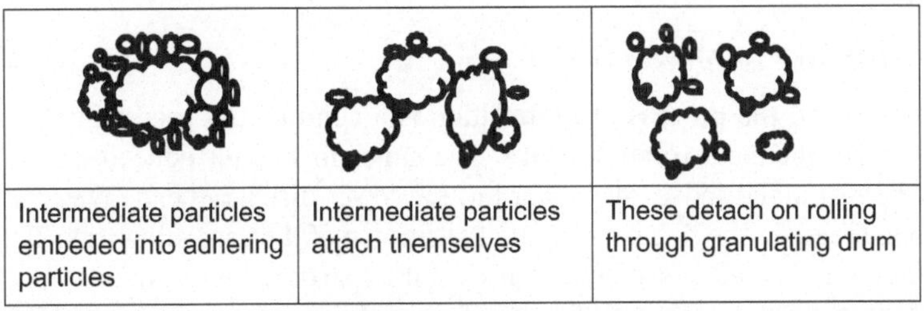

FIGURE 6.9 Schematic representation of formation and breakage of granules pertaining to intermediate particles

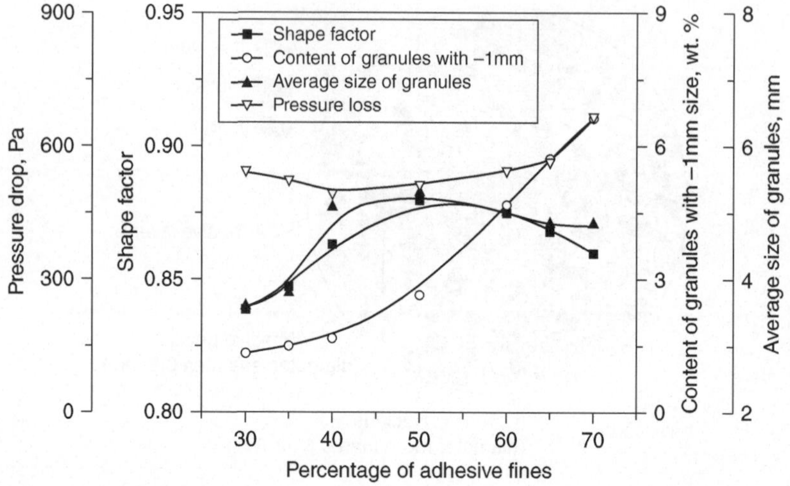

FIGURE 6.10 Effect of adhesive fine proportion on the granule characteristics (after Ref. [12])

Figure 6.10 shows how the characteristics of granules depend on the mass percent of adhesive fines in ore. The content of adhesive fines in the sinter mix does affect the particle diameter, Dp, and shape factor, φ, of granule positively before these falls. Besides, the pressure drop, ΔP, decreases initially and then increases. **Figure 6.10** also shows optimal values of these parameters at around the 40–50% level: both Dp and the φ achieve maxima, correspondingly ΔP has minimum value.[9] The optimal values of these indices at around 40–50% are attributed to the following:[12]

- The adhesive particle does not make a significant impact on the growth of granules at its lower content since the layer which surrounds the nuclear particles is thin. Therefore, the shape of granule is similar to that of ore and irregular; also the shape factor is smaller.
- On the contrary, at higher percentage of adhesive particles, several mini-pellets are produced which lack nucleus particles inside the adhesive fines. It will be difficult for these to grow further.

6.5.4 The Granulation Process

In a sinter plant, the process of granulation is a continuous process and is carried out in a long, inclined, rotating drum. The sinter mix is introduced into the high end of the drum and moves downward to the lower discharge end. Water is added close to the feed end as fine sprays onto the surface of the cascading mix. The water redistributes and manifests as small lenses at interparticle contact points. As the mix cascades in the drum, granules form and grow. In the process of granulation, fine adhering particles (below 0.25 mm) layer onto the surfaces of nuclear particles, also of intermediate particles, when sufficient water is present.

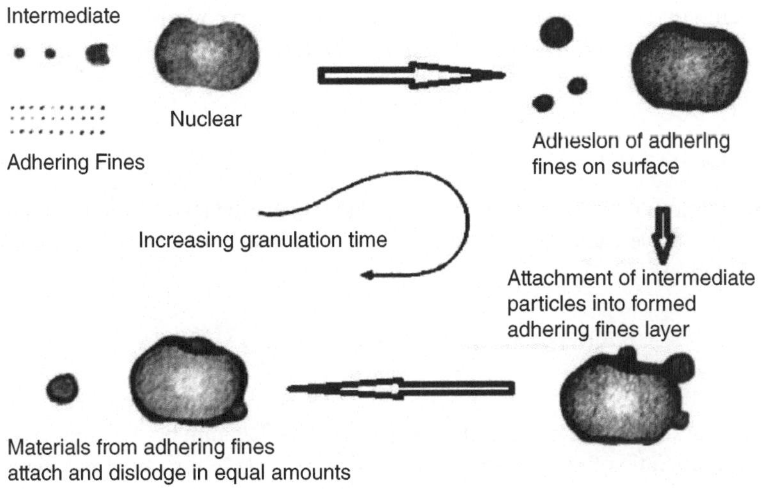

FIGURE 6.11 Schematic representation of the granulation scheme (after Ref. [15])

Granulation of iron ore sinter mixes is a continuous process. Steps involved in the formation of granules are (**Figure 6.11**):

a. Coating of fines material onto the surface of large particles
b. Incorporation of intermediate particles into the formed adhering fines layer
c. Dislodging of poorly adhering material

When (b) is equal to (c), a steady-state condition is achieved.

Different adhering fines (below 0.25 mm) have differences in their moisture absorbing capacity and apparent density;[15] hence they behave differently in granulation. The particles of intermediate size tend to pick up little adhering material and remain unattached (see **Figure 6.9**). These intermediate particles, on further addition of water, are embedded into the adhering fines layer. However, these are not incorporated into the layer securely because of their sizes; hence, they detach easily when the granules roll over each other in the cascading mix. In view of this, the proportion of mid-sized particles in blending should be controlled; that is, the proportion of middle-size particles (0.25 mm to 1 mm) should not be high.

As the materials cascade, a steady-state exists for a defined drum and water addition level where the rates of granule growth and degradation are balanced. Further increase in residence time in the drum does not result in an additional increase in the size distribution of the granulated mix.

The mixing and granulation of the sinter mix are affected by their pattern of movement in the drum mixture. In the low degree of filling, the frictional force between the drum and the sinter mix is large enough to overcome the granulating force of the sinter mix. This results in a cascading effect resulting ungranulated/partially granulated sinter mix in the drum.

Particles are held together by the surface (tension) forces of the bridging water lenses. The size distribution of the discharged granules is strongly dependent on the amount of water added. When the sinter mix contains porous particles, increased water addition is also necessary to maintain the same level of interparticle surface water.

6.5.5 Granulation Index

The granulation of sinter mix can be assessed by using the term "granulation index." Hara et al.[19] have used the term ($GI_{-0.25}$), defined as Eq. 6.11, for the estimation of the granulation.

$$GI_{-0.25}[\text{mass}\%] = [1-(B_{-0.25}/A_{-0.25})]*100 \tag{6.11}$$

where:

$A_{-0.25}$: mass ratio of raw material having size below 0.25 mm before granulation [%]
$B_{-0.25}$: mass ratio of mix materials having size below 0.25 mm after granulation [%]

What is meant is that the granulation improves with the increase of $GI_{-0.25}$.

The adhering particles in ore fines, defined in **Sec. 6.5.3,** relate to below 0.25-mm-size fractions ($A_{-0.25}$), undergo granulation, and hence decrease to result in a lower size fraction ($B_{-0.25}$). The middle-size particles do not take part in granulation. Hence, their proportion should be controlled; that is, their proportion should not be high.

Nippon Steel Corporation defines the G index for ore granule size classification, from the point of sintering behavior, using the expression:[20]

$$G = MF/N \tag{6.12}$$

where:

N = mass % above 0.7 mm size,
F = mass % below 0.2 mm, and
M = intermediate fraction.

The aforementioned definition, like the previous one, is based on the fact that the ore of size fraction greater than 0.7 mm does not take part in granulation, whereas the weight of below 0.2-mm-size fraction decreases as it forms granules of intermediate size.

The term "granulation index" can be defined in several ways depending on its applicability and the nature of the ore, though the basic concept is the same. The hematite-grade iron ore fines, typically from the eastern part of India, are heterogeneous in size as they are not normally distributed. In this case, the granulation parameters could be worked out differently. Typically, the identified parameters for the assessment of granulation are as follows.[21]

Harmonic mean diameter ($Dp_{[total]}$) of the sinter mix
Harmonic mean diameter ($Dp_{[-1.68]}$) of the size fraction less than 1.68 mm in sinter mix
Weight %: the size fraction of the sinter mix having size less than 1.68 mm and more than 0.6 mm ($wt.\%_{[-1.68+0.6]}$)

Harmonic mean diameter (Dp) is computed as follows: $Dp = 100/\Sigma\,(wi/di)$, where wi is the weight fraction of the material with particle size, di.

It has been observed that these parameters correlate well with the permeability during the sintering vis-à-vis its speed and other sintering indices.[21]

6.5.6 Granulation: Mechanism and Factors Affecting

The process of granulation initiates with "auto-layering" that involves the initial formation of an inner layer of ultrafines coating the surface of a nuclear particle. This is followed by an outer layer of intermediate-sized particles embedded in a matrix of fines.[18]

Capillary forces and liquid bridging are the principal bonding mechanisms in granulation. In the course of granulation, the agglomerates undergo a continuous gradual compaction, at least in the initial period of growth. As a result, the void spaces become increasingly filled with liquid. The liquid bridges formed are the principle cohesive forces that hold the particles together and favor granulation. For a given ore blend, sufficient nuclei are required to seed granule formation. The particles are held together by liquid bridges (liquid trapped) and surface tension of water is crucial in this respect.

The wet agglomerate undergoes different states of granulation as the material is saturated with liquid, during its rolling (see **Sec. 12.2.5** and **Figure 12.10**). Pendular state is formed at low saturation when water presents just at the particle contact point. This state is the suitable regime for water content with regards to the granulation of iron ore in the process of sintering. In this regime, two forces contribute to the tensile strength of the bond in an additive fashion; the pull is due to surface tension at the solid-liquid-gas contract line directed along the liquid surface and the negative capillary pressure existing in the liquid bridge (see **Figure 12.11**)[22]

Granulation is a complex process in which many input variables influence a multitude of product properties. Granulation is also material specific as different types of iron ores in the sinter base mix behave differently with regards to granule growth and moisture requirement for optimum bed permeability and void age.[18,23]

The characteristics of granules are determined by how the granulation proceeds with different sized raw materials. Factors that do influence the granulation process, also the characteristics of a granule are include:

a. Material properties of the sinter feed/nature of ore:
   These include particle size and size distribution, particle shape, density, microstructure and mineralogy, internal porosity and surface roughness of the ore, initial moisture content of sinter feed, moisture absorbing capacity of ore, extent of laterization, and loss on ignition (LOI) content of iron ore.
   - The particle sizes of the mix ingredients including ores, solid fuel, and fluxes, and also their distribution patterns in the mix, influence granulation during balling and transfer of heat from the hot gases to solids and vice versa.
   - The diameter of granules and porosity following granulation varies widely for the sinter mix with types of ores: whether they are laminated or softer ores (see **Sec. 6.3.3**).

- The presence of gangue minerals, including clay mineral (kaolin), also has significant influence on the granulating properties of the mix. High LOI ores, also goethite ores, make a positive contribution to granulation in view of their high moisture retention capacity.
- Therefore, the granulation of sinter mix improves when massive to hard-laminated ores (with high Fe) are blended with high LOI ore. The blending of different types of ores might be essential in order to incorporate higher amounts of ultrafines and also to improve the bulk density of ores.
- Internal porosity of ore determines the amount of moisture available for the granulation (see **Sec. 6.4.2**). The voids in the dry and wet zones could be prevented from being filled up with proper agglomeration of the fines at the mixing and balling stages.
- The granulation process is reflected through the mass fraction of adhering (smaller-sized) particles in the sinter mix that report to the coarse granules after granulation and also their sizes. The adhering behavior of different types of fines present in a typical sinter base mix differs and necessarily is required to be explored through experimentation (see **Sec. 6.5.3**). The granulation process is also modified by coating coke breeze and limestone on the surface of quasiparticles that have been granulated in the primary part of the drum mixer.
- Inherent moisture content of ore and the rate of granule growth with increasing feed moisture content are important parameters in affecting the granulation of sinter mix. Moisture in the sinter mix promotes a balling effect on the particulates that undergo granulation. Consequently, the amount of very fine particles in the mix decreases. For a given ore blend, with sufficient nuclei to seed granule formation, the granulation is optimized by varying the granule moisture. In general, the strength of the granules increases with increases in the water content during granulation. The amount of optimum moisture content in green mix required for granulation is, however, decided by material properties of the raw materials used in sintering.

b. Process variables

These include the types and amount of sinter mix ingredients and content of moisture: the specific consumption of lime in the mix, recirculation of sinter return fines, temperature of green mix, addition of binders in the mix, if any, and their type and dosage.

- **Moisture content**. The presence of moisture in the sinter mix has an important role in affecting the granulation of the sinter mix. Besides, the removal of moisture and its rate of removal is very important as the sintering progresses. This is in view of the fact that the rate of sinter formation depends on the shifting of the drying zone. Therefore, for having an optimum rate of sintering, not only the laws of gas dynamics but also those of evaporation of moisture have to be taken into consideration.

- **Lime (active CaO)** Burnt lime is slaked into hydrated lime when it reacts with water in the sinter mix in the mixing and granulation stage. Lime in contact with water (in the sinter mix) promotes the improvement in its surface tension (see **Sec. 7.4.5** also) resulting in its enhanced binding properties. The improved process of micropelletization provides the balls (and hence the bed structure) with more stability during the sintering process. The underlying mechanisms for increased strength of individual balls, with the addition of lime in sinter blends, are: (i) The contact between the reactive charge components of balls is enhanced, and (ii) the granules are hydration hardened caused by solution-colloids formation crystallization.
- **Lime and moisture** The addition of lime increases the moisture requirement as well as the granulation time. This is because the diffusion of moisture is not so easy when colloid $Ca(OH)_2$ is formed from the hydration of burnt lime, which has a huge specific surface area and strong hydrophiliciy. The optimum granulation time is 5 mins. This produces larger and stronger granules.
- **Surface tension** The ball-ability of the sinter mix improves when water is mixed uniformly throughout the sinter mix. This can be done by increasing the flow-ability of water when its surface tension is reduced.
- **Selective granulation** The ore having a particle size less than 3 mm is separated and granulated in advance. Then, granulates are added to the residual raw mix: The granulation is thus improved. The high alumina fraction of iron ore fines is separated out and pelletized to a size to act as nuclei particles. These high-alumina nuclei particles are then fed into the remaining base mix for further granulation. The adverse effect of high alumina content in sinter is, thus, neutralized (see **Secs. 7.4.3, 10.8.1,** and **10.8.2**, **Figures 10.10, 10.11, and 10.12**).
- **Return fines** The addition of return fines in the sinter mix also promotes granulation and supports the bed material to prevent it from collapsing.[9] However, increasing the amount of return fines in sinter mix has limitation.
- **Mix temperature** The temperature of sinter mix, increased by exothermic reaction of hydration and granulation, improves by the binding effect of $Ca(OH)_2$, which can ease moisture condensation in the bottom layer of the sinter bed and improve granule strength.

  Temperature of the green mix can be increased above its dew point with steam preheating or with the addition of hot sinter return fine (see **Sec. 7.5.3**). The mix temperature is increased by exothermic reaction of hydration and granulation improved by the binding effect of $Ca(OH)_2$, which can ease moisture condensation in the bottom layer of the sinter bed and improve granule strength (see **Sec. 7.4.5**).
- **Basicity of sinter (mix)** The increasing proportion of flux (nonferruginous) fraction in the sinter blends with increasing basicity is important in contributing the granulation of the sinter mix. Fluxes have a higher water absorption capacity than iron ore.

c. Operational conditions.
These include machine parameters related to mixing and granulating equipment, residence time in these equipment, and the stage of feed component addition. The granulation of sinter mix is reported to be improved with modification of machine parameters of feeders and balling units (see **Sec. 7.4.4**).

6.5.7 Moisture Content of Sinter Mix: Effect on Granulation and Permeability

The iron ore granule grows up with increasing the water content added into the mix. The mass fraction of the particles with smaller size decreases with increasing the water content. The water requirement for optimal granule growth varies according to the ore.

The controlling parameter of permeability and strength of the bed is the amount of moisture in the green mix during the granulation, which is decided by material properties of the raw materials used in sintering. Moisture has an important role in the preparation of the charge for sintering. It has two functions, namely,

- Improvement in balling. The amount of very fine particles in the mix decreases due to the balling effect and, consequently, the gas permeability increases.
- Heat transfer in sintering process. The presence of moisture sharply accelerates the heat transfer from the gas to the charge; due to this, the heat is concentrated at a small layer and helps in achieving the high temperature in the zone of sintering at a compensatively lower fuel rate.

Moisture in the sinter mix increases the gas permeability and decreases the amount of very fine particles of the mix. This is due to its balling effect on the particulates that undergo granulation. The effect of water addition to a sinter mix is schematically shown in **Figure 6.12**. Insufficient moisture prevents proper agglomeration of fines, which then fill the voids in the dry and wet zones. The agglomeration

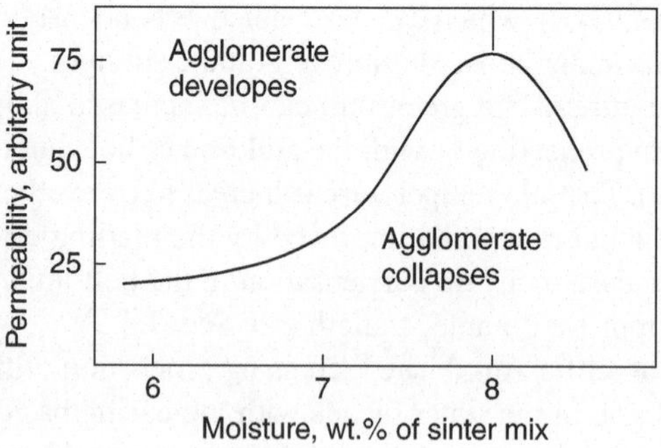

FIGURE 6.12 Schematic representation of the agglomeration of mix material with its water content

of the sinter mix is severely affected at higher moisture content. Excessive moisture promotes condensation in the bottom of the bed, which fills the voids with free water and may cause the collapse of the bed structure in the wet zone.

The moisture content of the sinter mix is to be kept slightly less than that corresponding to the mix. The pressure drop across the bed increases with higher moisture content of the sinter mix. This is attributed to a decrease in bed void-age due to:[9]

a. Accumulation of condensed water in the interstices between the granules
b. Swelling of the granules with a consequent increase in mean diameter of the mix material.

The sinter mix undergoes agglomeration at the mixing and balling stages before it is charged into the sintering machine. An important criterion of proper agglomeration is attributed to the fact that the voids in the dry and wet zones should be prevented from filling up with water. This is essential for effective gas-solid contact during the sintering through improved bed permeability.

Moisture Absorbing Capacity Not all the moisture added to the mix takes part in granulation as some part of it is absorbed through pores. The moisture available for granulation is the total moisture added during granulation less the moisture absorbed within the ore.[18] The ability of the ore particles to absorb or hold the moisture is termed their moisture absorbing capacity and is an inherent material property. The moisture absorbing capacity of different constituents of the sinter mix is different and increases in the order: Iron ore fines < limestone < return fines < coke breeze.

The mixes containing iron ore fines as adhering particles require lesser moisture (around 5 wt.%) to reach the peak bed permeability due to their lower moisture absorbing capacity. The sinter mix containing limestone, return fines, or coke breeze as adhering particles may require higher moisture (6–6.3 wt.%) to achieve peak bed permeability

The total moisture available for granulation depends on the number of pore openings exposed to the liquid. More water at the surface is available for granulation when absorption of liquid on solid decreases. This can be achieved by:

- Decreasing the ore porosity
- Increasing the contact angle of the ore

6.6 THERMAL CHARACTERISTICS DURING SINTERING

6.6.1 Heat Transfer and Sintering Reactions

The sintering process consists of the passage of a heat and reaction front through a packed bed of solids and occurs in layers. An important feature of sintering is that at each moment coke particles present in a narrow layer, 30 to 40 mm burns out. The heating of the sinter mix at each elementary layer takes place both due to heat transfer from layers lying above (mainly by convection) and due to the heat generated during combustion of fuel. The heat accumulated in these layers is spent in preheating and melting particles of the sinter mix and also in the occurrence of endothermic reactions (decomposition of hydrates and carbonates).

The air that enters the sintering bed is heated by the hot finished sinter until it attains its peak temperature in the combustion zone. The hot sinter is subsequently cooled while the gases following through the bed continuously carry the heat from the flame front further along the bed. Besides the heat transfer, the sintering process involves numerous chemical reactions and mineralogical changes that take place in heating, sintering (combustion), and cooling zones. The reaction zone is shifted continuously toward the grate as air is drawn through the bed.

To attain a high temperature, the heat generated by combustion in the upper layer must travel down the bed at the same rate as the combustion zone and so be available at the right time for raising the temperature of this zone.

6.6.2 Heat Front and Flame Front

The thermal characteristics of sintering are described by the formulation of the "heat front" and "flame front" concepts proposed by Voice and Wild.[4] Besides, the front, due to various reactions during the sintering process, decides the temperature profile. The measurement of temperature in the bed at various levels enables the movement of the combustion zone, where the (majority of) reactions take place. According to Voice and Wild:[4]

Heat Front. Heat front has been defined as the level at which the bed temperature first commences to rise uniformly and is taken as 100°C.

Flame Front When the solid fuel is added, the term flame front comes into the picture. It is described as the position of the rapid rise of bed temperature caused by combustion of the fuel.

In addition, there is:

Velocity of Heat Transfer A heat front moves through the sintering bed with a constant velocity that is related to the gas velocity across the sintering bed:

$$\text{The velocity of heat transfer front} = h_g V/h_s\,(1-\varepsilon) \tag{6.13}$$

where:

h_g = the volumetric heat capacities of the gases (Kcal/m^3),
h_s = the volumetric heat capacity of solids (Kcal/m^3),
V = the volumetric flow of the gases (m^3/m^2-min.), and
ε = void fraction of the bed.

The previous relationship shows that the velocity of the heat transfer front is directly proportional to the gas velocity. The specific volume of air needed for heat transfer has a close similarity with that for sintering. This suggests that the volume of air required for sintering is decided by the heat transfer needs rather than the fuel combustion needs.

Flame Front It is the region of coke combustion and is defined by the speed at which the maximum temperature thickness moves down through the bed. It is the combination of heat front and reaction front. The flame front speed can be determined by the time required for particles to reach combustion temperature.[24]

6.6.3 Speed of Flame Front

The flame front speed (FFS) is defined as the rate of travel of the combustion zone. On an average, the FFS can be calculated from dividing the bed height (mm) (of sinter mix in green stage) by the time (s) taken for waste gas temperature to peak indicating burn-through.

FFS is a function of convective heat transfer through the bed air flow rate and the temperature differential between the flame front and unfired bed.[24]

The convective heat transfer and air flow rates are closely related.

$$\text{Heat transfer flux per unit area of layer below } Q(J/s) = U\,\Delta T \tag{6.14}$$

where:

U = heat transfer coefficient (J/s/K)
ΔT = temperature differential between the flame front and un-fired bed

The heat transfer coefficient between solid particles and combustion gases increases with the decrease in particle diameter, which results in an increase of the average mass temperature of the particle. The coarser grains act as a heat sink as the heat content of the waste gas cannot be sufficiently absorbed by the solid.

FFS and Process Permeability The main factor that drives the flame front downward is the convective heat transfer from the flame front to the region below. The phenomenon of heat transfer is determined by the post-ignition permeability, which is well correlated with the FFS (**Figure 6.3**).[7]

The flame front is governed by

- **Amount of solid fuel** in the **mix** and **its combustion efficiency**. The combustion efficiency of the solid fuel is influenced by its reactivity and size.[25] Wood charcoal happens to be one of the most reactive fuels, whereas dense graphite electrodes are among the most unreactive that burns more completely to CO_2. The finer grinding of coke increases its reactivity.

  Figure 6.13a shows a plot of flame front speed (FFS) versus fuel reactivity ([r]). The relationship is given by the following empirical equation:[26]

$$\begin{aligned} FFS &= 0.2014.\ \ln\,[r] + 4{,}039, \\ R^2 &= 0.999 (R = \text{correlation coefficient}) \end{aligned} \tag{6.15}$$

  Figure 6.13a suggests that the rate controlling step shifts from a combination of chemical and diffusion control to being principally controlled by diffusion as fuel reactivity increases.

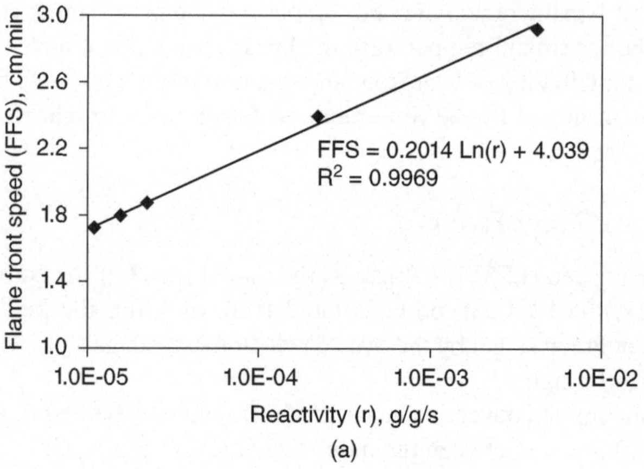

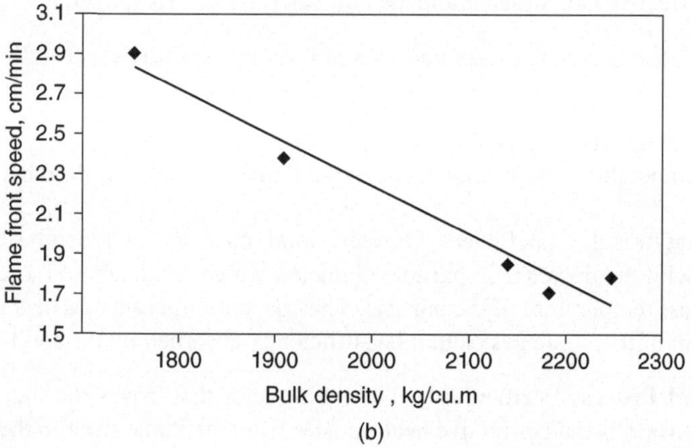

FIGURE 6.13 (a) Relationship between fuel reactivity and flame front speed (after Ref. [26]) (b) Effect of bulk density on flame front speed of sinter bed (after Ref. [26])

- **Bulk density** of **sinter mix. Figure 6.13b** shows the relationship between the bulk density of sinter and flame front speed. The flame front speed would decrease when bulk density of the bed increases.[26] The dense material acts as heat sink and affecting the heat transfer and flame front speed.
- **Composition of mix**. The particle size and the distribution particles in the mix affect the flame front speed.[27] With the decrease in particle diameter, the heat transfer coefficient between solid particles and combustion gases increases.

6.6.4 Gas Dynamics vis-à-vis Heat Patterns

The heat pattern is represented by the bed temperature profile during the sintering process. This is characterized by the thickness of high temperature zone and the maximum bed temperature. **Figure 6.14a** shows a sintering bed having an embedded thermocouple. As the flame front approaches the thermocouple and then traverses it,

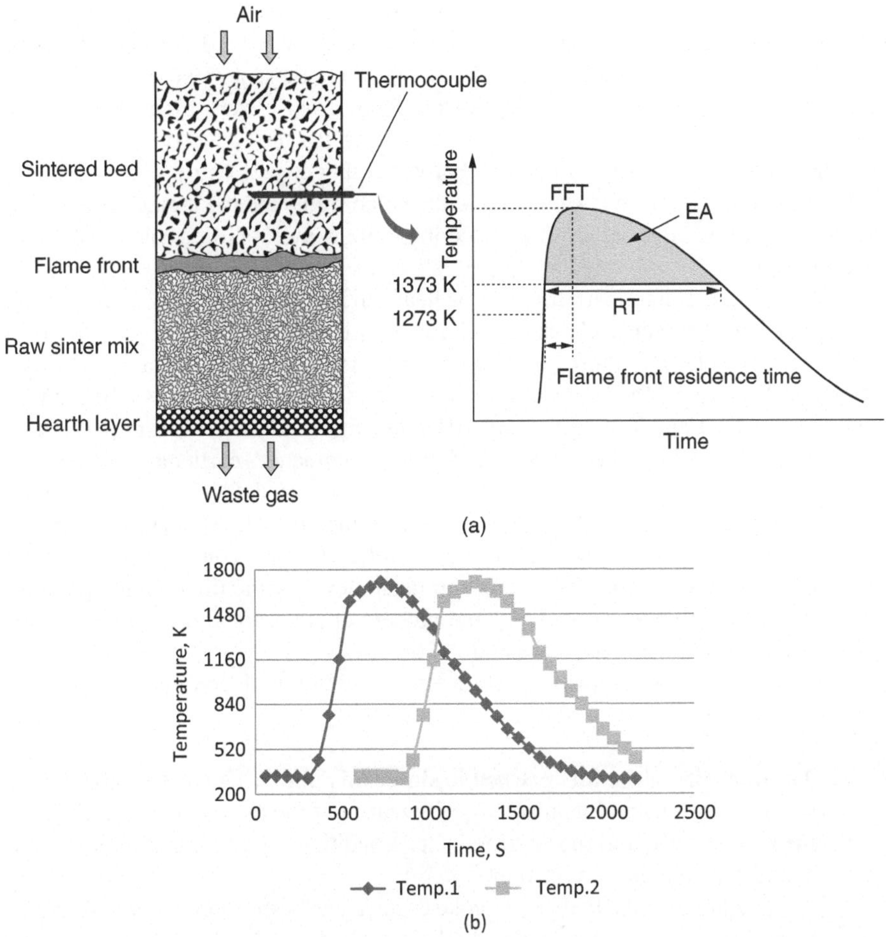

FIGURE 6.14 (a) Schematic of flame front location in the sintering bed and temperature profile (after Ref. [15]) (b) Temperature profiles as the sintering commences at different bed heights (after Ref. [28])

a temperature–time profile is obtained. At the lower region of the bed, the maximum temperature can be as high as 1,350°C (1,623 K). The temperature-time profile has a characteristic shape of a very steep temperature gradient at the leading edge and a much more gentle downward sloping section at the trailing edge. The three critical flame front properties are: flame front temperature (FFT), flame front width (FFW), and flame front speed (FFS)—the latter two determines the (flame front or heating) residence time (RT).[15] The left portion of the shaded area denotes the flame front area, while the right portion is the reaction area. The temperature-time profile shown in **Figure 6.14a** has two important parameters which influence the sinter quality. These are (a) residence time (RT) above 1,373 K, and (b) enclosed area (EA) above the ordinate = 1,373 K line. The area is proportional to the total heat imparted to the bed which causes the melt and greatly influences the sinter strength.[15]

Several endothermic reactions—namely, evaporation of water, transformation of goethite to hematite, and calcination of limestone—occur ahead of the flame front. The sharp temperature gradient, even before reaching these temperatures, indicates that the intensity of heat flowing out of the flame front is very high. As this heat flow intensity is maintained till 1,000°C, it is clear that the convection of heat from the flame front to the thin region ahead of its leading edge is extremely efficient. A solid–liquid–gas mixture is generated in the flame front, and the temperature gradient at the trailing edge shows that the cooling rate is much slower compared to the heating rate.

To better understand the transfer of heat during sintering, it is necessary to recognize the thermal cycle that is characteristic of the process. At any given time during sintering, there exists a temperature distribution that is typically shown in **Figure 6.14b**.[28] The curves denote the temperature profiles of the sintering bed, **Temp. 1** and **Temp. 2** at two different heights H1 and H2 from the bottom of the unit (H1 > H2). At height H1, the sintering commences earlier and the temperature reaches its peak around 720 S since the ignition started. By the time the section below it at height H2 is still above the ambient temperature, it is being heated by the incoming hot gas passing through the section H1 that is now (subsequently) cooled. The bed surface at H1 has cooled to a significant level by the time the temperature at section H2 attains its peak. Thus, the process of sintering continues downward through the various sections of the bed.

The bed temperature profile attained in a sinter bed containing a fuel is dependent upon:

a. **The quantity of the fuel present in sinter mix and its characteristic.** The size of solid fuel and its reactivity, the nature of the association of coke in the mix granule, and the environmental condition (partial pressure of oxygen) determine the combustion.

   A higher reactive fuel and/or coke of finer size burns more rapidly giving more CO to the exhaust gas and a broader bed temperature profile. Fuels containing low temperature volatiles are unsuitable for their use in the sinter mix not only because they are more reactive, but also the volatilities distill out ahead of the flame front resulting in a broader temperature profile. In addition, the volatilities could be condensed at the lower part of the bed and in the wind boxes and collecting mains. This means that the calorific value of coke is not effectively utilized.

   The effect of coke parameters on the thermal characteristics of sintering should not be seen in isolation as it is interinfluenced by the partial pressure of oxygen in the gaseous stream. The combustion rate should be such that the flame front travels with the heat front. This condition is called "matching" and gives the maximum bed temperature profile. In the case of coke, this occurs at 21 vol.% of oxygen. A much lesser oxygen content (approximately 9 vol.%) is required in the case of charcoal, which is much more reactive than coke. Any other condition is termed as "mismatching," which gives a broadened temperature profile. **Figure 6.15** represents these two extreme conditions: (a) coke and (b) wood charcoal under air and oxygen,

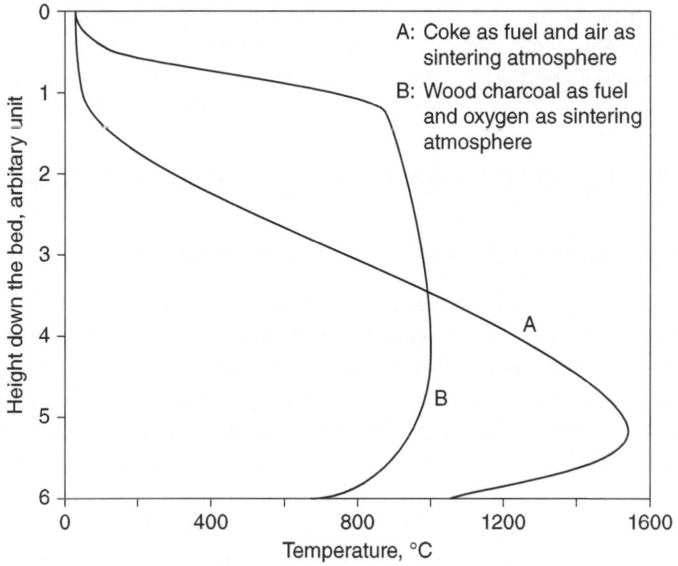

FIGURE 6.15 Bed temperature profile in the case of coke and wood charcoal in the mix, with air, and oxygen as the sintering atmosphere, respectively (after Ref. [29])

respectively, as sintering atmosphere.[29] These give to profile of normal bed temperature and much widened temperature profile, respectively.

The incorporation of coke breeze of a finer size (below 0.25 mm) as well as of a bigger size (above 3.15 mm) is not good for sinter making as incorporation of these fractions in the sinter mix causes a mismatch of flame front and heat front. In the former case, it burns faster and the burning front moves ahead of the sintering front; in the latter case, it lags. In both the cases, utilization of the heat is not proper

b. **The quantity of ignition heat input**. The influx of heat through ignition should be adequate to burn the solid fuel at the bed surface within the stipulated time. This is detailed in **Sec. 6.13**.
c. **The specific heat of bed material**. This determines the solid-gas heat transfer and consequently the requirement of air for sintering.
d. **The exothermic and endothermic reactions, exclusive of fuel combustion**. The sintering process experiences various reactions in the combustion zone and during cooling of sinter, besides the combustion of solid fuel. The reactions liberate heat or absorb heat influencing the temperature profile of the sinter bed.
e. **The temperature to which the solid is preheated prior to combustion**. The solid sinter mix, preheated to a certain temperature (using steam or from other sources) during its granulation process in the balling drum, releases its sensible heat to the sintering bed. This affects the bed temperature profile. The preheating of the mix reduces the requirement of solid fuel in the sintering process.

f. **The temperature and quantity of combustion gases after burning of fuel**. A higher temperature or/and higher amount of combustion gases indicates inadequate thermal efficiency. Additionally, a greater amount of CO in the exhaust indicates that the combustion of carbon in coke has not given its heat fully.
g. The use of a very high suction may result in lower thermal efficiency.

6.6.5 Heat Pattern in Sintering versus Sinter Quality

The heat pattern is strongly influenced by the mass flow rate of gas. The width of bed temperature profile reduces with the increasing flow rate: The higher the mass flow rate of gas, the sharper and shorter are the heating and cooling times. A higher gas flow rate increases the production of sinter while the strength of sinter decreases. A decrease in sinter strength is due to the fact that temperature of the solid particles changes more rapidly with the increase in flow rate; consequently, the retention time at high temperature zone (above 1,100°C) decreases.

The heat pattern could be tailored in such a manner that heating time is reduced while cooling time is lengthened at high temperature. This region results in lower FeO formation.

The desirable heat pattern for favored mineralogy of sinter and its improved quality is:

a. Maximum temperature in the bed (T max) should be below 1,300°C. This helps in suppression of formation of multicomponent magnetite and decomposition of calcium ferrite.
b. The temperature rise time (θ_1) in the high-temperature zone should be as short as possible because in this zone, partial pressure of oxygen is low on account of combustion of coke in the sinter mix (thus, generation of FeO is reduced).
c. The temperature fall time (θ_2) in the high-temperature zone should be long for the purpose of securing the strength of the sinter.

   Mill scale is used to increase the temperature fall time in the high zone. Addition of mill scale has been found to improve thermal conductivity, also the bulk density of high basic sinter, and it prevents the crumbling of sinter. However, if a very high amount of mill scale is added, it leads to breaking up granules in the mix and worsening of gas permeability.
d. With the increase in heat capacity of gas:
   - The speed of flame front increases.
   - The thickness of the high temperature zone becomes narrow.

   Water vapor has a higher heat capacity. Hence, the heat capacity of gas can be increased by the addition of water in air that is subjected into the sintering bed on suction.
e. The upper part of the bed suffers from inadequate heat, which results in weaker sinter. This problem can be addressed through:
   i. The enrichment of oxygen concentration in the induced air provides effective heat utilization required for the sintering bed, such as fast coke fuel combustion and high temperature at the subjected area. High

temperature should be maintained, especially in the upper part of the bed, in order to maintain proper strength of the sinter. Therefore, supplying oxygen into the projected region can produce better quality sinter.

ii. Besides, double layer charging of the sinter mix in the sinter strand: The top portion of the bed with the higher amount of coke and the lower portion with the lesser amount of coke compensates for the deficit of heat requirement in the upper layer and, consequently, stronger sinter is produced.

f. The combined use of closely sized limestone and coke in the sinter mix raises the maximum temperature in the sinter bed and shortens the high-temperature holding time. Further, the pore network of the sinter case is improved, the number of carbon pores in sinter is reduced, and the fine pores are increased in numbers and dispersed.[30]

6.6.6 Burn-Through Point

In a batch sintering setup, a thermocouple is inserted in the wind box to measure the temperature of exhaust gas. The temperature when plotted against the time as the sintering proceeds has the characteristic shape as shown in **Figure 6.16**.

In a continuous sinter plant, the exhaust gas temperature is measured by inserting the thermocouples at different wind boxes through the machine length from the beginning to the end of strand.[31] The profile, thus obtained, bears the characteristic shape similar to **Figure 6.16**.

- The temperature remains flattened for a major portion of the entire sintering duration and then raises suddenly. This point, temperature rise point (TRP) marks the position when the raw sinter mix zone in the bed just disappears. Ahead of this point, the heat associated with downdraft gases is absorbed by the wet raw mix.

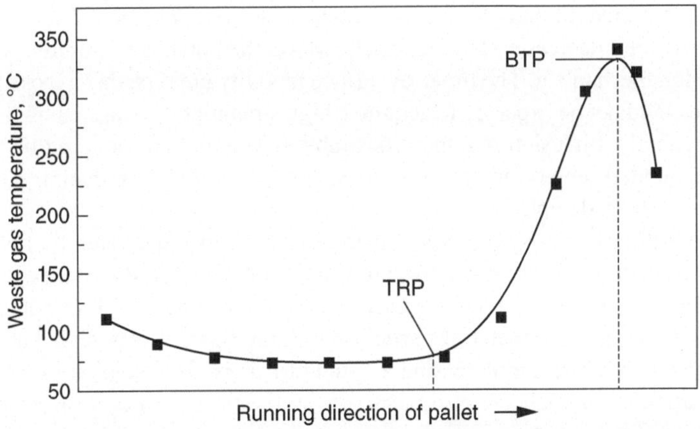

FIGURE 6.16 Profile of exhaust gas temperature showing the burn-through point (after Ref. [31])

- The burn-through point (BTP) refers to the position when the temperature of exhaust gas reaches a maximum and then decreases. This represents the completion of sintering (i.e., the entire of the bed is sintered); thereafter, cooling starts. The time to achieve the burn-through point is called "burn-through time." The BTP may be advanced or lagged depending on the bed permeability. This includes pre-ignition permeability and sintering bed permeability.

6.7 BONDING IN SINTER

Bonding is guided by the mineralogical changes that prevail during the sintering process. Theoretically, two types of bonding prevails:

- By diffusion: This is related to the process of recrystallization in the solid.
- By slag formation: The glassy melt forms a matrix in which the crystalline constituents are embedded.

Diffusion bonding sinter prevails due to recrystallization and crystal growth of hematite (magnetite). Hematite minerals have relatively high surface mobility at about 1,300°C and are the most important diffusing agent among the various constituting minerals. The criteria for diffusion bonding to take place effectively are:

a. Purity of the constituting minerals being sintered
b. Very close packing and good contact of particles

Such bonds are obtained in Fe-rich low silica ores since the impurities in lean ores form low-melting liquids before the "diffusion" temperatures are reached. The majority of the sinter mix contains significant quantities of low-melting constituents that form extensive slagging at temperatures well below those at which diffusion could begin to operate. In addition, very close packing and good contact of particles are necessary for diffusion to operate effectively. The surface-to-volume ratio becomes unfavorable with increased particle size; as a consequence, the number of contacts points between particles decreases. This adversely affects the bonding of different phases by the diffusion mechanism. The limits of surface mobility are such that a loosely packed aggregate could not be expected to acquire a high strength by diffusion bonding alone.

Theoretically, diffusion is a more desirable mechanism of bonding because the diffusion bonded sinters are more porous, are accessible to reducing gases, and hence are easily reducible.

Slag bonding prevails with the formation of (partial) liquid slag by fusion. The crystalline constituents are partially/completely embedded in the matrix of a fused glassy melt, the extent of which depends on the volume and wet-ability of the liquid phase. First, the gangue material participates in the formation of slag, but with the increasing temperature, an increasing amount of iron is also taken into solution. The crystallization of iron silicate also occurs when the melt reaches a sufficiently high temperature. On subsequent cooling, iron will come out of solution as secondary crystals and dendrites of magnetite. These secondary crystals, along with the remaining primary iron-bearing constituents, are firmly enveloped by the glassy slag.

The bond strength depends on the amount of glass as well as the amount and types of its constituents. The fuel rate and impurities that are present in the sinter mix or that are added, such as, SiO_2, CaO_2, CaO, MgO, and Al_2O_3, decides the amount of glassy phases in the sinter matrix.

6.8 ASSIMILATION AND COALESCENCE

6.8.1 Sintering of a Pseudo-Particle

The sintering reactions consist of melt formation and assimilation reactions among the sinter mix constituents. This is schematically shown in **Figure 6.17** in the case of a pseudo-particle that is formed during granulation.[32]

a. The basic structure of pseudo-particles formed during granulation consists of a large core particle (nucleus) and a coating layer of adhering fines and fluxes (stage a).
b. The iron oxide within raw materials reacts with nearby limes: Calcium ferrite is the first product that is formed by solid-solid reaction. At higher temperatures (approximately 1,200°C), the ferrite gets decomposed to produce the liquid melt. The liquid increases its volume by dissolving more iron oxide, also silica and alumina, as the temperature is raised (stage b).
c. Finally, the liquid melts starts reacting with the nucleus particles by dissolving them. This results in the formation of more melt (stage c).

Thus, the sintering proceeds through the assimilation. The assimilation reaction depends on the reactivity of melt and the solid nucleus. In case either/both of these components is/are more reactive, the assimilation is higher.

The bed structure of sinter following the process could be described by the physical phenomena shown in Figure 6.18[33]. Iron oxides reacts with nearby limes to produce liquid with the composition of eutectic point. Its volume increases by more dissolution as the temperature is raised (Stage 1).

The pores involved within the liquid-forming stage gather together to form larger-sized pore or escape from the liquid, which changes the pore ratio within the

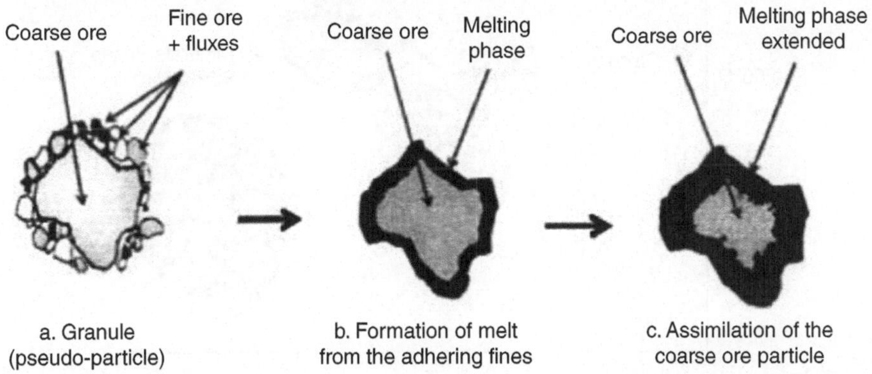

FIGURE 6.17 Schematic representation of the sintering of a pseudo-particle (after Ref. [32])

condensed phase. This is due to the fact that the diffusion of gas induced by the surface tension works on the pore wall. On the other hand, the liquid volume may increase to cover the whole surface of the pseudo–particles (stage 2).

The coalescence is initiated once a certain volume of melt of acceptable physical properties forms. The cohesion forces begin to work between two particles to contact to each other and make the pseudo-particles gather together. Later, the temperature stops increasing and strait to decrease as shown in stage 3. (**Figure 6.18**).[33] This is when the flame front approaches in a particular section of the sintering bed and is terminated with the departure of the flame front. The driving force of this phenomenon is reduction in free energy when the surface area of the three phase system: solid, liquid, and gas (voids) decreases. A dense cluster will approach a spherical shape since this configuration has the lowest surface energy level. Simultaneously, the porosity of these clusters decreases as the trapped bubbles in the liquid phase coalesce into larger bubbles.

Finally, when the temperature decreases, liquids begin to solidify and the sinters of the compound of minerals and pores form (**Stage 4**).

The manner in which these aforementioned processes is executed, and their intensities, affect the sinter bed.

The degree of coalescence depends on:[34]

- Properties of solids: size, distribution, and density. The porosity and minerals present in the solids also affect the coalescence, directly or/and indirectly, by influencing the density of solids.
- Properties of liquid: kinematic viscosity and surface tension.
- Ratio of liquid, solid, and void in the sinter mix.
- Size distribution of voids in the melt.

In the case of fluxed sinter, iron ore and fluxes constitute major material components of the sinter mix. Hence, their assimilation characteristics do have much influence on sintering.

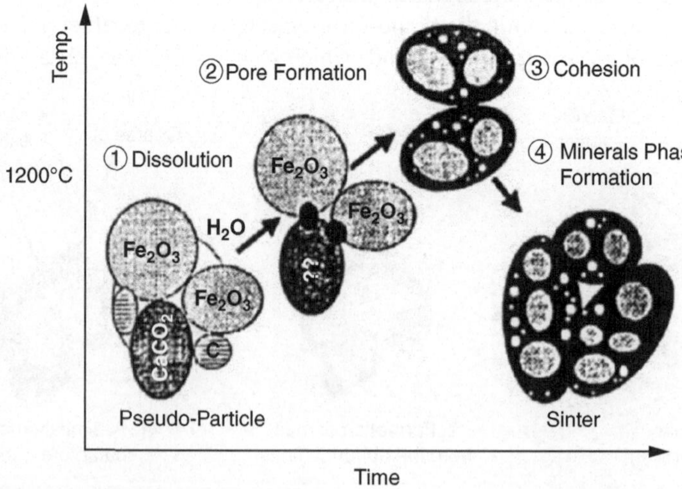

FIGURE 6.18 Schematic diagram of iron ore sintering process (after Ref. [33])

6.8.2 Assimilation of Iron Ore

The assimilation characteristics of iron ore has important roles: to control the reactions in the sintering bed and to obtain the target mineral structures. The rate of assimilation of ore nuclear particles depends on:[35]

- Solid-liquid interfacial area
- Temperature
- Chemical composition of primary melt

A conceptual model for the rate of solid dissolution is given by the expression:

$$dm/dt = K * Ar * \Delta C^n \quad (6.16)$$

where:

K = proportionality constant,
m = mass of solid assimilated,
t = reaction time,
Ar = area normal to the direction of mass transfer,
ΔC = the concentration driving force, and
n = a constant; the value is 1 if assimilation is diffusion controlled and greater than 1 when the dissolution reaction is also influential.

The amount of liquid phase during the assimilation of ore is an index to determine the assimilation characteristics of ore. However, it is difficult to evaluate the amount of liquid phase during the assimilation of ore. Hence, the degree of volume shrinkage (ε) is taken as a measure to evaluate the extent of sintering reactions. The volume can be calculated from measured buoyancy of the samples when immersed in mercury.[36]

$$\varepsilon(\%) = \left[1 - \left(V_f / V_o\right)\right] * 100, \quad (6.17)$$

where:

V_f = sample volume after sintering, and
V_o = sample volume before sintering.

The types of ores and their porosity do influence its assimilation as mentioned below.

Ore Type. The assimilation characteristic of iron ore depends on its type: whether the ore is hematite or magnetite. Besides, the assimilation of ore depends on its LOI and the impurities content. Broadly, rich ore has difficult sinter-ability.

The effect of individual ore components are described below:

a. **Hematite:** The fusion temperature of hematite ores differs depending on their sources, their contents LOI, and their impurities. The fusion temperature of the blend consisting of hematite decreases with the increase in limestone content and increases with the increase in alumina content.

Al_2O_3: The following relationship has been observed in the case of typical Indian ores within the range: wt.% Al_2O_3 = 1.5–5.5 and wt.% of admixed limestone (CaO = 50%) = 20–40:[37]

Fusion temperature = 34.4*(wt.% Al_2O_3 in ore - (3.6 wt.% limestone in ore) + 1322.

It is implicit that, with the increase in CaO content, the quantity of CaO-Fe_2O_3 melt increases due to lower liquidus temperature. The alumina gets dissolved into the CaO-Fe_2O_3 melt, resulting in an increase in its liquidus temperature.

Alumina in iron ore is refractory by nature. Hence, it demands higher heat input for going into sintering melt. In addition, the agglomerates with high alumina content increase the slag viscosity.

SiO_2: SiO_2 in the form of quartz assimilates with the calcium ferrite liquid phase that was formed by CaO-Fe_2O_3 (FeO) reactions in the first step of the sintering reaction. The assimilation causes secondary hematite and silicate slag to crystallize, resulting in a decrease in the amount of liquid phase in the system.

b. **Goethite:** Goethite (α-FeOOH) has high reactivity at high temperatures. The mineral undergoes dehydration around 573 K and is transformed into hematite (α-Fe_2O_3): $2FeOOH = Fe_2O_3 + H_2O$.

The resultant hematite has fine pores developed due to the release of water vapor, leading to its porous and brittle microstructure. As result, the specific surface area of the hematite mineral is extended, which accelerates the assimilation into the liquid phase. Consequently, the formation liquid phase of low-melting calcium ferrite is enhanced.

The pisolite ore when calcines become very porous as goethite transforms into fine hematite crystallites. Goethite ores assimilate much more readily than hematite ores. This is because the melts formed in case of goethite ore are very mobile and easily penetrate into the porous ore, which therefore has a very high reaction surface area.[35]

Noda et al.[36] have evaluated the assimilation of iron ores with 20% CaO using an electric furnace. The ores have different compositions of goethite (α-FeOOH), FeO, and SiO_2. In the sintering process, the reaction of Fe_2O_3-CaO takes place prior to other reactions. The composite formed at 20% CaO has the lowest melting point in the system; hence, the proportion of 20% CaO was selected in their study.

The following empirical relationship has been obtained:[36]

$$\varepsilon(\%) = 22.4 + 0.416 * [\alpha - \text{FeOOH}] + 0.679 * [\text{FeO}] - 0.770[\text{SiO}_2],\ (r^2 = 0.91) \quad (6.18)$$

c. Magnetite (FeO): FeO in the form of magnetite in iron ore affects the reactions in a CaO-Fe_2O_3 system where the formation of liquid phase proceeds more easily and facilitates the formation of calcium ferrite at around 1,200°C (1,473 K).

Porosity of Ores. Porous ores are assimilated very rapidly because melts are capable of penetrating into intraparticle pores. The effectiveness of a melt in penetrating the intrapores of a solid and its ability to accommodate the assimilated material is also a strong function of its chemical composition.[32]

6.8.3 Assimilation of Fluxes

The assimilation of fluxes with iron oxides results in CaO-Fe_2O_3 and CaO.FeO melts having melting points of 1,205°C and 1,120°C, respectively. The properties of melts, before these solidify, depend to a large extent on the chemical composition of the layers adhered to the granules and the assimilation of nucleus particles (see **Sec. 6.10.2**).

The assimilation of lime in the structure of sinter at higher basicity is not complete, which adversely affects its strength. This is, in particular, when the size of flux is higher. The coarser-sized flux is not calcined to be assimilated by iron oxide in the shorter heating and combustion duration of sintering[38,39] (see **Sec. 10.6.2**). Fluxed sinter loses strength during hydration of the unassimilated lime due to which an increase in sinter volume by 10% is observed. In view of this, limestone should be crushed appropriately.

Assimilation of Dolomite Dolomite decomposes to form CaO and MgO in stages according to the following reactions:

$$CaCO_3.MgCO_3 = CaCO_3 + MgO(s) + CO_2(g)(\text{at around } 1{,}070 \text{ K})$$
$$CaCO_3 = CaO(s) + CO_2(g)(\text{at } 1{,}170 \text{ K})$$

The CaO and MgO react with hematite individually and calcium ferrite and solid solution of magnetite by solid state diffusion:

$$CaO(s) + Fe_2O_3(s) = CaO.Fe_2O_3(CF)$$
$$MgO(s) + FeOx(s) = (Fe,Mg)O.Fe_2O_3(s)$$

The eutectic temperature of the system, CaO.2Fe_2O_3(CF2) – CaO.Fe_2O_3 (CF) - 2CaO.SiO_2 (C2S) in air is 1,465 K. Above this temperature, the calcium ferrite melt forms in the sinter and, subsequently, dissolution of CaO and MgO into the melt takes place.

The SiO_2 precipitates due to its low solubility in magnetite, leading to the formation of two types of melts with different chemical composition:

- One with high MgO, high CaO, and low FeOx content
- Another with low CaO, medium MgO, and high FeOx content

As the sinter is cooled, magnetite is precipitated from the second melt resulting in silico-ferrite of calcium and aluminum (SFCA) and slag structure. Hematite and slag phase is formed from the melt with high SiO_2 and low MgO content.[40]

6.9 SINTERING REACTIONS

6.9.1 Sintering: A Thermochemical Process

Sintering is a thermal–chemical process that is governed by the laws of heat transfer and reaction kinetics in packed beds of particles. The reactions involved in the sintering process are the result of a number of reactions that take place in the heating, sintering (combustion), and cooling zones (see **Sec. 6.9.3**). However, the fundamentals pertaining to reaction kinetics of a chosen geometry are difficult to apply to the sintering process. This is because:

- During the sintering process, heat transfer across the sintering bed, exothermic and endothermic chemical reactions, mineralogical and physical changes in the product, all occur simultaneously as the transmission of heat proceeds across the bed.
- The heat transfer is also complex through conduction and mainly convection.
- Sintering is a fast process, and it is unlikely that equilibrium will be reached during the process.
- The reaction mechanism followed and, subsequently, the sinter minerals formed are, however, dependent on the mix composition, including the sinter basicity and MgO content.

6.9.2 Mechanism of Sintering Reactions

The sintering process undergoes several reactions in its heating, sintering (combustion), and cooling zones. The sintering consists of melt formation and assimilation reactions among the sinter mix constituents (see **Sec. 6.8.1**).

a. Sintering reactions begin within the layer of adhering fines, where fine materials are in intimate contact. Each particle reacts only with the particles surrounding and in contact with it. The number of contact points between particles of lime and hematite exceeds that between lime and quartz.
   - During iron ore sintering, a low-melting-point phase is formed by a solid-state reaction between hematite and flux, such as CaO-Fe_2O_3 and Fe_2O_3-SiO_2. Although the affinity of CaO to SiO_2 is more than that of CaO to Fe_2O_3, a large extent of calcium ferrites is formed. This is because of the fact that hematite reacts easily with lime even in the solid phase, whereas quartz hardly reacts with lime and hematite in the solid phase. The minerals, calcium ferrites, are the first product to form by solid-solid reactions between hematite and lime.
      - Further temperature increases cause their incongruous melting to form the first melt.[32]
      - At increasing temperature, and above 1,100°C, the assimilation of solids into this primary melt occurs at a significant rate. The subsequent melt–solid reactions at the periphery of surrounding solids results in the formation of more melt. Melt volume continues to

increase as more and more solid is assimilated. The melt close to the lime particles has high basicity, the consequence of which is the crystallization of ferrites and silicates of calcium does occur.

- In most cases, complete assimilation of the solids is not obtained because the bed cools very quickly once the flame front has traveled down a particular section of the sintering bed.
- At approximately 1,200°C, calcium ferrites decompose and form the initial liquid melt, which can dissolve silica and alumina.
- Assimilation of quartz and melt of calcium ferrites proceeds, according to Matsuno,[41] in the following manner:
  a. Molten silicates with low basicity are formed around quartz and quartz dissolves into melt.
  b. Molten silicates react with molten ferrites and hematite precipitates. Reaction between the two melts continues until the assimilation is complete.

This is schematically shown in **Figure 6.19.**

The assimilation reaction depends on the properties of both the initial melt and nucleus particle. The more reactive the melt, the greater its ability to dissolve the solid nucleus. Similarly, the greater the reactivity of the solid nucleus, more of that amount will be assimilated.

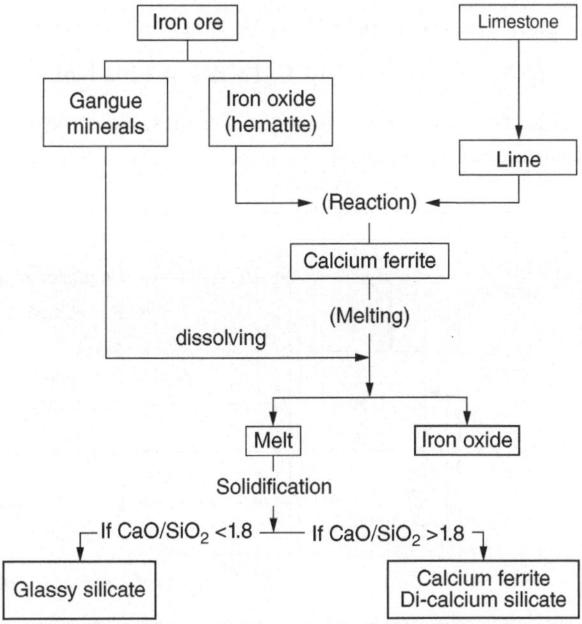

FIGURE 6.19 Assimilation of quartz and calcium ferrite melt: schematic representation according to Matsuno [41]

6.9.3 Chemical Reactions (Basicity Effect)

The product, sinter, is the result of various chemical reactions that take place in the zones of sintering (combustion) and cooling according to the material characteristics and its targeted chemical composition (**Figure 6.20**). The sinter basicity has a significant effect on the manner in which the reactions take place during the entire sintering process and the sinter minerals are thus formed.

a. In the case of siliceous (acidic) sinter magnetite, faylite and glass-containing iron oxides are the major mineral phases. The minerals are formed predominantly through the slag bonding. Mazenek and Jasienska[35] proposed the following reaction sequence:

$$\text{Hematite} \xrightarrow[1{,}050-1{,}200°C]{} \text{Magnetite} \xrightarrow[1{,}150-1{,}250°C]{FeO + Qtz} \text{Fayalite}$$

b. In the fluxed (with lime) sinter, the reactions undergo in a complex manner. The sequence of mineral formation at different stages of sintering in the case of hematite ore and the minerals formed in the final sinter product are schematically shown in **Figure 6.20.**

The reactions involved at different zones pertaining to different temperatures are given as follows:

Temperature (800–1,000°C) This refers to the heating zone.

- The calcination of carbonates proceeds to yield lime and MgO according to the following endothermic reactions:

$$CaCO_3\,(s) = CaO(s) + CO_2\,(g),\ \Delta\,H = +179\ \text{KJ/mol} \quad (6.19)$$

$$MgCO_3\,(s) = MgO(s) + CO_2\,(g)\Delta\,H = +102\ \text{KJ/mol} \quad (6.20)$$

- Iron oxide (hematite) gets crystalized. Besides, the sintering of hematite with quartz and lime initiates.

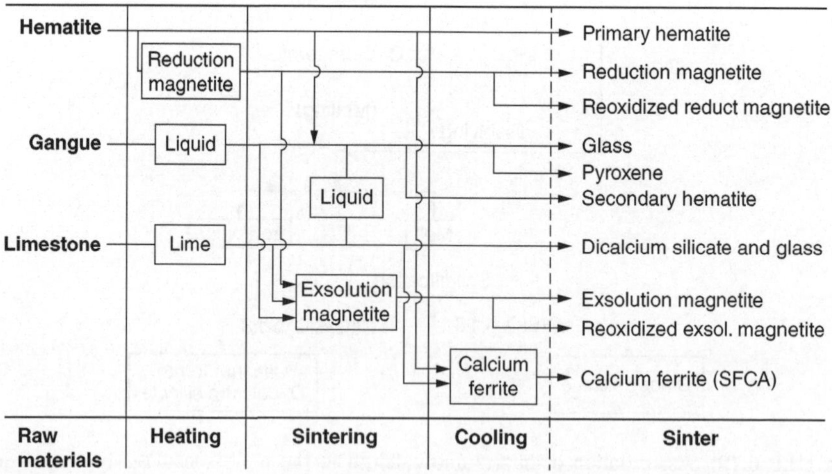

FIGURE 6.20 Development of minerals during hematite ore sintering (schematic) (after Ref. [42])

Temperature (1,050–1,200°C) The sintering reaction starts with the reduction of hematite.

- Formation of a low-melting liquid component, mono calcium ferrite ($CaO.Fe_2O_3$) also takes place at around 1,200° C by reaction with lime, which absorbs silica and more iron oxide.

$$\text{Part of hematite} + CaO = CaO.Fe_2O_3\ (\text{calcium ferrite})$$
$$+\ 2CaO = 2CaO.Fe_2O_3\ (\text{di-calcium ferrite})$$
$$SiO_2 + CaO = CaO.SiO_2\ (\text{calcium silicate})$$

- The melt so formed attacks the grain boundaries of hematite, and the grains dislodge into the melt forming $CaO.2Fe_2O_3$, $2FeO.SiO_2$, $CaO.FeO.SiO_2$, $2CaO.Fe_2O_3$, etc.

Temperature (1,250–1,350°C)

- Mono-calcium ferrite decomposes
- Magnetite + lime + silica = Calcium olivine
- More hematite along with gangue goes into the solution resulting in increased slag-bonded complex sinter structure SFCA.

6.10 SINTERING REACTION AND MINERALOGY

6.10.1 Stages of Chemical Reactions

The mineralogical and morphological characteristics of the sinter product are decided by the sum total of the chemical reactions involved in the sintering process that take place in the heating, sintering (combustion), and cooling zones. **Figure 6.20** shows the sequence of these chemical reactions.[42] The total sintering process consists of:

a. Heating and reduction stage: The combustion and gasification of carbon in coke provides sufficient heat and CO potential in the gaseous phase for reduction of hematite in the ore mix to magnetite.
b. Cooling and oxidation stage: This stage is initiated by the downward flow of air across the sintered bed on suction. Magnetite gets oxidized to hematite on oxygen potential.

FeO in sinter is associated with the reduction of hematite as well as oxidation of the reduced phases during cooling (**Table 6.2**). The conversion to wustite is never complete because the kinetics of oxygen dissolution into the melt are extremely slow.

TABLE 6.2
Sequence of Iron Oxide Reduction and Oxidation in the Sintering Process at its Different Stages

Sintering process	Minerals formed			
Heating and reduction	Hematite →	Magnetite →	Wustite →	(Metallic iron)
Oxidation and cooling	(Metallic iron) →	Wustite →	Magnetite →	Hematite

6.10.2 Sinter Mineralogy

The iron ore sinter consists of several pores of varying sizes and shapes in addition to the mineral phases[43,44] and is affected by the sinter basicity and its alumina content. The controlling parameters in this regard are:

a. The manner how the sintering reactions proceed in various stages of sintering
b. How the minerals get assimilated and their coalescence (also of pores) occurs

Effect of Sinter Basicity. The calcium ferrite phases are formed subsequent to the reaction between iron oxide and lime at low temperature (750–780°C). The stability of this (ferrite) phase depends largely on the thermal condition of the sinter bed and the cooling cycle. The basicity effect is also apparent at this stage: At lower basicity (CaO/SiO_2 less than 1.8), glassy silicates are formed, mainly through slag bonding; whereas, at higher basicity (CaO/SiO_2 more than 1.8), dicalcium silicates and calcium ferrite minerals are formed (see **Figure 6.19**).[41]

Figure 6.21 shows (part of) a phase diagram for a system containing calcium oxide (CaO, abbreviated as C) and hematite (Fe_2O_3 abbreviated as F). For a constant mixture of the two components, a fixed value for melting temperature can be picked out of the diagram.

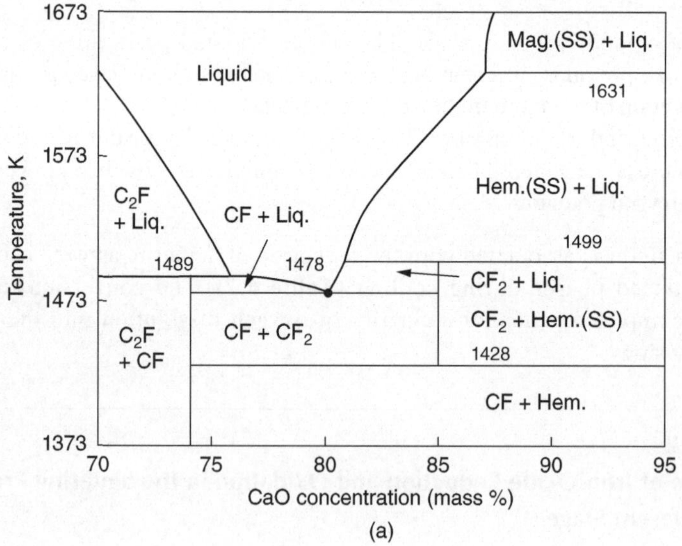

FIGURE 6.21 CaO–Fe_2O_3 system of importance in the sintering process, above 1,100°C (after Ref. [10])

6.10.3 Formation of SFCA

The melt, silica ferrites of calcium and aluminum, commonly known as SFCA is formed through the sequence of sintering reactions. The mineral monocalcium ferrite is formed initially in the sinter at high basicity, which is modified with the presence of Al_2O_3 and SiO_2 in iron ore to form melts whose composition in terms of their Fe_2O_3, CaO, Al_2O_3, and SiO_2 contents changes depending on the solids assimilated. The formation of SFCA (low Fe form) and SFCA -1 (high Fe, low silica form) is dominated by a solid-state reaction mainly in the system CaO-Fe_2O_3. Initially, hematite reacts with lime at low temperatures (750–780°C) to form calcium ferrite. The formation and stability of ferrite phase depends largely on the thermal condition of bed which undergoes sintering and the cooling cycle.[45]

Alumina and Silica Effect The presence of gangue minerals—namely, silica and alumina—does affect the SFCA formation. The calcium ferrite initially formed undergoes significant changes by diffusion of Al and Si species. Thus, SFCA is formed toward the end of the heating/reduction regime itself. With the high content of Al_2O_3 and SiO_2 in iron ore, there is little chance that ferrite is present as calcium ferrite. At sintering temperatures higher than 1,300°C, increased mobility of Al and Si into ferrite melt is a more likely proposition.

Cooling Zone With the passage of flame front, the cooling stage gets initiated whence hot sinter is (partially) oxidized by the oxygen in incoming air. At this stage, euhedral and skeletal crystals of magnetite and (predominating) hematite continue to exsolve from the slag. The increasing viscosity of the liquid phases during cooling and the passage of gas through this viscous material produce a porous texture in the sinter. It is possible for iron oxides (hematite and magnetite) to be reabsorbed by reaction with silica melt to form SFCA, particularly at a slow cooling rate. During cooling, the crystallites of SFCA exsolve from the melt, followed by the formation of dicalcium silicate. The cooling rate increases (the available time for cooling is substantially reduced) with lower bed height. Hence, increasing the flame front speed leads to decreasing SFCA content (see **Figure 8.4**).

When the temperature exceeds 1,300°C, SFCA starts to decompose into hematite, magnetite, and silicates. On cooling, various minerals and phases precipitate out of the melt. If the oxygen content of the flue gas is high, SFCA will transform into hematite; if the oxygen content is low, it will decompose to magnetite. Magnetite can also be oxidized to hematite during cooling.[45]

6.11 MASS BALANCE

6.11.1 Mass Balance in a Continuous Sinter Plant

The stream of material flow in and out of a sinter machine is shown in **Figure 6.22**. The zone of combustion and fusion, having the highest temperature,

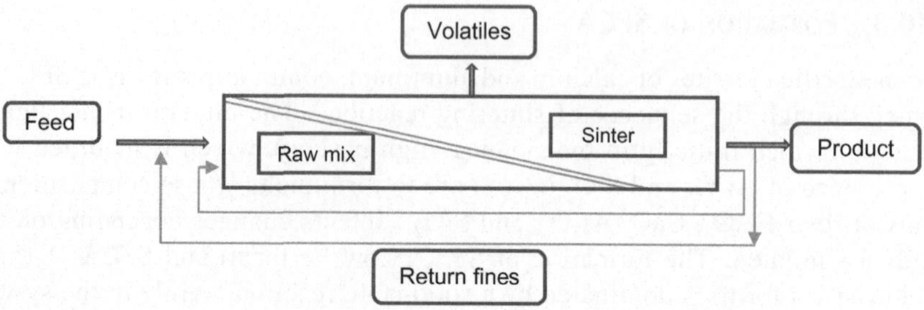

FIGURE 6.22 Schematic of mass balance in a continuous sintering plant

is also shown between two slant lines. The production of (sized) sinter, P (in T/h), can be expressed as:[46]

$$P = \left[H * W * \rho_s * (1-\omega_m) * (1-\varepsilon) * \nu - R\right] * (1-\lambda) \tag{6.21}$$

where:

H = bed height (m),
W = width of bed (m),
ρ_s = average density of the particles forming the bed (T/m^3),
ω_m = mass fraction moisture content (-),
ε = fractional void age (-),
ν = sinter machine speed (m/h),
R = return fines recirculated (T/h), and
Λ = mass fractional volatile (loss on ignition) content.

The mass balance in a continuous sintering plant according to **Figure 6.22** is:

Input of raw sinter mix + Return fines = Product (Sinter cake) + Volatilities + Return fines

The output of the blast furnace sinter, O, can be expressed by the following equation:

$$O = I(S - C) \tag{6.22}$$

where:

I = input of raw mix to the strand per unit time,
S = weight of sinter cake per unit weight of raw sinter feed, and
C = circulating load per unit weight of raw sinter feed.

The sinter cake output (S) per unit weight of sinter mix feed can be computed using the following equation:

$$S = F_O(1 - LOI_O) + F_F(1 - LOI_F) + F_L(1 - LOI_L) + F_C * as + F_{RF}, \tag{6.23}$$

where LOI_O, LOI_F, and LOI_L denote the loss on ignition (in wt.%) of ore fines, flux, and lime, respectively, and "as" (in wt.%) denotes the ash content of the coke.

F_O, F_F, F_L, and F_C denote fraction of ore fines, flux, lime, and coke in the sinter mix.

6.11.2 Charge Calculation

The mix that undergoes sintering is composed of various ingredients as mentioned previously. The composition of sinter mix varies according to the chemical characteristics of the ingredients constituting the sinter mix. The mass of these ingredients is calculated on the basis of mass balance. Besides ore fines, fluxes—namely, limestone and dolomite—are important components of the sinter mix. The ratio of Fe-bearing material to the lime and MgO-bearing material is guided by the chemistry of the product sinter. The chemical composition of sinter, in turn, is decided by the blast furnace requirement with respect to the slag chemistry. The basicity ratio of slag and its MgO content are the key determinants in this respect. The basicity ratio of sinter is taken as the basis in computing the sinter charge. That is, composition of the sinter charge is deduced on the basis of at what value the sinter basicity ratio is fixed. The basicity ratio is defined as: CaO/SiO_2.

Annexure VI.2 lists details of charge calculation.

6.12 HEAT BALANCE

The heat balance in the sintering process can be represented as:

$$\begin{aligned}&\text{Energy input} + \text{generation of energy due to exothermic reactions}\\&= \text{Energy output} + \text{losses due to endothermic reactions}\end{aligned} \tag{6.24}$$

The flow of energy consists of:

- Convective heat transfer
- Heat conduction in the solid phase
- Flow of energy due to radiation

The radiation effect within the packed bed from its top to bottom layers can be neglected.[47,48]

The major portion of energy input is contributed by the combustion of solid fuel. The sinter mass constitutes 45% of the total energy output. A significant portion of this sensible heat is transferred to the gases on cooling, which can be recovered.

In addition to these, the accumulation of energy within the system and losses are to be taken care of. These include:

- Generation of heat due to the exothermic chemical reactions including the combustion of coke and losses of heat due to the endothermic chemical reactions, such as calcination of carbonates, also from goethite to hematite roasting
- Energy flow due to the change in phases, such as melting and solidifying of solid material, vaporization, and condensation

Heat Balance across Combustion Zone The method for computing the heat balance across the zone of solid fuel combustion has been proposed by Wegmann.[49] According to Wegmann, the zone of solid fuel combustion has a thickness of 15–20 mm that moves down with the rate of VSS. For instance, with the sinter bed height of 400 mm and assuming the thickness of the combustion zone is 20 mm and the vertical speed of sintering 20 mm/min, the combustion zone will occupy in succession 20 positions for which 20 separate heat balances should be calculated. Minor variations in the parameters like sintering speed and widening of the combustion zone at the end of process as well as bed shrinkage have been accounted in the process. The solution of 20 equations of heat balance makes it possible to construct a theoretical curve of the least combustion of solid fuel in sintering of a given mixture in a bed of a specific height. By comparing this curve with the actual curve of carbon distribution across the height of the sintering bed, one can find ways for further reduction of the combustion of coke fines.

6.13 IGNITION

The granulated sinter mix is ignited initially when it is charged onto the moving strand via a roll feeder. As the bed travels under an ignition hood, coke particles on the upper surface are set alight to generate a narrow flame front. Mixed gas of the coke oven and blast furnace is used for its combustion in the ignition hood.

Types of Ignition Hoods Two types of ignition hoods are in use: (a) combustion chamber type and (b) nozzle/torch type. **Figure 6.23** shows a sectional view of the ignition furnace.

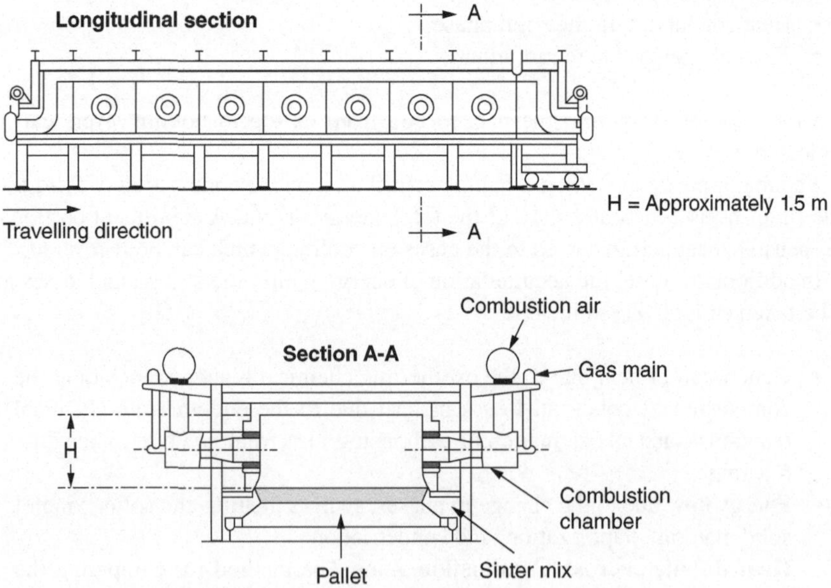

FIGURE 6.23 Sectional view of typical ignition furnace

It is important to have good ignition so that the coke in the sinter mix beneath the top surface undergoes adequate and effective combustion. Proper ignition is of great significance in the production of sinter of good strength and at the desired productivity level. A minimum of ignition intensity has to be maintained in this regard. This requires:

- Sufficient oxygen in the gaseous environment down on the sintering bed for combustion of the coke particles. As the coke particles are admixed with the noncombustible materials, the combustion of coke in the bed is affected by their presence.
- Temperature for ignition to be maintained above the flame temperature. The ignition temperature must be close to the mix sintering temperature. The ignition temperature is decided by the condition of gas combustion in the hood.

Ignition intensity is given by the following relationship:[50]

$$\left[\Delta H(CG) * V(CG) * L\right] / U(\text{J/sq.m}), \tag{6.25}$$

where:

ΔH (CG) = enthalpy of combustion product, J/cu.m. This is the product of the specific heat of combustion product (gas) [J/(cu.m*K)] and its temperature in K.
V (CG) = volume of combustion product (per square meter of the hood area per minute). It is decided by the amount of the combustible gas used for ignition and that of excess air required in the gaseous environment for combustion.
L = length of the hood (m), and
U = speed of pallet, m/min.

Controlling Parameters The controlling parameters in affecting the ignition are:

- Area of the ignition hood: In order to avoid sudden chilling of the top surface, post ignition, multisectional hearths with gradually decreasing the temperature are used. The length of ignition hood depends on the bed height. The sinter machine with a lower bed height requires a longer hood; whereas those with a higher bed height require a shorter hood. The sinter machine having a bed height greater than 600 mm has an ignition hood that is roughly 20% of the area.
- Calorific value of the combustible gases, coke oven gas, and blast furnace gas and their composition.
- Ignition temperature: The ignition temperature must be somewhat lower than the sintering temperature. This is to avoid slagging of sinter at the top of the bed.
- Volume percent of oxygen in the combustion product (flue gas) that is needed to burn the coke breeze in the sinter mix for sintering to take pace.
- Throughput of the raw mix on the sintering strand per unit time.
- Total duration (in minutes) for ignition The ignition duration should correspond to the overall time it takes to heat the top layer of the mix to the flame temperature of solid fuel and to the time of combustion of solid fuel. Coke needs to be burned for a certain duration: till that time the ignition is required.

Coefficient of Excess Air Two types of situations may prevail during ignition:

- If the volume of waste gases is balanced by the volume of gases drawn through the bed, then no excess air enters into the bed—a condition of inert ignition prevails.
- If the volume of the waste gases drawn through the bed is in excess of the combustion products, then coke at the top of the bed can burn in free oxygen—a condition that is termed as oxidizing ignition.

The presence of excess air in the environment has two opposing effects during ignition: When no excess air enters the bed (condition of inert ignition). CO_2 in the environment reacts with carbon (coke) in the sinter mix to form CO. A minor amount of heat is lost due to the endothermic carbon solution reaction: $C + CO_2 = 2CO$. With the presence of excess air in the environment, said reaction is minimized and the loss is compensated. However, the excess air, acting in the opposite direction, reduces the flame temperature.

With inert ignition, the coke particles, even red hot (due to high temperature), do not burn, and combustion zone moves down the bed extremely slowly. The machine area below the hood is not utilized, which results in inferior sinter quality. Due to these reason, a certain amount of excess air must be introduced through the ignition hood. Provision of excess air in the range 5–6 vol.% is suggested for proper burning of solid fuel in the sinter mix.

Leakage The mode of introduction of excess air into the bed also influences the sinter strength. Air can enter from the sides or also can be introduced through the burners. With air infiltration from the sides, complete mixing is not achieved and chilling of the bed top at the edges is possible. This reduces the sinter strength. It is, therefore, desirable that excess air is introduced through the hood burners.

Duration of ignition If the input of ignition ceases before the fuel particles have finished burning, the temperature in the combustion zone will fall as the heat is consumed for preheating the air. At the same time, the combustion of fuel particles will slow down and cease. The sinter at the top layer of the bed will be weak and may contain a considerable amount of residual carbon. For efficient sintering, it is essential that the combustion of solid fuel in the upper layer should be finished by the time the pallet comes out from the ignition hood.

Extended Ignition Hood The bed, which is being sintered, is subjected to sudden chilling by the incoming air when the sinter strand moves out of the ignition hood. In order to avoid this, the sinter machines are always designed with longer/extended ignition hoods, normally termed as multisectional hoods. The temperature in these extended hood zones is lowered stage wise. Normally, in section 1, the gas having a low calorific value is burned because no extra air is required in this section. While in the other two sections, gases of high calorific values are burned to provide excess air in the gaseous environment for combustion of coke and for efficient transfer of heat and the movement of the combustion zone.

Calculation of Ignition Hood Area The area of the ignition hood is calculated, taking into account the controlling parameters mentioned previously. As the width of the (sinter machine) pallet is fixed, this boils down to the calculation hood length. Besides the type and number of burners, the calorific value of mixed gases, etc., needs to be calculated. The ignition hood must satisfy the following criteria:

- Ignition temperature of 1,250–1,300°C, over 5% free oxygen.
- The duration of ignition should be adequate so that the coke admixed with the sinter mix at the top layer burns. Free oxygen in the combustion gaseous product is necessary for combustion of coke.
- The intensity of ignition for hematite type ore should be 67,000–75,000 (J/m^2) for satisfactory sintering. Beyond this, the sinter bed has excessive slagging.

Extended hoods are installed with stage-wise decreased temperatures to avoid sudden chilling of the top layer of the sinter once the pallet comes out of the hood.

Annexure VI.3 lists a typical calculation in this respect.

6.14 COMBUSTION OF SOLID FUEL IN SINTERING

Around 80% of the heat required in sintering is supplied by the combustion of solid fuel. The solid fuels used in sintering are primarily coke and, in some cases, anthracite, with each having different thermal effects depending on the chemistry (carbon content) and the combustion efficiency. The main combustible constituent of coke, used as a solid fuel in sintering, is carbon. As coke accounts for only 4–5 wt.% of the sinter mix, a small change in its properties (reactivity, size, and ash content) can have a large effect on the combustion behavior, heat generation, and sintering preformation.[51]

6.14.1 Effect of Combustion Behavior of Coke on Sintering Indices

The combustion behavior of solid fuel in the sintering bed along with the velocity of the descending gas (through the bed) determines the temperature, width, and speed of the transverse flame front, which in turn influences the quality of sinter, its yield, and productivity (**Figure 6.24**).

6.14.2 Combustion of Coke and Its Kinetics during Sintering

The combustion of coke is initiated when coke particles in the sintering bed react with O_2 present in the gaseous stream, which flows across the sintering bed under suction. The reaction is heterogeneous. Hence, not only the kinetics of reaction itself but also the mass transfer of reacting components are limiting factors in the combustion rate.

The combustion of coke in sintering, unlike that in a free oxygen environment, happens to be a much more complex phenomenon. The controlling parameters are:

a. The characteristics of coke, its association in the sinter mix before the sintering proceeds (pre-ignition condition), and—more importantly—the accessibility of coke by oxygen for combustion during sintering, which

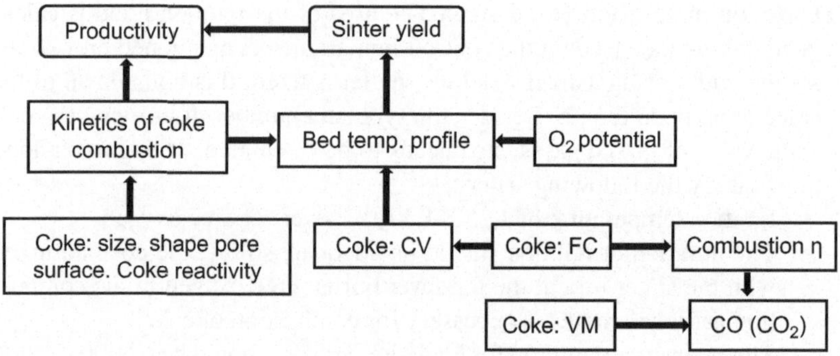

FIGURE 6.24 Schematic of the cause-effect of carbon characteristics vis-à-vis the sintering indices

decides the combustion behavior. In the latter case, the (burning) coke could be enveloped by the liquid slag.

- The combustion of coke in the case of sintering is neither that of a packed bed nor that of a single particle. This is because of the fact that coke particles in the sinter mix are disseminated among other sinter mix material. Hence, the nature of their association in the sinter blends, including the granulation process, is important. Larger coke particles may remain alone or could only have a thin layer of adhering fines.[52]
- The particle size of coke and its reactivity are important properties of coke that influence its combustion.
- The material being sintered passes through softening and melting stages. It is very unlikely that solid fuel particles can burn in the environment of a liquid melt since sufficient oxygen cannot enter the melt within the short duration of sintering.

b. The combustion of coke in the sintering process never goes to its completion. This is reflected by the presence of CO gas in the exhaust gas. Accordingly, the heating value of coke is decided. This is, however, material specific and depends on the process variables and operational conditions.

6.14.3 Combustion of a Single Coke Particle

The experiments conducted on combustion of a single carbon sphere in a laminar flow stream indicate that the burning rate of coke depends on the temperature region. Accordingly, the rate of combustion of coke is divided into three zones by the temperature region.[53,54]

Zone I (low temperature region): This is below 1,100 K. The reaction between carbon and oxygen on the surface of carbon material is controlled by chemical kinetics and is strongly temperature dependent.

The combustion rate of carbon materials at X (Rx [1/s]) is given by:

$$Rx = \{1/(1-X)\} * dX/dt; \text{ where } X = \text{is weight conversion on dry basis} \quad (6.26)$$

The combustion temperature of the coke and fine coke powders at the weight conversion X = 0.05–0.20 is between 600°C and 700°C. The activation energies for the combustion differ and depend on the types of carbon materials: 150–190 KJ/mol for the cokes, and 70–110 KJ/mol for carbon material with a large amount of volatilities.[55]

CO_2 appears as a primary product and in greater proportion with decreasing reaction temperature.

Zone II (middle temperature region): The combustion rate is influenced by the pore diffusion as well as chemical kinetics.

Zone III (high temperature region): This is above 1,400 K. The reaction rate is very fast. The reaction is diffusion controlled, that is, controlled by the transportation of the oxygen molecules through the external gas film around the solid particles.

Factors that influence the combustion are:

a. **Particle Diameter** Reaction rate decrease with the increase in particle diameter. With particles less than 100 μm, CO gas escapes sufficiently rapidly from the carbon surface and appears as a primary product and in greater proportion (in Zone III) with decreasing reaction temperature. CO_2 is formed by subsequent oxidation of CO in the surrounding gas phase.
b. **Nature of Carbon Source** In the internal diffusion (Zone II) and external diffusion (Zone III) rate determining steps, the differences in the combustion rate become small among the types of the carbon materials, and the apparent activation energies for the combustion of carbon materials becomes small (10–50 KJ/mol.).[55]
c. **Temperature** A certain mixture of CO and CO_2 is formed during the combustion of coke according to the following reactions.

$$C + 1/2\ O_2 = CO \text{ and } C + O_2 = CO_2$$

The ratio, CO_2/CO has the following temperature dependency:[56]

$$[CO_2]/[CO] = C * \text{Exp.}(6244/T)$$

The ratio CO_2/CO is mainly influenced by the particle temperature. The higher it raises, the more CO is formed. .

6.14.4 Association of Coke in Sinter Mix

Coke particles are disseminated among other sinter mix materials. The combustion rate should depend on how coke exists in the sinter mix. The coke particle encapsulated by a layer of fine noncombustible articles would burn at a very different rate compared to an equivalent uncovered particle.

Hida et al.[57] examined the granule structure of sintering feed and it classified into three types: "C," "P," and "S" (ref. **Sec. 6.5.3**).

- C type: The granules consists of nuclei particles having size more than 1 mm and fines (less than 0.5 mm) adhering on the nuclei. Large coke particles (more than 1 mm) exist mainly as free particles. These have very little adhering material. Free coke particles have good access to oxygen, which increases the combustion rate but decreases the combustion efficiency. The free coke particle emits the highest level of carbon monoxide.
- P type: This type of sintering feed is agglomerates of fines only. Fine coke particles (less than 0.25 mm) are well dispersed in the adhering fines layer (AFL) while the medium-sized coke particles were found mainly on the outside but attached to the surface of the AFL. The granule with fine coke interspersed between fine alumina in the adhering fines gives the lowest CO value.
- S type: This type sintering feed is lone particles of size 0.5–1 mm. Several reports have estimated the temperature distribution in the sintering process using the "S" type of state.

According to Hida et al.,[57] the combustion efficiencies for the three cases, as mentioned above, are very different. The rate of combustion in coke nuclear particles would depend on the thickness of the surrounding adhering fines layer. The combustion of coke with an "S" type distribution is normally faster than the real sintering process. This is because it ignored oxygen diffusion in the fine layer of quasiparticles.[58]

6.14.5 Profile of Oxygen in Coke Combustion

The resistance to combustion of coke, when assessed by O_2, during the sintering process depends on its particle size and its association in the granules. In the case of nuclear coke, the combusting particle has an additional resistance of coke ash that reduces the oxygen concentration at the coke surfaces, thereby reducing the combustion rate. The diffusion of oxygen inside a pseudo-particle, where smaller-sized coke particles are coated with the noncombustible matter, is different from that of a single (and separate) coke particle. This is schematically shown in **Figure 6.25**.[59] A correction factor to the fuel combustion rate is required to take care of the fuel distribution.

6.14.6 Characteristics of Coke and Upstream Processes that Influence Coke Combustion

The characteristics of coke and the upstream process of sinter mix granulation affect the combustion behavior of coke.

Important properties of coke that significantly affect the combustion of solid fuel are its particle size and reactivity, presence of mineral species in coke, and the manner in which sinter mix is granulated.

a. **Particle Size and Reactivity of Coke Particles** The wood charcoal, for instance, is porous and has higher reactivity than that of coke and anthracite; hence, the charcoal is heated in the sintering bed to a higher temperature in

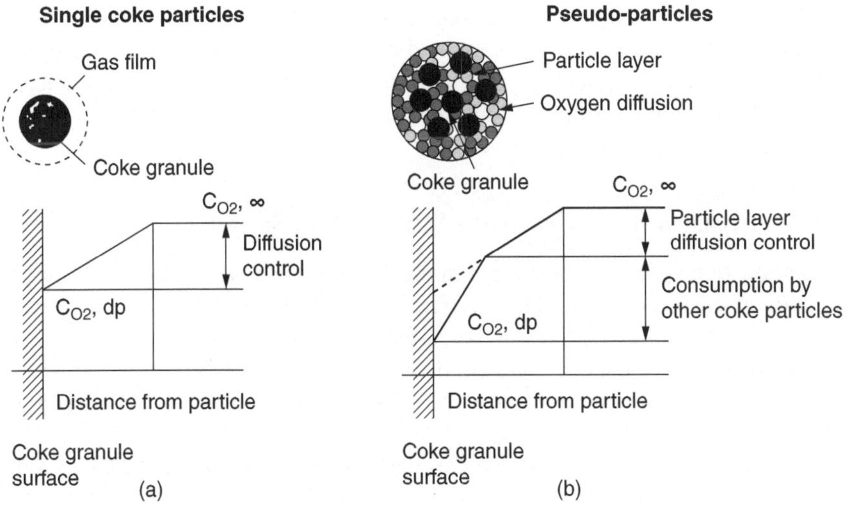

FIGURE 6.25 The profile of oxygen concentration in the neighborhood of a coke particle for the two situations: single coke particle (left side) and pseudo-particle (after Ref. [59])

a shorter time. A shorter holding time at a higher temperature (more than 1,200°C), where the main sintering reactions take place, produces the sinter with poorer strength and yield. The combustion increases with a decrease in particle diameter. With particles less than 100 micron the CO gas escapes sufficiently rapidly from the carbon surface.

b. **Presence of Mineral Species** The mineral species exerts a strong catalytic effect on the combustion reactivity of the solid fuel.[55] The char structure, including pore size distribution and porosity, plays a key role in determining particle burnout times. As the diameter of coke particles gradually shrinks with combustion, the void volume between non-coke particles grows. The reaction rate is thus affected in view of the accessibility of the internal surfaces with oxidizing gas under Zone III combustion conditions. Coke ash is either carried out of the sintering bed by the gas stream or bound to the non-coke sinter material. However, the effect can be neglected in view of its lower amount in the sinter mix.
c. **Granulation** The combustion of coke is affected by the manner in which the granulation proceeds. The size of coke particles besides its surface property (hydrophobicity) do have an important role in determining their positioning within a granule. The location of coke particles within the granule, and thereby their access to oxygen during combustion, is influenced by the granulation process.

The selective granulation practice, also the divided addition of coke, helps in the positioning of coke in the sinter mix. In the latter case, the entire coke is split into two fractions: the first one is put into the mixing drum along with the other mix ingredients, while the rest (the second one) is added in the second (balling) drum.

This facilitates positioning the coke particles for better combustion. According to Ishikawa,[52] the large coke particles should only have a thin layer of adhering fines to facilitate combustion, and fine coke particles should be located on the outermost part of the adhering fines layer of a granule for their better accessibility by oxygen to facilitate combustion.

6.14.7 Combustion of Coke with Melts during Sintering

The combustion process of solid fuel in the sintering bed takes place with the participation of the liquid phase. The material being sintered passes through softening and melting stages during the sintering process. It is very unlikely that solid fuel particles can burn in the environment of a liquid melt. This is due to the fact that sufficient oxygen cannot enter the melt within the short duration of sintering (its combustion zone).

The scheme proposed by Korotich[60] suggests that the formation of iron ore sinter begins once a melt appears around the burning fuel particles. As the temperature rises, the viscosity of the melt decreases and its wettability improves. The liquid phase under the influence of capillary forces is drawn into the lumps surrounding the coke fines particles. Simultaneously, the pore size increases. The scheme has a drawback as it considers the combustion of a free fuel particle taken separately outside the melt. In real situations, part of the fuel can be rolled into the granules.

The solid fuel differs sharply from the remaining mix with respect to its following characteristics:

- Its wettability by the melt; it is hydrophobic in nature with respect to the melts.
- Lower density of the fuel material compared with other sintering mix components.
- Its capability to react with ferruginous melt to form a gaseous phase.

The totality of these properties determines the possibility of the solid fuel becoming opened up (floating up) into a large pore or channel and burning on the melt surface.[61] The combustion of solid fuel, and consequently the formation of sintered material, takes place precisely by virtue of these properties in spite of the relatively low volumetric concentration (18%) of carbon in the bed.[62]

Mikhalevich et al.[62] have modified the scheme originally proposed by Korotich.[57] The combustion of solid fuel goes through different mechanisms as the sintering process is advanced through its different bed height (zones).

- During ignition, the fuel particles, which are free and coarse, burn at the top layer according to the scheme proposed by Korotich.[60] Fine fuel particles inside the granules may remain unburned.
- As the combustion zone moves down the middle of the bed:
  - The temperature level of the process rises.
  - The amount of melt that penetrates into the ever-increasing volume of granules increases.

The fine fuel rolled into the granules begins to react with the melt to form FeO. The smallest fuel particles can react fully.

In the middle of the bed, a considerable part of the fuel rolled inside the granules can burn in addition to free fuel particles. Coarse fuel particles, with a small amount of ore fines rolled on them, are freed from their envelope even more easily. In this case, the melt forms as drops that, owing to the hydrophobic behavior of carbon, roll from the fuel surface.

In the bottom part of the bed, the amount of melt formed increases. Around 70–90% of the mix goes through the melting stage. Part of the free fuel is covered with the liquid phase as the melt flows over and the flotation effect becomes important. The bottom part has a high potential of reducing agent also subjected to the excess heat. As a consequence, the hematite is reduced to lower oxides: the melts has a lower viscosity. The fine fuels undergo the reaction fully, while the coarser fuel floats up with the gas bubbles.

The stages of the aforementioned scheme are arbitrary as it depends on the amount of solid fuel in the mix. The increased fuel consumption shifts the previously mentioned stages to a higher level of the sintering bed. Conversely, as the fuel is decreased, the stages move to a lower level.

6.14.8 Heating Value of Coke Combustion

The heating value of coke is numerically equal to its standard heat of combustion but of the opposite sign. The gross heating value of fuel is the heat evolved in its complete combustion under constant pressure at 25°C when the combustion products are condensed to the liquid state. The net heating value is similarly defined, except that the final state of the product after combustion is taken as vapor at 25°C.

The theoretical amount of air required for complete combustion depends on the chemical composition of the fuel and the stoichiometric relations involved in combustion.

Wegmann[63] has formulated the thermal effect as follows:

For combustion of carbon in coke fines:

$$C + 1/2\ O_2 = CO + 10104\ \text{KJ/Kg C}$$

$$C + O_2 = CO_2 + 33685\ \text{KJ/Kg C}$$

For combustion of carbon in anthracite:

$$C + 1/2\ O_2 = CO + 10330\ \text{KJ/Kg C}$$

$$C + O_2 = CO_2 + 33911\ \text{KJ/Kg C}$$

The ratio CO_2/CO in the exhaust gas can be taken as a measure to compute the heat of combustion of solid fuel (Q_c) to CO and CO_2. For instance, if the ratio $CO_2/CO = 3$, then $Q_c = 0.75 * 33685 + 0.25 * 10104 = 27789.75$ KJ/Kg of C if coke breeze is used.

6.14.9 Combustion Efficiency in Sintering

Excess oxygen is always present in the sintering atmosphere. In spite of this fact, the combustion of coke never proceeds to completion as some carbon monooxide is present in the waste gases. The air flows down the bed and carries carbon monoxide produced at the surface of the burning fuel very rapidly into the cold zone of the bed. Hence, the oxidation of coke by atmospheric oxygen cannot proceed to completion. This explains the presence of CO in the exhaust gas. The fact that the ratio of CO/(CO_2+CO) in waste gas increases with an increased flow rate of gases down the bed due to increased suction supports the explanation.

Higher levels of carbon monoxide in exhaust gases and, consequently, lower sintering temperatures are achieved with increasing fuel reactivity, whether it is caused by the nature of fuel or by finer grinding of the same fuel.

Fine coke breeze combusts more rapidly and has a poorer CO_2/(CO_2+CO) ratio since the heat given off is not sufficient to heat the surrounding mix to sintering temperature. Coarse-grained coke breeze, on the other hand, causes local superheating and longer combustion time. This in turn leads to a broadening of the sintering zone. The excessively fine and also the excessively coarse particles in the mix require a higher fuel consumption to attain the same sinter characteristics while the sintering rate is simultaneously decreased.[25]

Higher CO results from increased suction, finer fuel size, and increased volatile content of the fuel. It would appear that as the accessibility to oxygen is reduced, the combustion efficiency gets improved possibly because of slower combustion rate. The bulk of the volatile matter escapes on suction and does not burn within the sinter bed and therefore may be considered as having no usable thermal contribution. **Figure 6.26** shows schematic depiction of the front at the middle of the bed. Coke is assumed to ignite at around 1,000°C. The temperature of solid (T_s) and gas (T_g)

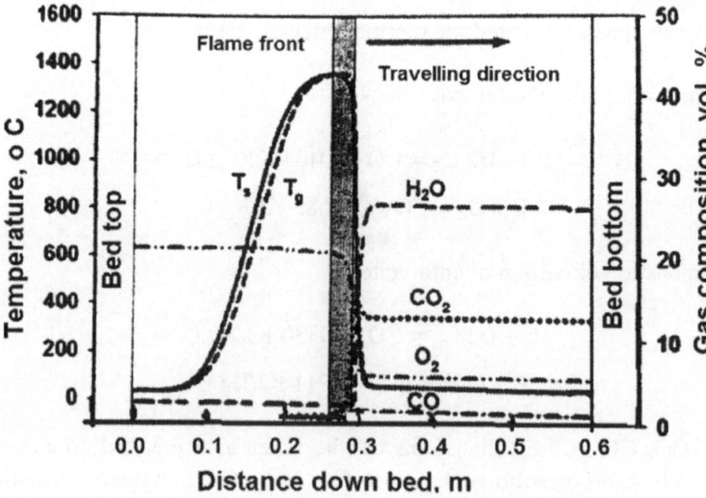

FIGURE 6.26 Composition of exhaust gas as a function of sintering time (after Ref. [64])

as well as the gas composition changes significantly around the flame front.[64] Within the flame front, a significant amount of heat is released by coke combustion, which results in a significant decrease in oxygen levels and increase in carbon dioxide levels. A small amount of carbon monoxide is also formed because of the incomplete combustion of the coke. A significant amount of the heat, that is released, is used to raise the temperature of the bed and also generate melt. This initiates at arounf 1,100°C and involves complex sintering reactions.

The CO_2/CO ratio decreases at a higher vacuum, with excessively fine fuel, or with an appreciable height of the sinter bed. According to Kotov,[63]

$$CO/CO_2 = 3.7 * 10^{-4}\left(V * D_m/\mu\right)^{0.7} * \left(Z/D_m\right)^{2.1} * \left(D_m/D_c\right)^{0.9} \tag{6.27}$$

where:

V = speed of air sucked through,
D_m, D_c = average particle size of sintered material and coke breeze,
μ = kinematic viscosity of waste gas g, and
Z = thickness of the zone of fuel combustion.

ANNEXURES

Annexures VI.1 to VI.3 to this Chapter are mentioned below. These, respectively, provide:

- Typical calculation of mean size and voidage of sinter mix granule using Ergun's equation vide **Sec. 6.3**
- Calculation of mass ratio of various sinter mix components at particular sinter basicity (CaO/SiO_2) and its MgO content by applying the mass balance equation vide **Sec. 6.11**
- Typical calculation of intensity of ignition of sinter mix using coke oven gas and blast furnace gas for a particular strand speed vide **Sec. 6.14**

ANNEXURE VI.1

TYPICAL CALCULATION OF GRANULE CHARACTERISTICS USING ERGUN'S EQUATION

Ergun's equation (vide Eq. 6.3)

$$\Delta P/h = A * \eta\, Vo(1-\varepsilon)^2/\varphi^2 d^2{}_p\varepsilon^3 + B * pV^2{}_o(1-\varepsilon)/\varphi\, d_p\varepsilon^3 \tag{A6.1.1}$$

can be rewritten as:

$$\Delta P = C\, V_o + D\, V^2{}_o, \text{or } \Delta P/V_o = C + D\, V_o \tag{A6.1.2}$$

where:

$$C = A\, \acute{\eta}(1-\varepsilon)^2 * h/\varphi^2 d^2{}_p\varepsilon^3, \text{ and}$$
$$D = B\, p(1-\varepsilon) * h/\varphi\, d_p\varepsilon^3 \tag{A6.1.3}$$

For air, the viscosity, ή, and density, p, at 25°C:

$$ή = 1750 * 10(E\text{-}8)Pa.\ s = 1750 * 10(E\text{-}8)(N/sq.\ m) * s$$
$$= 1750 * 10(E\text{-}8) * (0.102 * 9.81)\ kg/(ms)\left[1N = 0.102\,kgf = 1kg * 9.81m/(sq.s)\right]$$
$$= 1.75 * 10(E\text{-}5)kg/(m\ s)$$
$$p = 1.293\ kg/cu.m$$

With regard to typical hematite ore fines (100 wt.% below 10 mm in size) along with 1% lime, 30% return sinter fines, 7% moisture and fluxes (limestone and dolomite) so as to maintain sinter basicity at 2.5 typical observations on the velocities of the air infiltration through the bed of sinter mix, 0.3 m height were 0.835 m/s and 0.712 m/s, respectively, when subjected to the pressure drop, ΔP, across the bed (through a vacuum pump) 800 mm WG [78,480 kg/(m sq.s)] and 600 mm WG [58,860 kg/(m sq.s)]

$$D(\text{using Eq. A6.1.3}) = \left[(78{,}480 / 0.835 - 58{,}860/0.712)\right]/(0.835 - 0.712) = 91{,}060(kg/cu.m)$$
$$C(\text{using Eq. A6.1.3}) = \left[(78{,}480/0.835) - D * 0.835\right], \text{ or}$$
$$= \left[(58{,}860/0.712)\right] - D * 0.835\,] = 17{,}867\left[(kg / (sq.m * s)\right]$$
$$(0.07) * (C/D) = SSA * (ή / p), \text{ or}$$
$$(0.07) * (17{,}867/91{,}060)(m/s) = SSA(sq.m/cu.m) * \left[(1.75 * 10E(-5)/1.293)(sq.m/s)\right]$$
$$SSA = 1{,}014.8(sq.m/cu.m)$$

The void fraction, ε, can be found out using Eq. A6.1.4 (vide Eq. 6.9)

$$SSA/D = \left[6(1-\varepsilon)/\varphi * do\right]/\left[1.75\ p(1-\varepsilon) * h/\varphi\ d_p\varepsilon^3\right]$$

$$\text{Or } \varepsilon^3 = \left[1.75^* p * h/6\right] * [SSA/D] \qquad (A6.1.4)$$

$$\varepsilon^3 = \left[1.75^* 1.293(kg/cu.m) * 0.3(m)/6\right] * \left[1{,}014.8(sq.m/cu.m)/91{,}060(kg/cu.m)\right], [--]$$
$$= 0.0012608[--] \text{ or } \varepsilon = 0.108[--]$$

The value of shape factor, φ, varies along with the mix composition and nature of iron ore fines. Typically, for φ = 0.9:

The spherical equivalent diameter, d_p, can be computed from Eq. A6.1.5 (vide Eq. 6.10)

$$d_p = 6(1-\varepsilon)/\varphi * (1/SSA) \qquad (A6.1.5)$$

$$d_p = 6(1 - 0.108)/0.9 * (1/1014.8) = 0.0059\ m = 5.9\ mm$$

ANNEXURE VI.2

CALCULATION OF SINTER CHARGE

The mix that undergoes sintering is composed of various ingredients. The proportion of these mix ingredients is calculated using the overall mass balance as well as the balance of each of the chemical ingredients. The calculation is as follows.

Overall Mass Balance

If we consider the total of the ingredients, excluding moisture content, as one, then the equation for the mass balance of the sinter mix is

$$F_F + F_O + F_C + F_L + F_{RF} = 1 \qquad \text{(A6.2.1)}$$

where:

F_F = fraction of flux in the sinter mix,
F_O = fraction of ore in the sinter mix,
F_C = fraction of solid fuel in the sinter mix,
F_L = fraction of lime in the sinter mix, and
F_{RF} = fraction of return fines in the sinter mix.

Fe, CaO, MgO, SiO_2, and Al_2O_3 Balance

$$F_O * Fe_O + F_F * Fe_F + F_L * Fe_L + F_C * as * Fe_{as} + F_{RF} * Fe_{RF} = 1 * Fe_S$$
$$F_O * CaO_O + F_F * CaO_F + F_L * CaO_L + F_C * as * CaO_{as} + F_{RF} * CaO_{RF} = 1 * CaO_S$$
$$F_O * MgO_O + F_F * MgO_F + F_L * MgO_L + F_C * as * MgO_{as} + F_{RF} * MgO_{RF} = 1 * MgO_S$$
$$F_O * SiO_{2O} + F_F * SiO_{2F} + F_L * SiO_{2L} + F_C * as * SiO_{2as} + F_{RF} * SiO_{2RF} = 1 * SiO_{2S}$$
$$F_O * Al_2O_{3O} + F_F * Al_2O_{3F} + F_L * Al_2O_{3L} + F_C * as * Al_2O_{3as} + F_{RF} * Al_2O_{3RF} = 1 * Al_2O_{3S}$$

As the return fines are recirculated back to the sintering plant and have the same chemistry as that of sinter; the previous equations can be rewritten as:

$$F_O * Fe_O + F_F * Fe_F + F_L * Fe_L + F_C * as * Fe_{as} = (1 - F_{RF}) * Fe_S$$
$$F_O * CaO_O + F_F * CaO_F + F_L * CaO_L + F_C * as * CaO_{as} = (1 - F_{RF}) * CaO_S$$
$$F_O * MgO_O + F_F * MgO_F + F_L * MgO_L + F_C * as * MgO_{as} = (1 - F_{RF}) * MgO_S \qquad \text{(A6.2.2)}$$
$$F_O * SiO_{2O} + F_F * SiO_{2F} + F_L * SiO_{2L} + F_C * as * SiO_{2as} = (1 - F_{RF}) * SiO_{2S}$$
$$F_O * Al_2O_{3O} + F_F * Al_2O_{3F} + F_L * Al_2O_{3L} + F_C * as * Al_2O_{3as} = (1 - F_{RF}) * Al_2O_{3S}$$

where:

CaO_O, CaO_F, CaO_L, CaO_{as}, and CaO_{RF} denote the CaO content (in wt.%) of ore fines, flux, lime, ash of solid fuel, and return fines, respectively.

MgO_O, MgO_F, MgO_L, MgO_{as}, and MgO_{RF} denote the MgO content (in wt.%) of ore fines, flux, lime, ash of solid fuel, and return fines, respectively.

SiO_{2O}, SiO_{2F}, SiO_{2L}, SiO_{2as}, and $SiO_{2\,RF}$ denote the SiO_2 content (in wt.%) of ore fines, flux, lime, ash of solid fuel, and return fines, respectively.

Al_2O_{3O}, Al_2O_{3F}, Al_2O_{3L}, Al_2O_{3as}, and Al_2O_{3RF} denote the Al_2O_3 content (in wt.%) of ore fines, flux, lime, ash of solid fuel, and return fines, respectively.

Fe_s, CaO_s, MgO_s, SiO_{2s}, and Al_2O_{3s} denote the contents of Fe, lime (CaO), MgO, SiO_2, and Al_2O_3 in sinter, respectively.

When the return fines are recirculated back, they have the basicity ratio of that of sinter. Also, (it is assumed that) the ore fines do not contain lime. Hence, the basicity ratio of the sinter mix is contributed by the net basicity ratio that is contributed from the flux, lime, and solid fuel.

Then the basicity ratio B/R is given by the quotient (Eq.A6.2.3):

$$= \frac{\left[(1-F_O-F_C-F_L-F_{RF})(CaO_F - B/R * SiO_{2F}) + F_L * (CaO_L - B/R * SiO_{2L}) + F_C * as * (CaOas - B/R * SiO_2as)\right]}{\left[F_O * SiO_{2O}\right]}$$

Generally, the contents of lime and fuel in the sinter mix are assigned some fixed values. Hence, the fraction, F_O, can be computed from the previous equation, while assigning a value of return sinter fines (F_{RF}).

MgO Balance The MgO percent in the BF slag is maintained. In order to do so in the sinter that is charged into the furnace, it should have a desired MgO consistency that needs to be computed. The sources of MgO are limestone and dolomite (or pyroxene), which constitute the fluxes. Hence, the component F_F has two components: limestone and dolomite (and/or pyroxene).

The components CaO_F, MgO_F, and SiO_{2F} of the flux are decided by their individual proportions in the flux. If limestone in the flux is x, then dolomite (or pyroxene) in the flux is (1 – x). The chemical composition of flux can be computed as follows:

$$\begin{aligned} CaO_F &= CaO_{LS} * x + CaO_{Dolo} * (1-x) \\ MgO_F &= MgO_{LS} * x + MgO_{Dolo} * (1-x) \\ SiO_{2F} &= SiO_{2LS} * x + SiO_{2Dolo} * (1-x) \\ Al_2O_3F &= Al_2O_{3LS} * x + Al_2O_{3Dolo} * (1-x) \end{aligned} \qquad (A6.2.4)$$

The value of x can be computed through the iterative method. A certain value of x is assigned initially and then the composition of flux is computed using Eq. A6.2.4). The fraction of ore in sinter mix (F_O) is calculated using Eq. A6.2.3 at required basicity ratio and particular fractions of coke, lime and return fines. Following this, sinter MgO is computed (Eq. A6.2.1). If the (computed) value does not match the desired value of MgO in the sinter, then the value x is modified and the charge (F_O) is recalculated. Sinter MgO is recalculated and so on the computation continues.

ANNEXURE VI.3

TYPICAL CALCULATION OF IGNITION INTENSITY

In industrial practice, mixed gases of the coke oven and blast furnace are used for ignition of the sinter mix. **Table A6.3.1** lists the typical chemical composition of the coke oven and blast furnace gases. Also, the composition of mixed gases—namely, coke oven and blast furnace in the ratios 24% and 76%, respectively—are also shown in **Table A6.3.1**. The heat of combustion of the blast furnace gas is much lower than that of the coke oven gas. It is not possible to reach a temperature of over 1,100–1,150°C on combustion of blast furnace gas alone. This requires preheated air. To get an ignition temperature of 1,250–1,300°C with over 5% free oxygen, it is essential to use mixed gases of the coke oven and blast furnace.

The total amount of combustion product and free air is determined by a stoichiometric relation.

Basis: 1 m³ of mixed gas

Oxygen necessary for combustion of:

$$VO_2 = 0.01(0.5 * H_2 + 0.5 * CO + 2 * CH_4 + 3.5 * CnHn) = 0.336 \text{ m}^3$$
$$\text{Air required} = 4.76 * 0.336 = 1.6 \text{ m}^3$$

The components of these gases get combusted according to the following reactions

$$H_2 + 1/2O_2 = H_2O$$
$$CO + 1/2O_2 = CO_2$$
$$CH_4 + 2O_2 = CO_2 + 2H_2O$$

Hence, the amount of combustion products (in m³) on complete oxidation of mixed gases is:

$$H_2O = 0.17 + 2 * 0.062 = \mathbf{0.294}; CO_2 = 0.211 + 0.12 + 0.062 = \mathbf{0.393}; N_2 = 0.429$$
$$\text{Total} = 1.116(\text{from combustion product})$$
$$\text{Amount of } N_2 \text{ (associated with air)} = 1.6 * 0.79 = \mathbf{1.264}$$

TABLE A6.3.1
Chemical Composition of Gases

Gas	Chemical Composition							CV
	O_2	H_2	CO	CO_2	CH_4	N_2	C_nH_n	Kcal./m³
Coke oven	0.3	56.3	9.2	3.2	24.7	3.7	2.4	4,240
Blast furnace	0.1	3.9	25.2	14.9	---	55.9	----	867
Mixed gas (24% CO + 76% BF)	0.15	17.0	21.1	12.0	6.2	42.9	0,6	1,700

Hence, the total volume of combustion product (excluding excess air)

$$= 1.116 + 1.264 = 2.38 \sim \mathbf{2.4}\,\mathrm{m}^3$$

To ignite the coke particles in the mix, the oxygen content in the combustion product should be necessarily at minimum of 5 vol.%. That is,

$$0.21\mathrm{K} + (2.4 + \mathrm{K}) = 0.05\mathrm{K}, \text{ where, } \mathrm{K} = \text{amount of excess air.}$$

Solving the aforementioned, K = 0.75.
Total air required 1.6 + 0.75 = **2.35**.
Excess air = 0.75 m^3 ($O_2 = 0.75 * 0.21$, $N_2 = 0.75 * 0.79$)

Hence, the amount of gross combustion products:

$H_2O = 0.294$; $CO_2 = 0.393$; $O_2 = 0.1575\,(0.75 * 0.21)$
$N_2 = 0.429 + 1.264 + 0.5925\,(0.75 * 0.79) = 2.2855$
Total volume = 3.13 (incld. excess air) = 3.13 ~ **3.15** cu.m

As a cross-check:
Volume of combustion product = 2.4 + 0.75 = 3.15 cubic meter/m^3 of mixed gas
Specific heat of combustion products in J (mol *K) in the temperature range, 298 – 2500 K: CO (g) = 29.11; CO_2 (g) = 37.11; H_2O (g) = 33.58; N_2 (g) = 29.12; O_2 (g) = 29.35

Hence, the volume average specific heat of combustion product is:

$$(1/3.13)[0.294 * 33.58 + 0.393 * 37.11 + 0.1575 * 29.35 + 2.2855 * 29.12]$$
$$= 30.554\mathrm{J}(\mathrm{mol} * \mathrm{K}) = 30.554(\mathrm{J/mol.}) * 1/22.4(\mathrm{mol/liter}) * \left(1/10^3\right)\left(\mathrm{liter/m^3}\right)$$
$$= 1497.7\left(\mathrm{J/m^3} * \mathrm{K}\right)$$

Intensity of Ignition:

$\left[\boldsymbol{\Delta H(CG) * V(CG) * L}\right]\boldsymbol{/U(J/sq.m)}$, where:
$\Delta H(CG)$: Enthalpy of combustion product, J/cu.m
= Specific heat of combustion product (gas) * Temperature
$= 1497.7\left(\mathrm{J/m^3} * \mathrm{K}\right) * \mathrm{T(K)}(\mathrm{J/m^3})$
V(CG): Volume of combustion product (per sq. m of the hood area per min)
$= 3.15 * \mathrm{V(MG)}\left[\mathrm{cu\ m}/\left(\mathrm{m^2} * \mathrm{min}\right)\right]$

where:

V(MG) = volume of mixed gas required for ignition per m^2 of the hood area per min.
Q = productivity of charging the machine with raw mix : W * H * D * U * 60 T/h
W = width of pallet (m); H = bed height (m); D = bulk density of charge (T/m^3).

The average specific consumption of raw mix including the return fines is 2 T/T of sinter produced. The value of D is typically 1.8 T/cu.m. Speed of pallet, U, can be computed from the productivity figure.

If the productivity of the machine is P [T/(m^2 *h) and total strand area is A, then

$$P * 2 * A = W * H * D * U * 60 (T/h)$$

$$U = (P * 2 * A) / (W * H * D * 60) (m/min)$$

Hence, the intensity of ignition is given by:

$$1497.7 * T\left(J/m^3\right) * 3.15 * V(MG)\left[cu\ m/\left(m^2 * min\right)\right] * L(m)/U(m/min)\left[\mathbf{J/m^2}\right]$$

According to the literature, the intensity of ignition for hematite ore should be 66,988–75,362 (J/m^2) for satisfactory sintering. Beyond this, the sinter bed goes to excessive slagging, that is, considering at the higher value of it.

$$1497.7 * T\ (J/m^3) * 3.15 * V(MG)\left[cu\ m/\left(m^2 * min\right)\right] * L(m)/U(m/min)\left[J/m^2\right] = \mathbf{75,000}$$

REFERENCES

1. Yang, W., S. Choi, E. S. Choi, D. W. Ri, and S. Kim, "Combustion characteristics in an iron ore sintering bed-evaluation of fuel substitution," *Comb. Flame*, 2006, 145, 447–463.
2. Hass, K. P. and G. W. Watson, "Sintering," in *Blast Furnace—Theory and Practice*, Ed. J. H. Strassburger, New York: Gordan and Breach Sc. Publ., 1969.
3. Lovel, R. R., K. R. Vining, and M. D. Amico, "The influence of fuel reactivity on the iron ore sintering." *ISIJ Inter.*, 2009, 49, 195–202.
4. Voice, E. W. and R. Wild, "Importance of heat transfer and combustion in sintering," *Iron and Coal Trader Review*, 1957, 175, 841–850.
5. Loo, C. E. and R. D. Dukino, "Laboratory iron ore sintering studies, 2. Quantifying flame front properties," *Mineral Processing and Extractive Metallurgy (Trans. IMM ©)*, 2014, 123(4), 197–203.
6. Boranbaev, B. M., T. M. Srinivasan, S. S. Datta, N. K. Kakkar, R. P. Bhagat, B. R. Choudhury, S. Dharnipalan, D. S. Thakur, V. N. Tripathi, and N. Das. "Sinterability of various ore fines used in Bhilai Steel Plant," *Steel India*, 1982, 5(2), 115–123.
7. LOO, C. E., N. Tame, and G. C. Penny, "Effect of iron ores and sintering conditions on flame front properties," *ISIJ Inter.*, 2012, 52(6), 967–976.
8. Ergun, S. *Chemical Eng. Progr.*, 1952, 48(2), 948.
9. Rankin, W. J. and P. W. Roller, "The measurement of void fraction in beds of granulated iron ore sinter feed," *Trans. ISIJ*, 1985, 25, 1016–1020.
10. Mitterlehner, J., G. Loeffler, F. Winter, H. Hofbaur, H. Schmid, E. Zwittag, T. H. Buergler, O. Pammer, and H. Stiasny, "Modeling and simulation of heat front propagation in the iron ore sintering process," *ISIJ Inter.*, 2004, 44, 11–20.
11. Bhagat, R. P., "Modelling of air flow through sinter mix bed," *Proc. Inter. Conf. on Mathematical Modelling and Application to Industrial Problems (MMIP-2011)*, Org. by Dept. of Mathematics, National Institute of Technology, Calicut, Kerala, March 28–31, 2011, Eds. M. J. Jacob and S. Panda, New Delhi: Macmillan Publishers India Limited, 2012, 79–87.

12. Gan, M. X., H. Fan, Z. Y. Ji, X. L. Chen, L. Yin, T. Jiang, Z. Y. Yu, and Y. S. Huang, "Optimizing method for improving granulation effectiveness of iron ore sintering mixture," *Ironmaking & Steelmaking*, 2015, 42, 351–361.
13. Furans, C. C., Flow of gases through beds of broken solids, *US Bureau of Mines Bulletin*, 1929, 307, 32.
14. Kasai, E., W. J. Rankinand, J. F. Gannon, "The effect of raw materials properties on bed permeability during sintering," *ISIJ Inter.*, 1989, 29, 33–42.
15. Zhao, J. P., C. E. Loo, and R. D. Dukino, "Modelling fuel combustion in iron ore sintering," *Combustion and Flame*, 2015, 1019–1034.
16. Kapur, P. C., and D. W. Fuerstenau, "An auto-layering model for the granulation of iron ore fines," *International Journal of Mineral Processing*, 1993, 239.
17. Fuori, T., K. Sugawara, M. Kagawa, S. Uno, M. Kamazu, T. Fujiwara, and A. Sawamuru, "Article Title," *Nippon Steel Tech. Rep. Overseas*, 1977, 10, 36–46.
18. Litster, J. D. and A. G. Waters, "Influence of the material properties of iron ore sinter feed on granulation effectiveness," *Powder Technology*, 1988, 55, 141.
19. Hara M., T. Kawanguchi, M. Matsumura, and C. Kamijo, "Analysis of raw material flow with axial direction in continuous drum mixer for sintering," *ISIJ Inter.*, 2009, 49, 609–617.
20. Cores, A., M. Muniz, S. Ferreira, J. I. Robla, and J. Mochon, "Relationship between sinter properties and iron ore granulation index," *Ironmaking & Steelmaking*, 2012, 39, 85–94.
21. Bhagat, R. P., S. K. Pradhan, M. C. Goswami, B. Nandi, and T. Venugopalan, "Effect of LOI content of iron ores and other process variables on the reduction properties of sinter," *Proc. XXVI International Mineral Processing Congress 2012*, New Delhi, India, September 24–28, 2012, org. by Indian Institute of Mineral Engineers, Jamshedpur and Indian Institute of Metals, Kolkata.
22. Xuewei, D. Lv, C. Bai, G. Qiu, S. Zhang, and M. Hu, "Moisture capacity: Definition, measurement, and application in determining the optimal water content in granulating," *ISIJ Inter.*, 2010, 50(5), 695–701.
23. Ellis, B.G., C. E. Loo, and D. Witchard, "Effect of ore properties on sinter bed permeability and strength", *Ironmaking & Steelmaking*, 2007, 34(2), 99.
24. Loo, C. E., and D. J. Wong, "Fundamental factors determining laboratory sintering results," *ISIJ Inter.*, 2005, 45, 449–458.
25. Szekely, J., C. Evans, and H. Y. Solm, *Gas Solid Reactions*, New York: Academic Press, 1976, 185.
26. Lovel, R., R. Vining, and M. D. Amico, "The influence of fuel reactivity on the iron ore sintering," *ISIJ Inter.*, 2009, 49, 195–202.
27. Peters, K., H. Kersting, and H. Beer, "Ways and means of reducing the energy consumption in iron ore sintering," *Proc. 4th Inter. Symp. on Agglomeration Toronto, Canada 1985, Agglomeration 85*, Ed. C. E. Capes, Ontario Iron & Steel Society of AMIE CIM, 1985, 157–179.
28. CSIR-NML REPORT, *Study of the Factors Influencing the Sintering and Cooling of Super-Fluxed Sinterr (CaO/SiO2 1.8–2.5) on the Sintering Indices: Structure Properties Correlation*, TISCO Completion Report, Sept. 2014.
29. Ball, D. F., J. Dartnell, J. Davison, A. Grieves, and R. Wild, *Agglomeration of Iron Ores*, London: Heinermann Education Books Limited, 1973 (2), chapter 10.
30. Hostani, Y., N. Knno, J. Shibata, T. Sato, and H. Szuki, "Technology for granulating coke breeze by centrifugal roll type pelletizer and effect of granulated coke breeze on sintering operation," *ISIJ Inter.*, 1985, 35, 1340.
31. Chen, X., L. H. Fan, Y. Wang, H. M. Long, T. Jiang, J. Shi., Q. Y. Song, and X. D. Yang, "Control guidance system for sintering burn through point," *Ironmaking & Steelmaking*, 2009, 36, 209–211.

32. LOO, C. E., R. P. Williams, and L. T. Matthews, "Influence of material properties on high temperature zone reactions in sintering of iron ore," *Trans. Inst. Min. Metall.*, 1992, 101, C7–C15.
33. Hidueyuki, Y. and K. Takazo, "Development of a 3-D sinter process mathematical simulation model," *ISIJ Inter.*, 2005, 45(4), 522–531.
34. Loo, C. E. and J. Heikkinen, "Structural transformation of beds during iron ore sintering," *ISIJ Inter.*, 2012, 52, 2158–2167.
35. Mazaneck, E. and S. Jasienska, *JISI*, 1963, 201, 60–67.
36. Noda, H., H. Yanaka, R. Yamamoto, H. Kawata, and Y. Yamaoka, "Assimilation characteristics of iron ores and their influence on sinter quality," *Trans. ISIJ*, 1985, 25, 1103–1110.
37. Reddy, A. S., A. K. Roy, R. V. Ramana, and A. Kumar, "Managing the adverse effect of alumina in sintering process through selective granulation of high alumina iron ore fines," *Tata Search*, 2007, 111–117.
38. Bhagat, R. P., S. K. Gupta, and H. S. Ray, "Heat transfer consideration for improvement in the sintering indices," in *Proc. 48th Iron Making Conference, ISS–AIME*, Chicago, 1989, 481–490.
39. Bhagat, R. P., "Factors affecting return fines regime and strand productivity in iron ore sintering," *ISIJ Inter.*, 1999, 39, 889–895.
40. Umadevi, T., K. Nelson, K. C. Mahapatra, C. M. Prabhu, and M. Ranjan, "Influence of magnesia on iron ore sinter properties and productivity," *Ironmaking & Steelmaking*, 2009, 36, 515–520.
41. Matsuno, F., "Changes of mineral phases during the sintering of Fe_2O_3 – CaO-SiO_2." *Trans. ISIJ*, 1979, 19, 595–604.
42. Ostwald, J., "Mineralogy and micro-texture of Australian iron ore sinters," *BHP Technical Bulletin*, 1981, 25(1), 13–20.
43. Raju, K. S., Petrographic studies of ten samples of iron ore sinters, Report by Indian Bureau of Mines (IBM), Nagpur, Submitted to Steel Authority of India Limited (SAIL) Research & Development Centre for Iron & Steel (RDCIS), August 1981.
44. Lu, L., J. Holmes, and J. R. Manuel, "Effect of alumina on sintering performance of hematite iron ores," *ISIJ Inter.*, 2007, 47, 349–358.
45. Webster, N. A. S., M. I. Pownceby, I. C. Madsen, and J. A. Kimpton, "Silico-ferrite of calcium and aluminum (SFCA) iron ore sinter bonding phases: New insights into their formation during heating and cooling," *Metallurgical and Materials Trans.*, 2012, 43B, 1344–1357.
46. Loo, C. E. and M. F. Hutchens, "Quantifying the resistance to airflow during iron ore sintering," *ISIJ Inter.*, 2003, 43, 630–635.
47. Yong, R. W., "Article Title," *Ironmaking & Steelmaking*, 1997, 6, 321.
48. Dash, I. R., J. Rose Mitterlehner, G. Loeffler, F. Winter, H. Hofbaur, H. Schmid, E. Zwittag, T. H. Buergler, O. Pammer, and H. Stiasny, "Modeling and simulation of heat front propagation in the iron ore sintering process," *ISIJ Inter.*, 2004, 44, 11–20.
49. Wegmann, E. F., *Trans. Indian Inst. of Metals*, 1973, 6, 61–65.
50. Research and Development Centre for Iron & Steel Authority of India Limited (RDCIS-SAIL). Optimization of ignition regime of sinter charge in sinter plants of Bhilai Steel Plant, Completion and Implementation Report (No. R&D 12.o2.606.86) July 1986.
51. Kawaguchi, T. and M. Hara, "Utilisation of biomass for iron ore sintering," *ISIJ Inter.*, 2013, 53, 1599–1606.
52. Ishikawa Y., K. Sugawara, Y. Umezu, Latest development of sintering technology, *Proc. 2nd Inter, Symp. Agglomeration*, Vol. 2, Ed. K. V. S. Sastry, New York: Am. Inst. Mining, Metallurgical & Petroleum Engrs., 1977, 503–525.
53. Klose, W. and M. Wolki, "On the intrinsic reaction rate of biomass char gasification with carbon dioxide and steam," *Fuel*, 2005, 84, 885–892.

54. Andersen, J. A. and A. K. Andresen, Influence of mass transfer on the thermogravimetric analysis of combustion and gasification reactivity of coke. *Fuel*, 2010, 89, 1541.
55. Murakami, K. Sugawara, and T. Kawaguchi, "Analysis of combustion rate of various carbon materials for iron ore sintering process," *ISIJ Inter.*, 2013, 53, 1580–1587.
56. Arthur, J. R., *Trans. Farad. Soc.*, 1951, 49, 64.
57. Hida, Y., M. Saski, T. Enokido, Y. Umrzu, T. Lida, and S. Uno, Effect of the existing state of coke breeze in quasi –particles of raw mix coke combustion in the sintering of ores. *Tetsu-to-Hagane*, 1982, 68, 400.
58. Ohio, K., K. Noda, T. Nishoka, Maeda, and M. Shimzu, "Combustion rate of coke in quasi particle at iron ore sintering process," *ISIJ Inter.*, 2013, 53, 1588–1593.
59. Hou, P., S. Choi, E. Choi, and H. Kang, "Improved distribution of fuel particles in iron ore sintering process," *Ironmaking & Steelmaking*, 2011, 38, 379–385.
60. Korotich, V. I., "Principles of theory and technology of preparation of raw materials for blast furnace," *Metallurgiya, Moscow*, 1978, 208.
61. Ten, S. B., A. G. Mikhalevich, B. M. Boranbaev, B. I. Kolokol'tsev, and V. V. Dyabin, "Making fuel utilization in sintering more effective," *Stal*, 1980, 10, 868–870.
62. Voskoboinikov, V. G., S. B. Ten, B. M. Boranbaev, and A. G. Rusakova, "Mechanism of solid fuel combustion in sintering," *Steel in the USSR*, 1980, 50, 357–359.
63. Wegman, Eu F., *A Reference Book for Blast Furnace Operators* (English translation), Part IV, Moscow: Mir Publisher, 1984.
64. Zhao, J. and C. E. Loo, "Dependence of flame front speed on iron ore on iron ore sintering conditions," *Mineral Processing and Extractive Metallurgy*, 2016, (125), 165–171.

7 Sinter Productivity: Theoretical Consideration and Plant Practice

ABBREVIATIONS

LOI *Loss on ignition*
RI *Return fines input*
RO *Return fines output*
VSS *Vertical speed of sintering*

7.1 SINTER PRODUCTIVITY

7.1.1 Definition and Computation

The sinter productivity through continuous machine strand (see **Sec. 6.11.1; Figure 6.22**) is related to:

a. Tonnage of fresh materials (called sinter blends), which is processed per unit time
b. Amount of return fines recirculated
c. Amount of moisture and volatilities lost to the atmosphere due to the sintering process

The production of (sized) sinter, P (in T/h), can be expressed as[1]

$$P = [H * W * \rho_s * (1 - \omega_m) * (1 - \varepsilon) * V - R] * (1 - \lambda) \tag{7.1}$$

where:

H = bed height (m),
W = width of bed (m),
ρ_s = average density of the particles forming the bed (kg/m^3),
ω_m = mass fractional moisture content (–),
ε = fractional bed void age (–),
V = sinter machine speed (m/h),
R = return fines recirculated (T/h), and
λ = mass fractional volatile (loss on ignition) content (–).

The operational parameters—namely, H and W—are set to defined values for a sintering machine. Hence, in relation to a particular sinter blend having its fixed characteristics in terms of particle density, contents of moisture, and volatilities, the sinter productivity can be increased through either/or:

a. Increase in sinter machine speed
b. Decrease in the amount of return fines recirculated

7.1.2 Sintering Speed

The vertical rate of sintering is defined as the shifting of sinter formation in millimeters along the height of the layer of sinter charge over the grates in unit time (in seconds). The speed of the sinter machine is inversely proportional to the sintering time. The sintering time has a significant effect on the strand productivity as this factor goes into the denominator in case of batch sintering (ref. Eq. **7.2**).

The rate of sintering can be accelerated by increasing the rate of air filtration across the sinter bed. This can be increased by:

- Increasing the suction pressure. To increase the suction pressure, which is a design (and machine) parameter, has restrictions. An increase in suction pressure gives more opportunities for leakage through the pallet sides in an existing operating sintering plant.
- Decreasing the gas dynamic resistance of the sinter charge; that is, increasing the bed permeability during the sintering. The resistance of sinter mix to gas dynamics on sintering is fairly related to the material characteristics and process parameters, including the granulation of sinter blends. These aspects have ben detailed in **Sec. 6.3** and **Sec. 6.5.**

7.1.3 Return Fines Recirculated

Figure 7.1 is the schematic of sinter production and its stabilization in a continuous sinter plant. A portion of sinter, that is produced in the sintering machine remains unsintered/weak which disintegrates to fines and screened out. This forms the return fines and recycled back to the machine. The recirculation of return fines, R, affects the productivity of sinter, as is evident from Eq. 7.1. The input of fresh material on the sinter strand decreases with increasing the value of R.

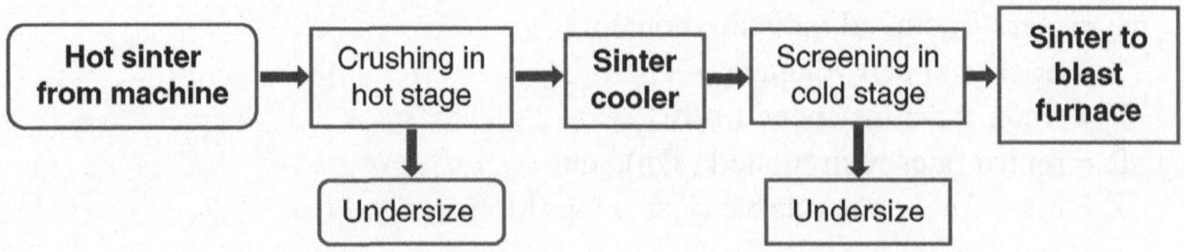

FIGURE 7.1 Flow scheme of sinter stabilization (crushing and screening)

The amount of return fines recirculated per unit time, R, in Eq. 7.1 increases when

- The yield of sinter produced is low, when the sinter is less strong in cold condition
- With higher degree of stabilization

Yield of Sinter The term, yield of sinter, denotes how much sinter, is produced per unit weight of sinter cake. The term is also referred as "sound' sinter" in the sintering plant operation. The definition of sinter is with regards to its size and hence changes depending on its acceptance by the blast furnace and hence is plant specific: whether it is above 6.3 mm or above 10 mm size. Obviously, the yield should be lower when above 10-mm-size fraction, compared to above 6.3 mm, is considered as sinter.

A lower sinter yield denotes the sinter is less strong in cold condition.

Degree of Stabilization The sinter produced on the sintering strand is subjected to crushing and screening at the hot stage followed by its cold screening and transportation to the blast furnace site. During the operation, the sinter passes through a number of transfer points: from its discharge stage through the sinter machine till it reaches the top of the blast furnace stack zone (see **Figure 5.1**; also **Figure 7.1**). Obviously, the sinter undergoes stages of mechanical treatment, often referred to as stabilization. The granulometry of sinter and its (generated) amount is decided by how the sinter undergoes stabilization. Accordingly, the figures pertaining to the yield of sinter and quantity of circulating load (return fines) changes as does the productivity.

The stabilization of sinter in the plant can be simulated in the laboratory by adopting the shatter test. This depends upon the blast furnace requirement with regards to size of sinter: whether it is above 6.3 mm or above 10 mm. The yield of sinter (also the sinter productivity) is calculated from the laboratory data following the shatter test (see **Annexure 2**). In the case of batch sintering, the strand productivity is computed from the weight of the sinter cake produced (weight of the sinter mix charged into the pot per batch minus the loss of moisture and volatilities), the yield of sinter after certain degrees of stabilization, the cross section of the pot, and sintering time (Eq. 7.2).

Strand productivity ($P_{+6.3}$/ P_{+10}) [T/(sq. m*d)]:

$$\frac{\text{Cake weight} * \text{Yield of sinter } (\% + 6.3\,/\,10 \text{ mm after n drops}) \#(\text{T})}{\text{Cross sectional area of pot (sq. m)} * 100 * \text{Time of sintering (d)}} \tag{7.2}$$

where: Yield of sinter ($Y_{+6.3}$/Y_{+10}) (%): Weight percent of +6.3 mm/+10 mm size fraction of sinter (cake) after "n" drops through shatter test on steel plate from 2 m height.

Treatment of sinter by allowing three or five falls from a height of 2 m is commonly practiced in the shatter test. The yield, thus obtained, reflects the productivity

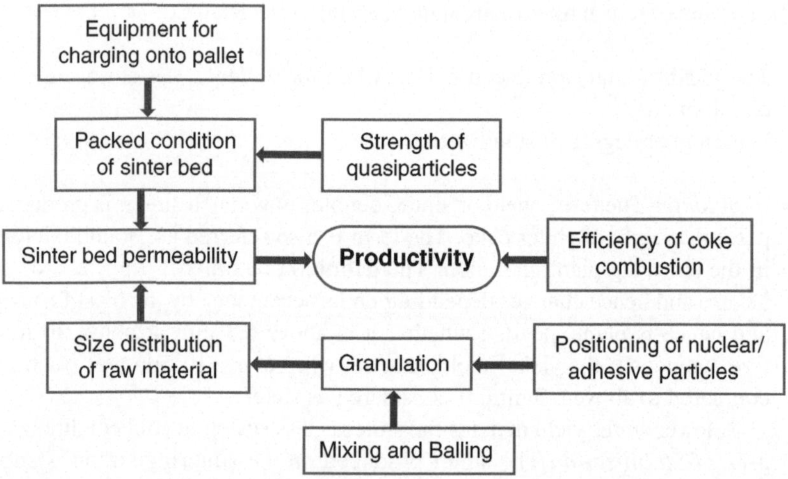

FIGURE 7.2 Dependence of sinter productivity on process and operational parameters

figure but not in absolute terms. The drop number (n) in the previous equations is to be simulated, keeping the layout of the sintering plant as well as the blast furnace (top of the stack zone) in view. For a larger setup steel plant where the blast furnace and sinter plant have more distance, the (drop) number, n, obviously becomes larger. The figure "n" in Eq. 7.2 can be decided through simulating the return fines balance (see **Sec. 7.8.2**).

7.1.4 Parameters Affecting Sinter Productivity

Figure 7.2 schematically relates the sinter productivity on a continuous strand with the process and operational parameters. The packed condition of sinter blends influences the flow of air and gases across the sinter blends before ignition and during the sintering process. The quasiparticles of sinter blends need to have sufficient strength; besides, the sinter blends should be charged onto the moving pallets with due care to ensure the optimum bed permeability. In addition, size distribution of the raw materials affects the bed permeability: the parameter is influenced by the manner the granulation proceeds. The granule characteristics of mix material, besides the feed characteristics, are affected by the operation of mixing and balling of raw materials.

7.2 BED PERMEABILITY AND BED STRUCTURE

Maintaining the optimum bed permeability before ignition (pre-ignition) and during sintering, as well as its strength are necessities for good sintering.

a. If the bed is less permeable, it would restrict the proper passage of the reducing gas and adversely affect the sintering speed, also the sinter quality.

b. On the other hand, if the bed is more permeable, the reducing gas may be channelized and may not come in proper contact with granules in the bed, and thereby adversely affecting the sintering process. Also, a much large decrease in aerodynamic resistance (higher bed permeability) of the mix, leads to an increase in the requirement of specific air consumption. The bed permeability is related to the fractional bed void-age ε (Eq. **6.3**). Hence, with increasing permeability, the throughput of the input material (the productivity) also decreases to some extent.

The structure of bed during sintering is important as it determines the bed permeability. The bed shrinks as the sintering proceeds (see **Sec. 6.4.1**). The structural changes of the bed that occur during sintering, are affected on account of the following:

a. The water vapor, initially generated at the top of the bed, gets condensed at the lower portion of the bed once the sinter mix is ignited, which decreases the bed permeability to a minimum for some time.
b. The structure of the sintering bed is influenced by the manner the sintering proceeds at its various stages (see **Secs. 6.9.2** and **6.10.1**) and the textural characteristics of sinter that appears following the dissolution of the minerals, and the formation of minerals and pores.

A sinter feed with good permeability from ignition until burn-through is required for efficient operation of a sinter plant. The permeability is greatly affected by these factors:

- the moisture content,[2] and
- the proportion of return fines in the sinter mix.[3]

In view of this, the sinter mix is required to have proper granulation. Kasai et al.[4] confirmed that pre-ignition permeability was related to the mean size of granules, which was, in turn, related to moisture addition.

This is more important for the iron ores with high ultra fines content that have low pre-ignition permeability.

In addition to the aforementioned, the bed of sinter should have adequate stability. In this regard, methods need to be devised to control and reduce the sinter cake load. The methods in this regard are:

- To provide adequate strength to the sinter mix granules through the modification in process parameters. For instance, the addition of binder(s), lime in particular, in the sinter mix provides the balls with more stability during the sintering process
- A magnetically- levitated sintering method has been devised to control and reduce the sinter cake load. The improvement in permeability resulting from applying the magnetic levitation are attributed to the formation of a homogeneous network with a pore size appropriate for gas flow as well as an increase in porosity of stem pores by about 8%.[5]

7.3 IRON ORES: PARTICLE SIZE AND CHARACTERISTICS

7.3.1 Particle Size of Ores

The particle size of ore and its distribution affects the sinter productivity in the following ways.

- Its direct influence on sintering speed: The bed permeability of the mix decreases when the percentage of very fine grains in the sinter mix is preponderant. On the other hand, the coarser particles do not take part in granulation.
- Indirectly through its influence on mix granulometry: The granulometry of sinter mix is decided by the size distribution of individual raw material components and the extent of their granulation under the influence of water and when subjected to mechanical forces in mixing and balling drums.

The ore with too much coarse fine is also harmful because it affects the granulation of the sinter mix in balling. The coarser fraction of iron ore does not take part in balling. The degree of balling improves significantly when ore fines below 5 mm size is used compared to those below 10 mm size.

The size of ore fines in some cases may not be normally distributed (see **Sec. 6.5.5**). Iron ore from a typical mines contain higher mass fraction of coarser size, at the same time, higher amount of finer sized fraction.[6,7] Accordingly, the bed permeability vary depending on how the size distribution gets skewed from the normal one.

7.3.2 Incorporation of Ultrafines

Blue Dust/Iron Ore Slimes The slimes are generated from the wet processing of iron ores fines. The values recovered from slimes make an additional source of iron ore for sintering. Another less-utilized source of iron ore is blue dust, which is rich in iron and low in SiO_2 and Al_2O_3. Both these materials are extremely fine and thus have limited usage in the sintering. Sintering with fine concentrate alone without any additive or special preparation is not stable due to low gas permeability. Studies carried out in various Indian research and development (R&D) laboratories[8-11] have shown that blue dust in the range of 20-40 wt. % of the total ore fines can be used without affecting sinter properties and the strand productivity. The sintering rate decreases with increasing blue dust in the mix. However, the strand productivity increases up to a limiting value as the sinter yield is enhanced.[12]

Magnetite Concentrate The magnetite concentrate, typically from Kudremukh, is rich in Fe (68 wt. %) with low gangue content and is extremely fine (60-70 wt. % below 44 micron). This had been a feedstock in pellet making. Its unfavorable grain size restricts its use through sintering.

Typical laboratory sintering experiments conducted at Bhilai[13] at sinter basicity ratio CaO/SiO_2, 2.5 show:

- The addition of 2 wt. % lime results in enhanced sintering speed compared to that when no lime was added. This is because of improved balling.
- Addition of the concentrate up to 20 wt. % of the ore part makes the sinter mix nonhomogeneous, and segregation takes place due to the difference in particle size between the concentrate and rest of the ore. The mix approaches homogeneity as the amount of concentrate is increased. The balling process becomes more favorable with the addition of lime and, consequently, strength of sinter improves.
- At any level of concentrate in the mix, the yield increased when the lime percentage was increased from 0% to 2%.

7.3.3 Ores: Textural Parameters and Chemical Composition

Ore Type The sinter mix constituting more massive to hard-laminated hematitic ore had lower permeability than that which consisted of a higher proportion of soft-laminated to goethite mineral. Typically with reference to Indian ores, this resulted in poorer productivity (in the case of hard-laminated ore), although a marginal increase in the sinter yield was observed.[2]

Silica and Alumina The sintering is delayed with the ores having high silica and/or high alumina content. The study at Chiba sinter plant (No. 4) shows that a high productive operation at 1.8 T/h*(m^2) without adding any binder(s) could be achieved when low SiO_2 operation was maintained. The total amount of return fine decreased from 400-450 to 250-300 kg/T-sinter.[14]

7.3.4 Blending of Ores

The ores from various resources have different sinter-ability in view of their differences in textural and size parameters. Also, these have different moisture retention capacities. This results in different sintering indices in terms of sintering rate and quality of sinter produced. These indices can be very well optimized by blending different verities of ores. Hard-laminated ore, when blended in higher proportion with laminated to soft-laminated ore, results in increased green permeability of the sinter mix; subsequently, an improved productivity of sinter has been attained.[2] The ore fines with high loss on ignition (LOI) content and those with high gangue minerals make a positive contribution to granulation in view of their high moisture retention capacity. Therefore, the granulation of sinter mix improves when massive to, hard-laminated ores (with high Fe) are blended with the high LOI ore.

The ores from the Indian resources, Noamundi and Joda of Tata Steel have different textural characteristics: the latter being more of a laminated (soft) to goethite type. These, when blended in 70:30 proportion, have resulted in improved sintering indices, including sinter productivity,[6] compared to the indices when these were individually sintered.[15]

7.4 GRANULATION

The mix granulometry plays an important role in deciding the sinter bed permeability and works better in its optimum, but limited, range[4,16,17]. **Figure 7.3** shows how the sinter productivity is influenced by granulation related parameters. The granulometry of sinter mix is decided by the size distribution of individual raw material components and the extent of their granulation under the influence of water and when subjected to mechanical forces in mixing and balling drums. The improvement in the granulometry of sinter mix aids the sintering efficiency through better solid gas contact and bonding: The topic is covered in **Sec. 6.5.**

7.4.1 Granule Size/Size Parameters of Mix Ingredients

The size of ores as well as that of solid fuel and fluxes does affect the flame front speed and transfer of heat from the hot gases to solids and vice versa. When the percentage of very fine grains in the sinter mix is preponderant, the permeability of the mix and, hence, the sinter output, decreases.[18] On the other hand, the heat transfer between the solid particles and combustion gases is enhanced with decreasing the particle size of mix materials,[19] which should accelerate the diffusion bonded assimilation and hence sinter productivity (**Sec. 6.5.3**). In view of this, the optimum value of the particle size needs to be worked out.

7.4.2 Bulk Density of Sinter Mix

The amount of mix material that undergoes sintering is the material volume multiplied by its bulk density. Since the strand has a constant volume (bed depth and

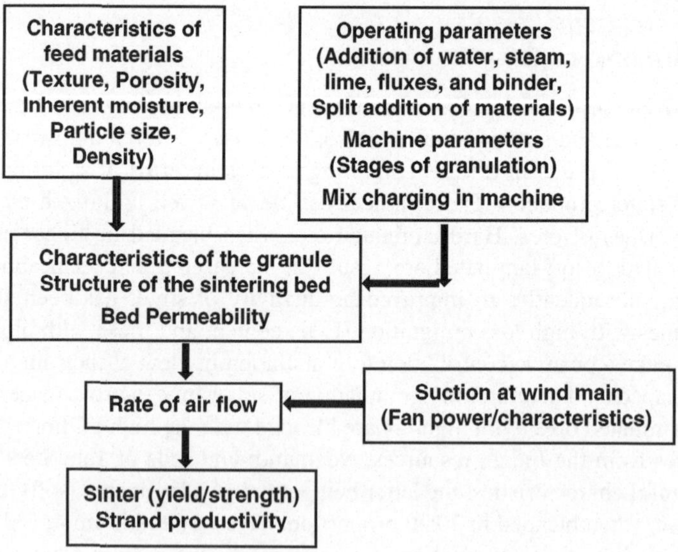

FIGURE 7.3 Factors involved in the granulation and sintering of iron ores and sinter productivity

width), the production of sinter is directly proportional to the bulk density of the sinter bed material (Eq. 7.1).

The increase in bulk density of the mix has the following effects on the sintering indices:

a. Flame front speed. The flame front speed, as determined by the time required for particles to reach combustion temperature, will decrease when bulk density of the bed increases (see **Figure 6.13b**). The dense material acts as a heat sink, which affects the heat transfer and flame front speed (see **Sec. 6.6.3**).
b. Thickness of high temperature zone becomes narrower.
c. Maximum temperature of sintering bed increases.

Sinter yield may increase as the sinter is subjected to higher densification due to the combined effect of the aforementioned phenomena in (b) and (c).

The bulk density of sinter mix can be increased through:

a. The addition of the surface active agents.
b. Reduction in coke consumption: Sintering beds have around 3.5-4.5 wt. % coke particles on a dry basis. Coke particles have a density of 1,100 kg/m^3, whereas the packed bed for sintering has a bulk density of around 2,000 kg/m^3. The bed has around 10% coke on a volume basis.

7.4.3 Selective Granulation

A process called the "selective granulation" has been developed. The pisolite ore having a particle size less than 3 mm is separated and granulated in advance. Then, granulates are added to the residual raw mix (**Figure 7.4**). The process has made the

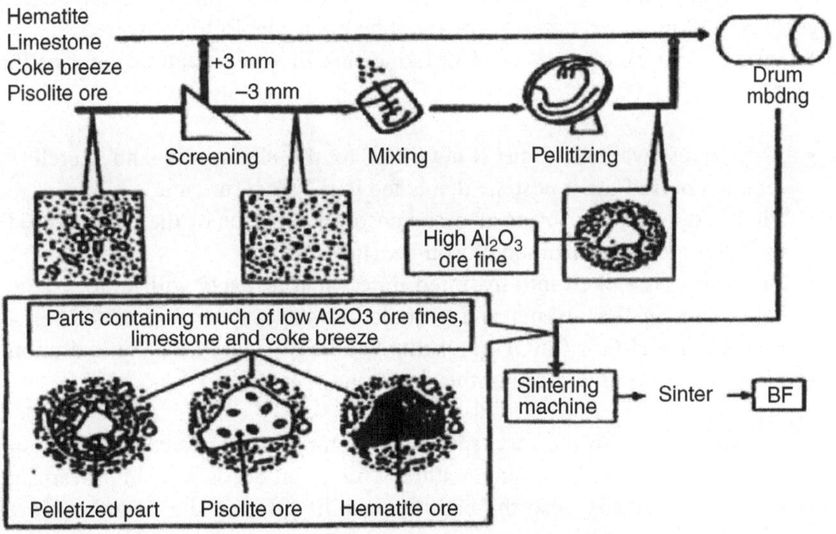

FIGURE 7.4 Flow of selective granulation process (after Ref. [20])

operator able to accommodate larger amounts of fines, from 36% in 1995 to 56% in 2002 at Oita works.[20,21] (Also see **Secs. 10.3.3, 10.8.1, 10.8.2; Figures 10.3, 10.10, 10.11, and 10.12.**)

The finer fractions of iron ore fines, in general, contain high alumina and form the adhering layer during the granulation process. In conventional granulation, alumina particles directly go into the surface layers of the granules. The adhering layer, which becomes almost uniform in the mixing drum, mainly participates in the melt-forming reaction during sintering. This results in a high fusion temperature and poor bond formation during sintering.

7.4.4 Modification in Operational and Machine Parameters

The granulation of sinter mix is reported to be improved with the modification of machine parameters, especially with respect to the application of newer kinds of feeders, such as the slit wire type,[22] the feeder with a magnetic device[23] including the intensified shifting feeder[24] to promote the particle-size segregation of raw mix on the sloping chute of a sinter plant (see **Figure 5.10; Sec. 5.6**).

The provision of a bigger sized lifter decreases the carbon variation in the sinter mix and increases the chemical consistency of product sinter. The modification in the parameters related to balling units such as water spray point, speed of mixing drum, and machine length also contribute to the improvement in granulation.

7.4.5 Usage of Lime

The hot metal, produced in many of the blast furnaces, typically from Southeast Asia, is high in silicon and phosphorus contents and hence requires high-purity soft burnt lime to the extent of or even more than 130 kg/T for basic oxygen furnace (BOF) steel making.[25] Around 20-30% of the total lime produced degrades to generate –10 mm-sized fines that cannot be fed to the BOF converter.

Sintering is an excellent way of utilizing this lime (dust) on account of these factors:

- It provides available lime (CaO-SiO_2) to the sinter mix and therefore replaces part of the limestone that is fed to the blast furnace.
- Lime acts as a flux, causing an easier absorption of iron by the slag followed by extensive crystallization of iron-bearing minerals.
- Burnt lime is slaked into hydrated lime when it reacts with water in the sinter mix in the mixing and granulation stage according to the reaction: $CaO + H_2O = Ca(OH)_2$ + heat. The heat released during hydration increases the temperature of the sinter mix.

Lime, when added in the mix, speeds up sintering and reduces the time spent by the sinter mixture at high temperature. This is on account of improvement in pre-ignition permeability, also the bed permeability, during the sintering process. Nyquist[26] discussed the reason for improvement in sintering efficiency with the addition of lime in the sinter mix. High surface tension and low viscosity of the melt, combined with short solidification time and a high temperature coefficient of viscosity have been found to lead to rapid sintering, particularly in relation to the combustion zone.

The underlying mechanisms for improvement in bed permeability (with lime addition) are as follows.[25,27]

- Lime (active CaO) in contact with water (in sinter mix) promotes the improvement in its surface tension resulting in its enhanced binding properties. This provides the balls with more stability during the sintering process.
- The contact between the reactive charge components is enhanced with the addition of lime.
- The granules are hydration hardened, which is implemented through a solution– colloidization–crystallization scheme.

These result increased strength of individual pellets and hence more stability to structure of the bed.

The lime should be finer in size for complete hydration during the mixing stage itself. This is because the hydration of lime, being a chemical process, is guided by its kinetics. The resistance for the diffusion of moisture increases as the reaction proceeds and the formed product (slaked lime), which covers around the lime particle surface, becomes thicker and thicker. Hence, more time is needed for moisture to penetrate the layer. The coarser the burnt lime size, the longer the time required for burnt lime to get hydrated. For instance, hydration time for burnt lime having size below 3 mm is about 8 min. Obviously, the size should be much less.

The laboratory scale experiments using the Bhilai (India) ores show that the specific productivity of sinter increased by 13.5% with the addition of 2 wt.% lime in the sinter blends.[28] The contribution to the productivity was by virtue of:

a. Increase in the vertical speed of sintering (VSS) by 11.3%
b. Increase in the yield of sinter (Y_5) by 5.2%

Further increase in the consumption of lime from 2 wt. % was not useful, although more available lime (CaO-SiO_2) could be incorporated. When lime consumption increased to 4 wt. % of the mix, the specific productivity decreased marginally from the level obtained at 2 wt. %.

7.5 MOISTURE CONTENT OF THE SINTER MIX

The presence of moisture in the sinter mix has an important role in affecting the sinter productivity. The removal of moisture and its rate of removal as the sintering progresses is important in this regard. This is in view of the fact that the rate of sinter formation depends on the shifting of the drying zone: how quickly the moisture is removed. Therefore, to have an optimum rate of sintering, not only the laws of gas dynamics but also those of evaporation of moisture have to be taken into consideration. The moisture content of the sinter mix is to be kept slightly less than that corresponding to the mix with the maximum pre-ignition permeability.

7.5.1 Role of Moisture in Granulation

Adequate moisture is required for proper agglomeration of fines, whereas excessive moisture promotes condensation in the bottom of the bed, which fills the voids with free water and may cause collapse of the bed structure in the wet zone.[29]

Optimum Moisture: In this regard, over-moistening at the lower zone of the bed should be avoided because it gives maximum resistance for air and gas flow through the bed due to partial destruction of agglomerates and packing of mass. In addition to that the condensation of water vapor at the lower level of the sintering bed happens to be a major problem in the sintering process. This decreases the bed permeability and often results in the collapse of the bed. Water is carried out from the upper part of the sintering bed in vapor form that gets condensed at the lower part. Recondensation occurs due to saturation of water vapor pressure in the waste gas passing through the lower horizontal sinter bed.

The optimum moisture content of the sinter mix should be as low as possible. This can be achieved through:

- **Initiating the granulation operation with low moisture addition:** The flow-ability of water, prior to its addition in the balling drum, is increased following the reduction of its surface tension. This is achieved while passing water through a magnetic conditioner. This attempt has resulted in improved ball-ability of the sinter mix.
- **Raising the mix temperature:** Increase of the sinter mix temperature improves the process of balling and the strength of granules in wet and dry state also improves. The phenomenon of recondensation of water vapor in the lower level sinter bed is suppressed by raising the temperature of the mix above the dew point, which is around 60°C (333 K). Hence, if the temperature of the mix is raised above this temperature, recondensation of water vapor can be avoided. Prevention of water vapor to recondensate will result in improvement of the bed permeability through preservation of granular structure, which results in the increase of machine speed.

Temperature of the sinter mix can be raised externally by:

a. Use of hot water, in place of normal water (at room temperature), in the sinter mix during granulation raises the sensible heat of the sinter mix temperature.
b. Steam injection: Low-pressure steam is injected into the sinter mix, which undergoes balling. The temperature of the sinter mix is raised when sensible heat of steam is transferred to the sinter blends. The low-pressure

steam can be generated through recuperation of sensible heat of (hot) sinter during its cooling.

c. Recirculation of hot sinter fines in the sinter mix increases its temperature to some extent by transfer of its sensible heat.
d. In addition, lime, when admixed with sinter mix, due its exothermic hydration by water [$CaO + H_2O = Ca(OH)_2$], releases heat, which increases the mix temperature. The mix temperature is increased by the exothermic reaction of hydration and granulation improved by the binding effect of $Ca(OH)_2$, which can ease moisture condensation in the bottom layer of the sinter bed and improve granule strength.

7.5.2 Granulation with Low Moisture

The possible way to improve the granulation of the sinter mix is to reduce the moisture content of the sinter mix. Low moisture addition in the granulating drum could be initiated by increasing the flow-ability of water. The increase in flow-ability of water can be achieved by reducing its surface tension. The electric and magnetic fields have been reported to lower the surface tension of water by up to 8%. The surface tension of water can be decreased from 72 to 65 $mN.m^{-1}$ while passing water through a magnetic conditioner. The water with increased flow-ability can then be added to the balling drum, in place of normal water, for granulation. In this way, water is mixed uniformly throughout the sinter mix, and this helps in increase in ball-ability of the sinter mix. This approach can lower the thermal requirement of a sintering plant. The possible reasons, in this respect, are:

a. The bed permeability gets improved.
b. Recondensation of water vapor at the bottom of sinter bed gets decreased.

The possible accrued benefits, while adopting this methodology, are as follows.

- A reduction in moisture content.
- Decrease in specific coke breeze consumption.
- The content of below 5 mm-sized fractions in the skip sinter is reduced.

Although the surface tension of water decreases when water is passed through a magnetic field, as mentioned previously, at a higher magnetic field, the effect is opposite. Recently, high magnetic fields (10 T) have been reported as increasing the surface tension of water by almost 2%. The surface tension of water increases linearly with an increase in the square of the magnetic field. The surface tension increased by 1.32 +/– 0.13 mNm^{-1} or 1.83 +/– 0.18% at 10 T. One possible explanation for this is that the stabilization of hydrogen bonds increases the bulk Helmholtz free energy, which thereby increases the surface tension.[30]

The surface tension of water increases with an increase in dissolved salts, also with a decrease in temperature. However, it has been reported that the measurement of the surface tension of water is sensitive to impurities.

7.5.3 Preheating of Sinter Mix

Sinter plants across the world, and in Russia in particular, have developed and adopted different ways of preheating the sinter mix.[31]

a. *Magnitogorsk Steel Plant:* Preheating of the sinter mix in the secondary mixing drum is carried out by the combustion product of coke oven gas. The gas burner, using coke oven gas having calorific value (CV) 17,640 KJ/Nm^3 (4,200 Kcal/Nm^3), has been installed on the discharge side of the mixing drum. The combustion products are withdrawn from the drum by a fan on the charging side. The entire length of the mixing drum is used to preheat the sinter mix. The results of trials show: (a) an increase in sinter productivity by 1.4% and (b) a reduction in solid fuel consumption by 4.2%.
b. *Krivorgaski Steel Plant:* Steam is supplied into the sinter mix in the balling drum through the steam outlet. Temperature of the sinter mix is increased from approximately 35 to 55°C. The sinter productivity increased by 7% and the content of below 5-mm fractions in skip sinter decreased by 2.5 wt. %. Typical arrangement is shown in **Figure 7.5.**
c. *Donestsk Research Institute for Iron & Steel* has developed another method of steam injection. Steam is supplied in the lower part of the bunker before the balling drum. The mix temperature is raised from 35°C to 45°C and productivity increased by 3.4%.
d. A steel Plant at *Fukuyama Japan* has developed a system of preheating the mix. Hot water is added in the balling drum. Temperature of the mix was raised from 30°C to 45°C. The sinter productivity was raised by 7%.
e. An industrial trial was conducted at the Bhilai sinter plant of Steel Authority of India Limited (SAIL), India, using the specially developed injection system. Steam was injected in the intermediate bunker before the drum feeder, which discharges the mix onto the pallet before ignition. Temperature of the sinter mix was raised from approximately 35°C to approximately 60°C. The benefits achieved were: (a) specific coke consumption decreased by 2 kg/t, and (b) sinter productivity increased by 2.6%.[32]

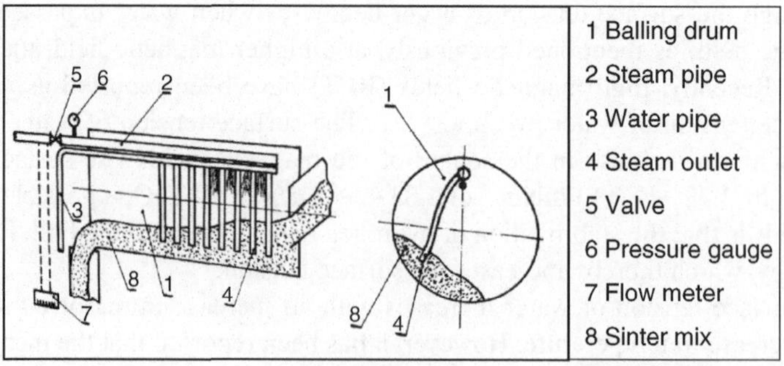

FIGURE 7.5 Schematic diagram of steam injection system in balling drum

7.6 SINTER BASICITY AND MGO CONTENT

7.6.1 Sinter Basicity

The basicity of sinter has considerable influence on the productivity.[33] The sintering speed increases with increasing basicity because the latter improves the permeability[34] and reduces the melting temperature.

An increase in sinter basicity reduces the sintering time due to:

- Reduction in specific air requirement on account of increased air utilization efficiency.
- Improvement in the pre-ignition and average (bed) permeability during sintering. This is on account of improved granulometry of the sinter mix.
- At higher sinter basicity, the proportion of flux (nonferruginous) fraction in sinter blends increases, which contributes to the improvement in granulation.[35] This is because the fluxes (limestone and dolomite) have a higher water absorption capacity than iron ore (see **Sec. 6.5.7**).

Joda (Tata Steel) Ores The experiments on granulation of sinter mix (at its different basicity) using Joda ores of Tata Steel show that the mean diameter of the sinter blends, Dp_{total}, also the weight fraction of below 3.3 to above 1.68 mm size, increases linearly with increasing sinter basicity (CaO/SiO_2) in the 2.1-2.5 range. This is at the cost of decreasing weight fraction of below 1.68 to above 0.6 mm size following granulation (**Table 7.1**).

7.6.2 Sinter MgO

The sintering time increases with the increase in sinter MgO.[36] In the case of Joda (Tata Steel) ores, the sintering as well as cooling of sinter happened to be faster when sinter MgO was lowered to 2 wt. % from 3 wt.% at sinter basicity (CaO/SiO_2) 2.0.[6] Further to that, when sinter basicity was raised to 2.5 and MgO content was further reduced to 1.5 wt.%, significant improvement in the sintering time was noticed.

TABLE 7.1
Size Analysis of Sinter Mix before and after Granulation

Sinter basicity	Dp of total mix material, mm		Wt.% −3.36 + 1.68 mm fraction of mix		Wt.% −1.68 + 0.6 mm fraction of mix	
CaO/SiO_2	Before	After	Before	After	Before	After
2.1	0.560	3.21	9.52	12.45	7.79	13.21
2.3	0.562	3.73	9.60	13.51	7.88	9.77
2.5	0.565	4.03	9.68	24.61	7.98	2.68

7.7 COKE AND FLUXES: CONTENT, NATURE, AND PARTICLE SIZE

7.7.1 Nature of Solid Fuel and its Content

A much higher amount of coke and/or the addition of solid fuel with higher reactivity result in broadening of the bed temperature profile (see **Section 6.6.4, Figure 6.14a**). As a result, the thermal effect of solid fuel is not fully utilized.

A lower ash content of coke increases the sinter productivity marginally. Stoichiometrically, an increase in coke ash content decreases the Fe content of sinter with a simultaneous increase in silica and alumina content.[37] The addition of carbon content in the mix through coke has limitations with respect to the sinter yield and sintering speed.[38]

The fact that the yield should go up with an increase in the amount of coke in the mix holds good within a limited range of coke percentage. A higher amount of coke may result in over-fusing of the sinter, which may result in more glass formation.[19] This fact needs to be considered in sintering practices. The pot sintering results with typical hematite ore show that the sintering time had a minimum value at about 4.2 wt. % of coke and is prolonged with an increase in coke content.

7.7.2 Particle Size of Coke and Flux

Use of sized (below 3 to above 0.25 mm) coke or limestone resulted in improved permeability before and during sintering.[19] Bhagat et al.[39,40] have shown that the VSS increased when below 0.5 mm -sized fractions in limestone were eliminated compared to the base. Hosotani et al.[41] in their studies have also observed that the sintering time was shortened by using limestone from which below 1 mm -sized fractions were removed. Coke size also affects the speed at which the flame front descendsed down the bed. The presence of finer fraction of coke (size below 0.25 mm) in the sinter mix had also detrimental effect on sintering time and hence productivity.[42] This is on account of improper combustion as the size of coke particles decreased resulting in higher levels of carbon monoxide in exhaust gases and, consequently, lower sintering temperature. When coke was forced to the surface of granules through its split addition, its combustion gets enhanced and consequently sinter productivity increases.

7.7.3 Split Addition Flux and Coke

Improvement in the granulation of sinter blends is not only affected by their moisture content but also by the mode of mixing and balling. The granulation process is modified by coating coke breeze and/or limestone on the surface of quasiparticles that have been granulated in the primary part of the drum mixer. This is achieved in the plant by injecting coke breeze and/or limestone at high speeds from the end of the drum mixer. **Figure 7.6** shows schematic of such arrangement. These materials segregate in the quasiparticles; hence, the process controls the excess melting reaction between mix iron ore and limestone.[43]

In the normal granulation process, the coke is added to the sinter blends in bulk; the fine coke particles remain embedded into the granules, and their combustion rate

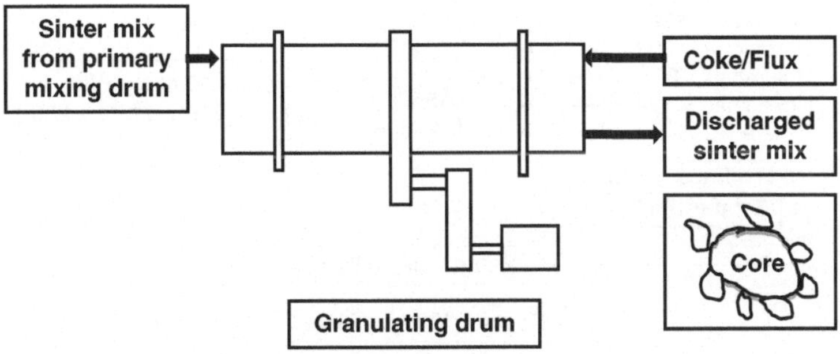

FIGURE 7.6 Scheme of injecting the fines of coke and fluxes in the drum mixer and the product of quasiparticles coated with these fines

is decreased. As a consequence, the width of the coke combustion zone broadens resulting in lower sinter productivity. In contrast, the late addition of coke improves sinter productivity[44] (see **Sec. 10.8.1; Figures 10.10 and 10.11**). The operative data of the JFE Kurashiki sinter plant in Japan show that the productivity increased through the implementation of this technology (see **Sec.10.8.1**).

Marginal reduction in the sintering time could be achieved with an improved method of mixing and balling. With two-stage, sintering time decreases by around 8% compared to one-stage mixing and balling.[38]

7.8 RETURN FINES

7.8.1 Return Fines Regime

Return fines are produced inadvertently in a sinter plant and are recirculated back to the plant along with the fresh material. A sinter plant maintains a return fines regime; the amount of return fines constitutes a certain percentage of the input raw sinter blends. A much higher amount of recirculated return fines is indicative of poorer performance of the sinter plant as the sinter mass has greatly disintegrated with a lower yield of sinter. On the other hand, in case the recirculated amount is less, the sinter that is produced on the traveling grate strand gets over-fused and hence could have a deteriorated reduction quality parameter.

The return fines regime is guided by:

a. The process parameters, including the content of coke in the sinter blends
b. The blast furnace requirement for sinter with regards to:
   - Its size: whether above 6.3/5 mm or above 10 mm
   - Amount (in wt.%) of the undersized fraction (below 5 mm) that is accepted in the furnace

It is possible to control return fines' recirculation in the plant. The techno-economic indices of sinter production are decided based on the way the return sinter regime is chosen. Hence, it is very important to understand the phenomenon of disintegration of sinter under the action of mechanical forces on one hand and to determine the influence of the factors related to the sintering technology on its mechanical strength.[45]

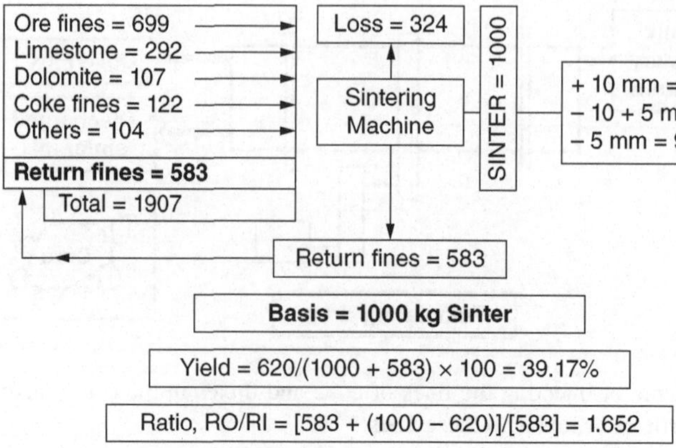

FIGURE 7.7 Typical mass balance around a sinter plant (after Ref. [37])

7.8.2 Return Fines Balance

Continuous Sinter Plant In the case of a continuous sinter plant, the return fines are balanced and regulated by the process parameters, including coke. **Figure 7.7** shows a typical mass balance around a sinter plant.[37] It appears that the return fines get balanced as the generated return fines are recirculated back to the plant. But in the true sense, the return fines output (RO) and return fines input (RI) are not balanced. A part of finer -sized fractions (below 10/6.3 mm size) is also associated with the sinter. This also constitutes a part of the RO in true sense.

Pot Sintering Unit In the case of a batch unit, return fines in a fixed weight percent of the mix are (initially) charged. Output of the return fines will depend on the process and operational conditions as well as the treatment of sinter after its cooling: how much the sinter is stabilized following the shatter test. The amount of return fines output increases when sinter is subjected to a higher number of drops. In this case, the sinter becomes stronger for the obvious reason. Typical result is shown in **Table 7.2.**[35]

TABLE 7.2
Generation of Sinter and its Strength following Shatter Tests of Three, Five, and Eight Drops (Height of Sinter Fall 2 m)

	+10 mm-sized sinter (wt. %)		Tumbler index, %
	Immediate	After 24 h	Immediate
After 3 drops	85.79	83.08	69.3
After 5 drops	80.09	77.56	72.0
After 8 drops	73.77	71.44	

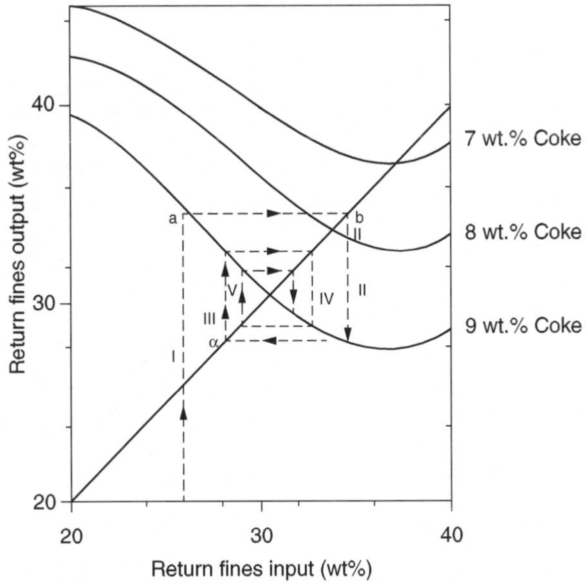

FIGURE 7.8 A typical return fines balance in the sintering process (after Ref. [33])

Figure 7.8 typically shows the relationship between return fines output and return fines input at three levels of coke in the mix: 7 wt.%, 8 wt.%, and 9 wt.%. The data were obtained from the pot sintering tests of 300-mm bed height. The parameters fixed were: (a) sinter basicity (CaO/SiO_2) = 3.25 and (b) content of moisture in mix = 7.25 wt. %. The straight line Y = X cuts the curves at three different points. Obviously, these points are the levels of balanced return fines at three levels of coke in the mix. Any change in the proportion of return fines from the balanced one would result in a difference between the return fines output and the return fines input. However, the difference decreases during subsequent sintering and recirculation of the return fines till the system reaches the balanced return fines regime. This is typically illustrated in **Figure 7.8** starting from a hypothetical sintering experiment (Exp. I) using 9% coke and 26% return fines in the mix.

7.8.3 Ratio (RO/RI)

In a continuous sinter plant, the ratio, RO/RI can be computed on the basis of sieve analysis data of the skip sinter. It is apparent from **Figure 7.7** that the sinter that has come out of the plant consists of some portion of below 5 mm and a significant portion of below 10 mm. These fractions are also return fines since, in this case, the +10 mm-sized fraction of sinter is accepted by the blast furnace as feedstock. As the sinter is transported from sinter plant to blast furnace, it degrades generating fines (below 10 mm size). Other reason for having below 10 mm-sized

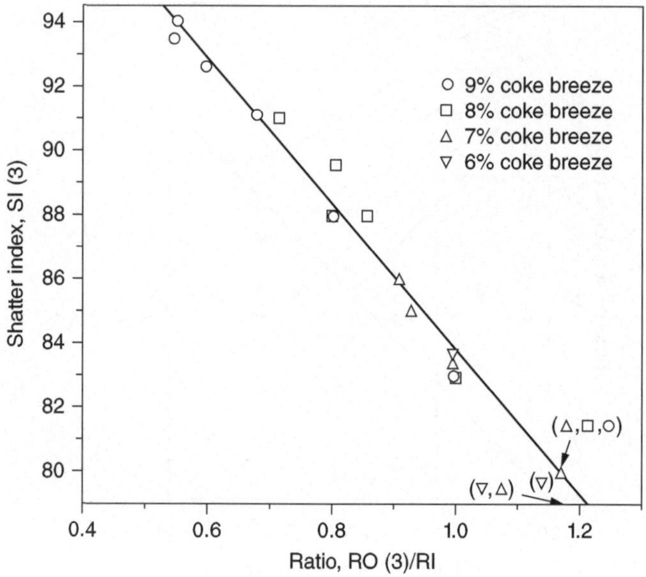

FIGURE 7.9 Relationship between strength of sinter and RO (3)/RI (after Ref. [46])

fraction in the sinter is the effectiveness of the screen. In the laboratory, the screen effectiveness has been cent percent (perfect); whereas in case of a plant setup, this could not be the case, that is, some undersize materials also report to the sinter directed to the furnace.

The index, RO/RI, is indicative of the "balance of return fines" and can be a useful tool in correlating the stabilization sinter at various transfer points (different stages of sintering plant operation and till the blast furnace charging point).

A close correlation could be established between the ratio RO/RI with the sintering variables and so with the shatter index at the different degrees of stabilization (**Figure 7.9**).[46] Such correlation has been substantiated from the reported data of Rauch and Cappel.[47]

7.8.4 Simulation of Return Fines Balance

For quality control purposes, the index, RO/RI, obtained through the plant operation could be simulated with that generated through pot sintering studies. That is, the value of "n" in Eq. 7.2, which is required to be followed in the control laboratory for the shatter test, can be arrived at while computing the ratio, RO/RI, of plant data at the discharge point of the sintering plant (stage) and at the blast furnace stage.

Figure 7.10 shows the value of RO/RI following the shatter test (height 2 m) of the sinter generated through pot sintering experiments at different numbers of drops, n (i.e., at different degrees of stabilization).[46]

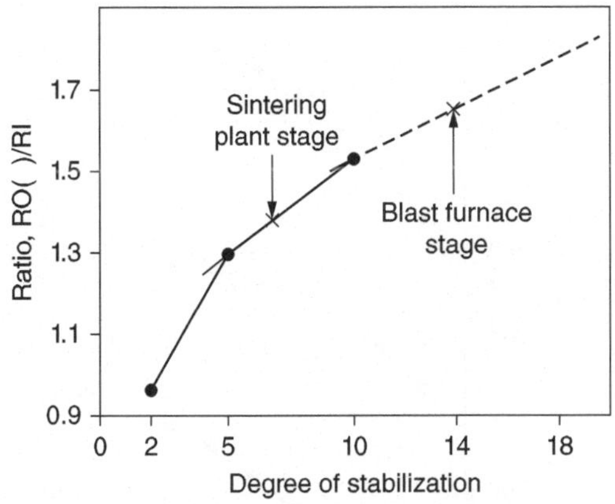

FIGURE 7.10 Empirical simulation of return fines balance in plant with laboratory data (after Ref. [46])

The pot sintering tests were carried out using the sinter raw mix from the plant and following other sintering conditions, including bed height, as those used in the plant.[46] The values of RO/RI equivalent to sinter plant and blast furnace stages are shown. Accordingly, the degrees of stabilization can be worked out. In this case, these are around 7 and 14 at the sinter plant and blast furnace stages.

7.8.5 Optimum of Return Fines

Although return fines are produced inadvertently in a sinter plant, they have a significant role in the iron ore sintering.

- The addition of higher return fines in the mix reduces the input of fresh material. However, part of the loss in strand productivity, due to the recirculation of return fines, gets compensated by a simultaneous increase in the yield of sinter[48] because of better fusion and bonding.[49] Hence, the cutoff point in relation to the amount of return fines in the mix could be arrived at (shown in the next paragraph).
- Return fines in the sinter mix improve the bed permeability resulting in more even distribution and transfer of heat with the lower shrinkage of sinter mix.[47] Consequently, the heat losses decreases, reoxidation heat is liberated, and the shrinkage of the mixture is lowered.

Figure 7.11 shows the influence of return fines input in sinter blends (wt. %) on the sinter yield, Y (5) (+10 mm-sized fraction) (%) following shatter test of five drops (2-m height) and (inverse of) sintering time, 1/t (h^{-1}). The data were obtained from the pot sintering tests of 300-mm bed height. The parameters fixed were: (a) sinter basicity (CaO/SiO_2) = 3.25; (b) content of coke (ash = 29.5 wt. %) in mix = 8 wt. %; (c) content of moisture in mix = 7.25 wt. %.[33,50]

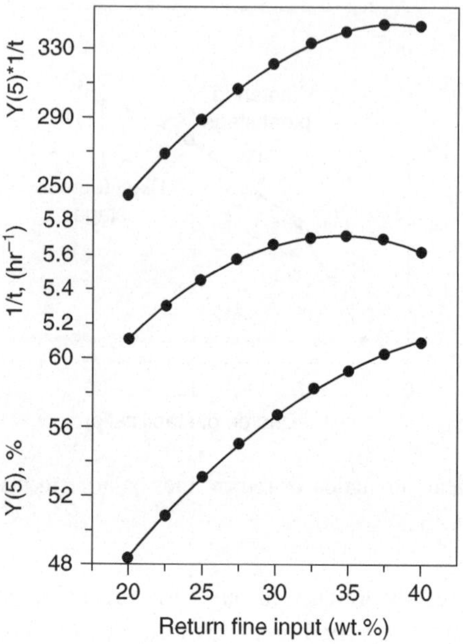

FIGURE 7.11 Influence of return fines input on the sintering indices (after Ref. [33])

The product, (Y (5) * 1/t), which is proportional to the sinter productivity, increases with the percent of return fines in the mix but flattens beyond 40 wt. %. The additional amount of return fines occupies some volume of the sinter strand, which decreases the input of fresh materials in the form of the sinter mix. The cutoff point in this case is around 36 wt. %. A decrease in the index, 1/t, beyond 35 wt. % is possibly due to the poor thermal efficiency resulting from poor absorption of heat from incoming hot gases to the sintering mix.[51,52]

7.8.6 Carbon Equivalent of Return Fines[33]

The sinter yield increases with increasing return fines and/or increasing coke in the sinter mix. This holds well within their limited range of variation in the sense that a higher amount of coke may result in over-fusing of the sinter, which may result in more glass formation. Though the sinter yield can be increased by increasing the coke content or increasing the return fines in the mix, the latter has a more favorable effect on the sinter yield in view of the aforementioned reason.

Theoretically, the variability in return fines generation and the gap between the return fines output and the return fines input could be reduced by controlling the content of solid fuel in the sinter mix. This is because of the fact that a reduced heat is required to re-sinter return fines as against the normal heat requirement for sintering the fine ore. Hence, the carbon equivalent of return fines can be worked out. That is, the balanced amount of return fines, recirculated to the plant, can be worked out at different levels of coke breeze in the sinter mix.

From **Figure 7.8**, it is apparent that the decrease in the balanced return fines when the coke breeze increases from 7 wt.% to 8 wt.% is 3.2 wt.%.; whereas that when the coke breeze increases from 8 wt.% to 9 wt.% it is 2.8 wt.%. Obviously, the effect of increasing the coke consumption on the yield should not be as pronounced at the higher level of coke breeze as that at the lower level of coke in the mix.

From the aforementioned, it is apparent that a five increase in the amount of return fines recirculation in the mix requires the wt.% of coke to be decreased by: (5/3.2 =) 1.562 in the range of 33-36 wt.% of the mix, and (5/2.8 =) 1.786 in the range of 30-33 wt.% of the mix and vice versa.

7.9 SINTER MEAN SIZE[50]

The mean size of sinter produced in many sintering plants, for instance those in India, is lower than that of lump ore.[34] This incompatibility adversely affects the permeability of the stack zone and also makes a restriction to use a higher percentage of sinter in the burden (at the same coke size) (see **Sec. 3.6.3; Figure 3.8**). Hence, improvement in the sinter mean size is essential.

An improvement in the mean size of sinter can be achieved through:

a. Mechanical means
b. By optimizing the process variables

Usually the mean size of sinter is regulated through mechanical means by screening the sinter material at the sinter plant itself through 10-mm/6.3-mm screens and recirculating the undersize to the plant as "sinter return." However, depending upon the effectiveness of the screen as well as the subsequent generation of fines with falls and abrasion between the sintering plant and the blast furnace, the undersize fraction also associates with the sinter, which decreases the mean size. The sinter mean size can be improved by screening out the fine fraction from the sinter at the blast furnace skip point and recirculating the same to the sintering plant as return sinter fines. A higher recirculation decreases the throughput of the fresh material into the sintering strand. However, a part of the loss in strand productivity gets compensated by a simultaneous increase in the yield of sinter because of better fusion and bonding.[34,53]

REFERENCES

1. Loo C. E. and M. F. Hutchens, Quantifying the resistance to airflow during iron ore sintering, *ISIJ Inter.*, 43 (2003), 630–635.
2. Boranbaev, B. M., T. M. Srinivasan, S. S. Datta, N. K. Kakkar, R. P. Bhagat, B. R. Choudhury, S. Dharnipalan, D. S. Thakur, and V. N. Tripathi, Narayan Das, Sinterability of various ore fines used in Bhilai Steel Plant, *Steel India*, 5(2), (1982) 115–123.
3. Wendeborn H. and F. Cappel, Limits of the physical aspects of the sintering process, in: *Agglomeration*, W. A. Knepper (Editor), Wiley Interscience, NY, 1962, 1041–1065.
4. Kasai, E., W. J. Rankin, and J. F. Gannon, The effect of raw mixture properties on bed permeability during sintering, *ISIJ Inter.*, 29, (1989), 33–42.

5. Inazumi, T., M. Fujimoto, S. Sato, and K. Sato, Effect of sinter-cake load reduction by magnetic force on iron ore sintering, *ISIJ Inter.*, 35, (1995), 372–379.
6. Bhagat, R. P., S. K. Sil, U. S. Chattoraj, M. C. Goswami, D. P. Singh, and B. K. Das, Pot sintering studies for improvement in reducibility and RDI values of sinter, *Tata Search*, (2006), 127–134.
7. Bhagat, R. P., S. K. Pradhan, M. C. Goswami, B. Nandi, and T. Venugopalan, Effect of LOI content of iron ores and other process variables on the reduction properties of sinter, in: *Proc. XXVI International Mineral Processing Congress 2012*, New Delhi, India, September 24–28, 2012, org. by Indian Institute of Mineral Engineers, Jamshedpur and Indian Institute Of Metals, Kolkata.
8. Wadhwani, H. et al., Mixed grade iron ore concentrate with ultra-fines for sinter making, in: *Proc. International Symposium on Beneficiation and Agglomeration*, 1986, 423.
9. Banerjee, G. N. et al. Sintering of beneficiated iron ore slime of Barsua Iron Ore Mines (Orissa), in: *Proc. International Symposium on Beneficiation and Agglomeration*, 1986, 431.
10. Maulik, S. C., R&D on beneficiation and sintering of iron ore at NML, Australian IMM Conference on Iron Ore Resources and Reserve Estimation, *Perth*, September 25–26, 1997.
11. Murthy V. G. K., A. De, A. Chatterjee, and V. S. Rao, Reduction of alumina in iron ore classifier fines and its influence on sinter produced, *Tata Search*, (1994), 7–13.
12. Chatterjee A., A. De, and S. S. Gupta, *Monograph on sinter making at Tata Steel*, January 1993, 69–71.
13. SAIL R&D Centre for Iron and Steel, SAIL, *Effect of Kudremukh concentrate addition on sintering parameters*, Report No.12. 02.268.82, 1982.
14. Obata, H., H. Takahashi, M. Nakamura, T. Natsumi, and K. Komamutra, High productivity operation at Chiba No.4 sinter plant, *ISIJ Inter.*, 31(5), (1991), 478–486.
15. Yadav, U. S., V. M. Korane, and B. K. Das, Role of different iron ore fines on sintering, *Tata Search*, (2005), 101–109.
16. Yang Y. H. and N. Standis, Fundamental mechanisms of pore formation in iron ore sinters and pellets, *ISIJ Inter.*, 31, (1991), 468-477.
17. Kawaguchi, T., K. Kuriyama, S. Sato, K. Takata, and T. Miyake, in: *Proc. 6th International Iron and Steel Cong.*, Nagoe, Japan, 1990, Iron and Steel Institute of Japan, 40–47.
18. Yang L., X. Yang, and L. Davis, Assimilation and mineral formation during sintering for blends containing magnetite concentrate and hematite/pisolite sintering fines, *ISIJ Inter.*, 39 (1999), 239–245.
19. Kasai, E., W. J. Rankin, R. R. Lovel, and Y. Omori, An analysis of the structure of iron ore sinter cake, *ISIJ Inter.*, 29, (1989), 635–641.
20. Shabata D. et al., Installation of selective granulation equipment at Oita No. 2 Sintering Plant, *Camp-ISIJ*, 10, (1997), 804.
21. Okuno, Y., *New approaches in research & development of iron making technology in Japan*, Conf Proc. Asia Steel International Conference, Jamshedpur, India, Vol. 1, T. Mukherjee and A. S. Dhillon, Eds., Indian Institute of Metals and Zhongguo-Jinshu-Xuehui, Organisers, Indian Institute of Metals, Kolkata, Publisher. April 9–12, 2003.
22. Ishikawa, Y. et al., Control of charging on sintering bed by segregating slit wire and its effect on productivity, *Camp-ISIJ*, 15, (2002), 710–713.
23. Matsuoka, H., et al., Raw mix packed structure control by ISF and sinter cake structure control by load control using stand sintering, *Camp-ISIJ*, 15, (2002), 698–701.
24. Oyama, N. et al., Development of charging technology for raw mixture for the improvement of packed structure in the sintering bed, *Camp-ISIJ*, 15, (2002), 694–697.

25. Jahn, F., in: *Proc. Inter. Symp. on Modern Developments in Steel Making*, Jamshedpur, 1981, 239–249.
26. Nyquist, O., Effects of lime on the sintering of pure hematite and magnetite concentrates in: *Agglomeration*, Ed. W. A. Knepper Interscience, NY, 1962, 809-864.
27. Savelev S. G., R. D. Kamenev, O. G. Fedorov, and N. M. Panchoshnyi, Using lime of different degrees of calcination in sintering mix, *Steel In The USSR*, 10, (March 1980), 123–125.
28. Boranbaev, B. M., Datta, S. S., Srinivasan, T. M., Bhadoria, D. K. S., Bhagat, R. P. and Aswath, H. S., Investigation of the process of sintering during the production of high basic sinter by the method of balanced sinter return fines regime, in: *Proc. the Institute for Briquetting and Agglomeration (18th Biennial Conf.)*, Colorado, August 1983, 295–314.
29. Wendeborn, H., and F. Chappel, Limits of the physical aspects of the sinter process, in: *Agglomeration*, W. A. Knepper (Editor), Wiley Interscience, NY, 1962, 1041–1065.
30. Hosoda, H., H. Mori, N. Sogosi, A. Nagasawa, and S. J. Nakabayashi, *Phy. Chem.*, 108, (2004), 1641.
31. Gebenkin, G. A., S. S. Datta, R. P. Bhagat, and B. K. Chatterjee, Preheating of sinter mix, in: *Proc. All India Seminar on Extraction of Iron*, Bhilai, org. by India Institute of Metals and Institution of Engineers, Feb. 1984., II.19 –Il. 24.
32. Preparation of sinter mix including secondary mixing and steam pre-heating for the sintering plant at Bhilai (BMP 2.), Steel Authority of India Limited, Ranchi/Bhilai, August 1986.
33. Bhagat, R. P., Factors affecting return sinter fines regimes and strand productivity in iron ore sintering, *ISIJ Inter.*, 39 (1999), 889–895.
34. Dongshim, J., A. Ohaba, and M. Tanaka, *Trans. National Research Inst. for Metals* (Japan), 22, (1980), 31–37.
35. CSIR-NML studies to increase reducibility and decrease RDI of iron ore sinter (CLP 0029), Sponsor: The Tata Iron & Steel Co. Ltd., Jamshedpur, March 2004.
36. Pangrahy, S. C., M. Rigaud, and P. Dilewijns, Effects of replacement of limestone by dolomite and other MgO bearing materials on production rate of iron ore sinter, *Ironmaking & Steelmaking*, 11(5), (1984), 246-252.
37. Bhagat, R. P., H. S. Ray, and S. K. Gupta, Effect of some process parameters on the performance of a sintering plant at Bhilai Steel Plant—a statistical analysis., *Trans. Indian Institute of Metals*, 41, (1988), 465–474.
38. CSIR-NML: Pot test for iron ore in relation to Tata Metalik Ltd. 40.5 sq.m. linear sinter plant, March 2013.
39. Bhagat, R. P., S. K. Gupta, and H. S. Ray, Heat transfer consideration for improvement in sintering indices, *Proc. 48th Iron Making Conference*, ISS-AMIE, Chicago, (1989), 481–490.
40. Bhagat, R. P., Studies on calcination of limestone and its assimilation by hematite during the sintering using thermo-gravimetric and X-ray diffraction analyses, *Scandinavian Journal of Metallurgy*, 21, (1992), 246–250.
41. Hosotani, Y., N. Konno, K. Yamaguchi, T. Oromoto, and T. Inazumi, Reduction properties of sinter with fine dispersed pores at high temperature of 1273 K and above, *ISIJ Inter.*, 36, (1996), 1439–1447.
42. Loo, C. E. and L. T. Matthews, Assimilation of large ore and flux particles in iron ore sintering, *Trans. Inst. Min. Metall.* Section C, Mineral Processing & Extractive Metallurgy, 101, (1992), 105–118.
43. Sato, S., T. Kawaguchi, M. Kato, *ISIJ Inter.* 28, (1988), 705.
44. Teo, C. S., R. A. Mikka, and C. E. Loo, Positioning coke particles in iron ore sintering, *ISIJ Inter.*, 32 (1992), 1047–1057.

45. Boranbaev, B., M. Yu, S. Yusin, T. M. Srinivasan, D. K. S. Bhadori, and S. K. Gupta, Bf & Sp. Operating Committee Meeting Rdcis, Sail. Preprint Indian Iron Making Conference Ranchi, January 10–11, 1989.
46. Bhagat, R. P., H. S. Ray, and S. K. Gupta, A relationship between strength of sinter and the ratio of return fines output to return fines input, *NML Tech. Jour.*, 34, (1992), 20–27.
47. Rauch, H. and H. Cappel, *Stahl Eisen*, 81, (1961), 96; (Loc. Cit. Ref. 29)
48. Pimenta, H. P. and V. Seshadri, Characterization of structure of iron ore sinter and its behavior during reduction at low temperature, *Ironmaking & Steelmaking*, 29(3), (2002), 169–174.
49. Ostwald, J., Mineralogy and micro-texture of Australian iron ore sinters, *BHP Technical Bulletin*, 25 (1981), 13–20.
50. Bhagat, R. P., Effect of screening and recirculating fines from the blast furnace sinter on the strand productivity vis-à-vis improvement in mean size of sinter, *SEAISI* Qty, (October 1992), 47–53.
51. Ravat, Y.F. and A. Chatterjee, *TISCO Tech. Jour.*, 26, (1979), 51–60.
52. Chatterjee, A., Y. F. Rawat, and A. K. Dey, *Trans. IIM*, 35(6), (1982), 601–613.
53. Grebe, K, K. Heinrich and S. K. Peter, Study of factors influencing granulation and strength of sinter, *Stahl.Eisen* (German), 100 (22), (1980), 1279–1285.

8 Sinter Mineralogy

ABBREVIATIONS

FFS	*Flame front speed*
RDI	*Reduction-degradation index*
SFCA	*Silico ferrite of calcium and aluminum*

8.1 PREAMBLE

Iron ore sinter is the result of various physical-chemical reactions that take place among the constituting minerals of the sinter mix in the zones of sintering (combustion) and cooling according to the material characteristics and their targeted chemical composition. The mineral phases that precipitate out of melt are influenced by the variables maximum temperature reached, chemical composition of the melt, oxygen partial pressure, and cooling rate. Hematite, magnetite, and fayalite as well as silicates are the major mineral phases in the case of acidic sinter, whereas hematite, calcium ferrites, magnetite, and mixed calcium iron silicates are the major mineral phases in the case of fluxed sinter.

The nature and quantum of these phases do influence:

a. Yield of sinter from sinter plant
b. Sinter mean size
c. Quality parameters of sinter in terms of strength, reducibility, low-temperature breakdown behavior during reduction, as well as softening and melt-down characteristics

The porosity of sinter, including the size, shape, and volume of pores, is important in determining the structural strength of sinter. Sinter can be treated as that of polycrystalline brittle materials with reference to porosity, which along with pore size distribution, affects the sinter strength. The inherent strength of the bonding phases present and the structure of the pores are crucial in determining these sinter qualities.[1]

Many of the important bulk properties of sinter are related to how easily oxygen is removed from the iron oxide by the reducing gases, CO. For this reason, small open pores in the sinter are preferable to large pores that can cause tumble strength to decline. The coalescence process in the reaction zone determines the mean size and size distribution of the product sinter.

The effect of sinter mineralogy on the quality parameters of sinter is described in **Chapter 9**.

8.2 MAJOR CONSTITUENTS AND DESIRED MINERALOGY OF SINTER

According to Cohen,[2] a sinter is regarded as consisting fundamentally of three types of materials:

a. Original mineral constituents of the raw ores that have remained unaltered throughout the sintering process.
b. Original mineral constituents that have readjusted their structure and shape by recrystallization in the solid state. These do not lose their chemical or physical identity at any time during or after the sintering process.
c. Secondary constituents that result from material that is fused or dissolved during sintering. Two groups may be classified here:
   - Constituents that remain mutually dissolved and consolidated as a glass
   - Constituents that crystallize out of solution as dendrites or as skeletal or fully developed crystals

Cohen[2] has enumerated the theoretical constituents of a desirable sinter, which can thus be stated as follows:

a. The sinter should contain a large proportion of unaltered grains of iron oxide.
b. The proportion of secondary crystals should be small.
c. The crystalline constituents should be embedded in the minimum amount of slag compatible with sufficient strength.
d. The slag should be free from iron.

8.3 MINERALOGICAL TERMINOLOGY

The mineralogical phases in relation to fluxed sinter are identified following their typical characteristics. The minerals can be grouped into:

a. **Crystalline Phase:** This includes all types of hematite, mentioned previously, magnetite (inclusive of magnesio-spinels), calcium ferrites, and magnesiowustite.
b. **Slag Phase:** This includes vitreous glass, precipitates of dicalcium silicates, olivines, and pyroxenes besides the relicts of flux constituents.

Various important terminologies in this regard are as follows:[3]

Primary Hematite This form of hematite mineral appears in undigested form in sinter. These minerals have not changed their identity throughout the sintering (and cooling) process. These are quite distinctive in appearance due to their internal porosity.

Secondary Hematite These hematite minerals are recrystallized from the melt. Another group of secondary hematite minerals is formed as a result of solid-state reoxidation (either partial or total) of magnetite during the cooling of sinter. It is normally seen around some pores or on the edges of sinters and also appears in lamellae form on magnetite surfaces.

Primary Magnetite This involves those magnetite minerals that appear in undigested form in sinter. Like primary hematite, these minerals have not changed their identity throughout the sintering (and cooling) process. Obviously, this phase in sinter is not expected in cases where the ore mix does not contain any magnetite

Secondary Magnetite These magnetite minerals occur as a result of reduction of hematite and are crystallized in a slag matrix background. The magnesio-spinels are also included in this category as they are optically unidentifiable from the magnetite.

Calcium Ferrites This category involves all calcium ferrites that belong to binary and ternary systems. These ferrites are a form of silicon-ferrite of calcium and aluminum (SFCA) (see **Sec. 6.10.3**). Normally, these minerals are formed following the diffusion process in solid state during the sintering process.

Magnesium Ferrites These are ferrites having the stoichiometric or approximate composition of $MgO.Fe_2O_3$. The ferrites of such a composition are generally not observed in the sinters.

Glass The glass in sinters is a group of minerals that have differing chemical composition and properties. This is noncrystalline in nature and is referred to as vitreous glass. This is formed after slag bonding during sintering.

Dicalcium Silicate It appears as a crystallized product in the glass and has a needle or dot-like appearance.

Olivines They also occur as crystallized products of glass but have more orderly dendritic patterns and are dull in appearance. This includes both calcium- and magnesium-bearing olivines.

Pyroxenes They occur in dendritic patterns having a widely varying composition and are not easily identifiable from the olivines.

Pores They appear as dark areas and are easily distinguished from the background slag matrix.

Magnesium-Wustite These round precipitates, dull in appearance, normally are seen near dolomitic relicts and are undissolved dolomite and relicts from other MgO-bearing sources.

8.4 SINTERING REACTION AND MINERALOGY

The mineralogical characteristic of sinter is decided by the sequence of various chemical reactions that take place in various zones including the cooling one (see **Sec. 6.10**), also the manner the minerals are assimilated. The melt, silica ferrites of calcium and aluminum, commonly known as SFCA, is formed through the sequence of sintering reactions. The composition of melts, thus formed, in terms of its Fe_2O_3, CaO, Al_2O_3, and SiO_2 contents, changes depending on the solids assimilated. The solid-state reaction, mainly in the system CaO-Fe_2O_3, is responsible toward the formation of SFCA. Initially, hematite reacts with lime at a low temperature (750–780°C) to form calcium ferrite. Depending upon the Fe content, two types of SFCA are formed (see **Sec. 8.5.2** also):

a. SFCA (low Fe form)
b. SFCA-1 (high Fe, low silica form)

Acicular SFCA increases when the temperature is below 1,220°C. At higher temperatures, columnar SFCA forms.[4] When temperature exceeds 1,300°C, SFCA starts to decompose into hematite, magnetite, and silicates. On cooling, various minerals and phases precipitate out of the melt. If the oxygen content of the flue gas is high, SFCA will transform into hematite; if the oxygen content is low, it will decompose to magnetite. Magnetite can also be oxidized to hematite during cooling.

8.5 COMPOSITION AND MINERALOGICAL CHARACTERISTICS OF FLUXED SINTER

8.5.1 Mineralogical Composition

Hematite, magnetite, and fayalite as well as silicates are the major mineral phases in the case of acidic sinter; whereas hematite, calcium ferrites, magnetite, and mixed calcium iron silicates are the major mineral phases in the case of fluxed sinter.

The fluxed sinter includes the following phases.[5,6]

- **Iron oxides** (hematite and magnetite): This constitutes about 40–70 vol. % of the total minerals present in the sinter, depending on the composition of the mix (sinter basicity). These minerals may be the primary one or be secondary. Their formation is guided by sintering and cooling conditions.
- **Calcium ferrites:** Different types of ferrites are formed. This constitutes about 20–50 vol. % of the total mineral constituents. These minerals are formed through, primarily, diffusion bonding. The ferrites crystallize as interlocking lathes and form a strong framework to bind neighboring grains of ore together. Formation of these mineralogical constituents starts at a lower temperature, even at solid state, and these have high reducibility. The calcium ferrites usually contain some silica and alumina and are thus referred to as silico-ferrites of calcium and aluminum and abbreviated to SFCA. They form the main bonding phase in sinter.[7]
- **Magnetite:** Small quantities of magnetite crystals are also found in irregular skeletal shapes in a calcium ferrite background. Formation of these mineralogical constituents starts at a lower temperature, even at solid state, and these have high reducibility.
- **Glasses:** These minerals are formed as a result of slag bonding. Their composition and quantity may vary according to the sintering condition and the amount of solid fuel present in the sinter mix. A proportion up to approximately 10 vol. % is accepted for good strength. These are the main bonding phases among the minerals in the sinter matrix.

- **Dicalcium silicates:** These also constitutes up to approximately 10 vol. %. The formation of silicates releases iron to form ferrites or retains its original composition as hematite rather than the formation of fayalites and, this is good for reducibility.
- **Minor phases:** These may include quartz, sulfides, pyroxenes, and lime, but the total of these rarely accounts for more than a few percent of the total sinter volume.

8.5.2 Mineralogical Characteristics

Characteristics of the major mineralogical constituents of fluxed sinter, namely, hematite, magnetite, and SFCA are described as follows.[8,9]

Hematite (Fe_2O_3): Hematite is grayish in color and is the brightest among the minerals present in a sinter sample under reflected light. It is distinctly anisotropic and shows weak bireflectivity under reflected light.[9]

Hematite occurs in three different forms:

a. ***Original/primary hematite,*** which is bonded by SFCA but that has been little altered. This is primary or relict hematite and is typically quite pure Fe_2O_3 (**Figure 8.1**).
b. ***Secondary hematite,*** which exsolves as crystallites. Secondary hematite results from the high temperature reoxidation of ex solution magnetite. The natural slagging constituents of the ore may take into solution much of the finer ore hematite and exsolve these as crystallites of secondary hematite (**Figure 8.2**). Such hematite is often aluminous and titani-ferrous with these elements in the hematite crystal lattice.

Magnetite (Fe_3O_4): Magnetite is major constituent in the sinters of medium basicity (2.75–3.0) and becomes minor phase in the sinters of higher basicity

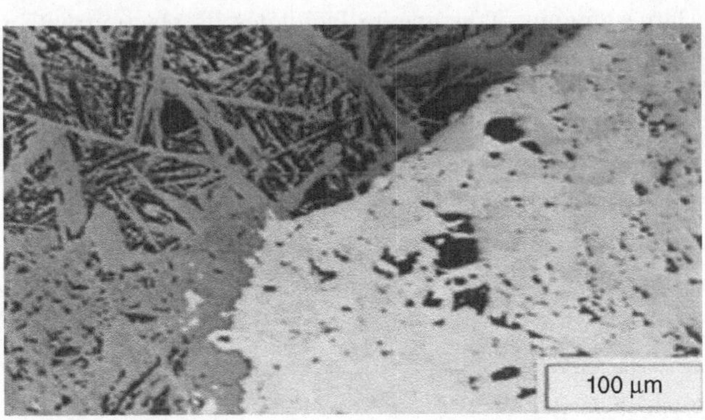

FIGURE 8.1 Photomicrograph showing primary hematite (white) bonded by prismatic SFCA (gray); interstitial cavities and dicalcium silicates are black (after Ref. [9])

FIGURE 8.2 Photomicrograph showing ex-solution crystals of secondary hematite (white) and SFCA crystals (pale gray) in glass (dark gray) (after Ref. [9])

(3.5–4.0). Magnetite is isotropic in nature, and it is light gray with a brownish tint under reflected light. Its reflectivity is more than calcium ferrite and less than hematite.[10] The magnetite mineral in sinter exists in different forms, such as:[10]

- Fine, rounded grains having a size 0.01–0.03 mm
- Medium-sized (0.03–0.1 mm), subhedral to euhedral or octahedral grains
- Partly fused grains in the form of massive crystals of irregular shape
- Irregular skeletal shaped grains
- Very coarse (0.1–0.3 mm), eutectic grains undermixed with dicalcium silicates

The morphology of magnetite is distinctly seen in sinters of low basicity (CaO/SiO_2 = 1.0–1.2) that are composed of ex-solution magnetite grains in glass that exhibit a range of forms from euhedral skeletal to anhedral crystallites. The crystal shape is a function of the rate of cooling of the sinter bed. In addition to this, ex-solution magnetite may also be formed by the essentially solid-state reduction of hematite.

The large amount of magnetite facilitates its reaction with silica and lime. Therefore, the sinter made from magnetite is already partially fluxed at 1,150°C and contains a large quantity of olivine.

Ferrites of Calcium: Fluxed iron ore sinters contain crystals composed of major calcium and iron oxides. Depending on the molar ratio of CaO and Fe_2O_3, they are classified as follows.

Mineral	Chemical Formula	Abbreviated as
Monocalcium ferrite	$CaO\text{-}Fe_2O_3$	CF
Dicalcium ferrite	$2CaO\text{-}Fe_2O_3$	C2F
Tricalcium ferrite	$3\ CaO\text{-}Fe_2O_3$	C3F
Calcium diferrite	$CaO\text{-}2\ Fe_2O_3$	CF2
Calcium aluminoferrite	$CaO\text{-}(FeAl)_2O_3$	CFAL

FIGURE 8.3 Photomicrograph showing typical field of prismatic SFCA (gray); interstitial cavities and dicalcium silicates are black (after Ref. [9])

Monocalcium ferrite is more common in these samples and forms a predominant phase in the sinter samples of higher basicity (3.7–4.0). Calcium ferrites usually occur as needle- or plate-shaped interlocked crystals. Monocalcium ferrite is light gray in reflected light, and it is deep reddish in transmitted light. Its reflectivity is lower than magnetite and higher than silicates. Calcium ferrite is usually prismatic and occurs as lath-shaped or needle-shaped grains (**Figure 8.3**).

Silico-Ferrite of Calcium and Aluminum (SFCA): The calcium ferrite, initially formed, undergoes significant changes by diffusion of Al and Si species. Thus, SFCA is formed toward the end of the heating/reduction regime itself.

The formation of SFCA is more likely in cases where:

a. The sinter consists of high content of Al_2O_3 and SiO_2 in iron ore.
b. At sintering temperatures higher than 1,300°C, which cause the increased mobility of Al and Si into ferrite melt (see **Sec. 6.10.3** also).

The SFCAs are of two different types: SFCA and platy SFCA-1. These are of a low Fe form and a high–Fe low silica form, respectively. The SFCA-1 appears needle-like or fibrous in a cross-section structure,[7] which is distinct from the columnar or blocky form, SFCA. SFCA-1 phase is the most desirable bonding phase in iron ore sinter since microstructures composed entirely of SFCA-1 (fine and fibrous structure and resistant to crack propagation) show higher physical strength and higher reducibility than microstructures composed predominantly of SFCA.[11]

Dicalcium Silicates Dicalcium silicates are the darkest phase in reflected light. They occur as small and elongated needle-like crystals mostly present as interstitial interlocked grains with ferrites.

8.6 PROCESS VARIABLES AND SINTER MINERALOGY

8.6.1 Flame Front Speed (FFS)

Figure 8.4 shows the distribution of mineral phases under different flame font speeds (FFSs). The content of hematite increases, whereas that of SFCA and magnetite decreases with increasing the FFS.[12] The residence time above 1,100°C becomes shorter with increasing FFS; as a consequence, the assimilation drops and SFCA decreases. The structure of sinter, in terms of ferrite and silicate, gets finally modified during solidification on cooling and is influenced by its rate. The content of magnetite in sinter decreases with increasing FFS because of a faster cooling rate and less reduction reaction. When FFS is increased, air flow through the bed is greater and the sinter-bed temperature drops. As a consequence, the oxidation of magnetite increases, leading to an increase in total hematite.

8.6.2 Cooling Rate

During the cooling stage, various minerals and phases precipitate out of the melt: the types of minerals formed, however, depend on the oxygen potential of the gaseous environment. If the oxygen content of the flue gas is high, SFCA will transform into hematite; if the oxygen content is low, it will decompose to magnetite. Magnetite can also be oxidized to hematite during cooling [4] (see **Sec. 6.10.3** and **Figure 6.20** also). The SFCA is formed in the cooling cycle when iron oxides (hematite and magnetite) are reabsorbed by a reaction with silica melt, particularly at a slow cooling rate.

A fast cooling rate will result in the formation of finer precipitated grain. The amount of melt generated in the reaction zone will determine bonding phase volume and composition. The cooling rate increases (the available time for cooling is substantially reduced) with lower bed height. Hence, the increasing FFS leads to decreasing SFCA content.

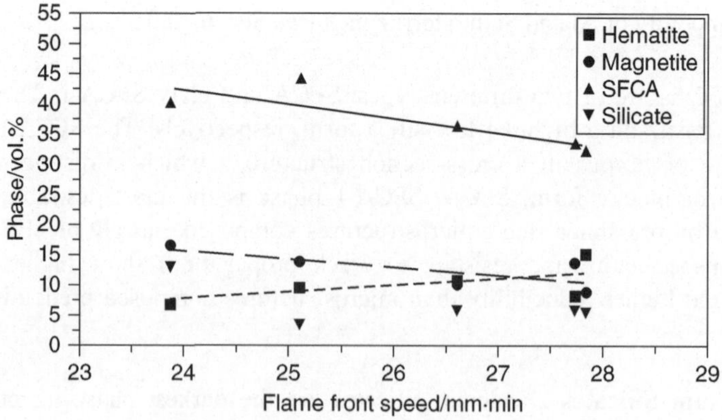

FIGURE 8.4 Relationship between mineral phase and flame front speed (after Ref. [12])

8.6.3 Bed Height

The mineral phases of the sinter samples change with the bed height: more calcium ferrites and magnetite phases are formed as the bed height increases whereas the proportion of hematite decreases.[13–15] With lower bed height, the FFS increases, and so does the cooling rate (the available time for cooling is substantially reduced). Hence, the increasing FFS leads to decreasing SFCA content.

8.6.4 Return Fines

The input of return fines in sinter blends has significant influence on the sinter mineralogy and pore structure. In the case of fluxed sinter at 3.25 basicity ratio (CaO/SiO_2) when the return fines in the mix was lower (20 wt. %), dicalcium ferrite mineral was of a fine needle-type compared to that at higher return fines (40 wt. %).[10] The pores happen to be smaller and spherical at lower return fines input.

8.7 SINTER CHEMISTRY AND ITS MINERALOGY

8.7.1 Sinter Basicity

The sinter basicity is one of the most important factors affecting the mineralogy of sinter. Stoichiometrically, an increase in the basicity decreases the iron content of the slag phase by replacing part or all of the FeO by CaO or MgO to form silicates as well as ferrites. **Figure 8.5** shows the variation of mineralogical compositions of sinter with respect to its phases, hematite, magnetite, glass and calcium ferrites, with changing basicity (CaO/SiO_2).[16]

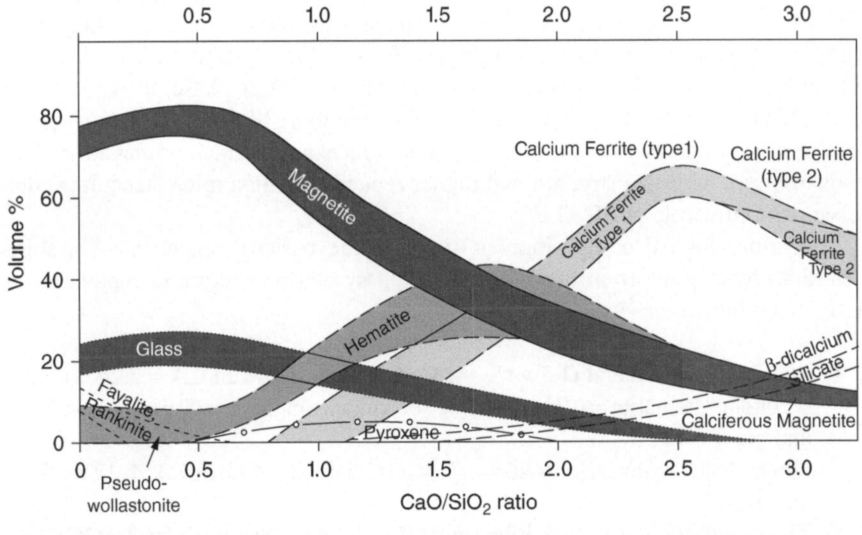

FIGURE 8.5 Change in the mineralogical composition of sinter with its basicity (after Ref. [16])

At lower basicity (CaO/SiO_2 less than 1.8), glassy silicates are formed, mainly through slag bonding. As the basicity increases above 1.4, the glassy matrix is replaced by calcium silicate and ferrites[17] (see **Figure 6.19** also). The needle-like ferrites precipitates take over as a bonding phase (hold the ore grain together).[18,19] The ferrite mineral also has higher reducibility.[20] In addition, iron calcium olivines of composition $Ca_xFe_{2-x}SiO_4$ are formed with the value of x increasing from 0.25 to 1.75 as the basicity is increased.[21]

Kawaguchi et al.[22] have observed that the volume of secondary hematite decreases on sintering of pseudo-particles with different CaO content. One with high CaO (limestone, mill scale, and Fe ores) and another with low CaO (return fines/coke breeze and Fe ores) are processed. This results in a decrease in sinter reduction–degradation index (RDI).

8.7.2 Fe Content in Sinter

The Fe content of ore has a major influence on the formation of SFCA (SIO_2-Fe_2O-CaO-Al_2O_3) constituents in sinter (see **Secs. 8.5.2** and **6.10.3**).

- Lower grade (Fe: less than 62 wt. %) ores and concentrates typically form SFCA as a part of the final structure.
- Medium-grade (Fe: 62–65 wt. %) ores form a mixture of SFCA and SFCA-1.
- High-grade (Fe: more than 65 wt. %) ores will largely form SFCA-1.

8.7.3 Sinter Alumina

Alumina in sinter affects the formation of SFCA and its amount and type. The phase, SFCA-1 is favored when the total alumina and silicate content in the bonding phase is low and the hematite content is high (see **Secs. 6.10.3, 8.4, and 8.5**). Therefore, although alumina tends to stabilize and promote the SFCA phase, it may actually reduce the amount of SFCA-1 produced by forming more SFCA. The SFCA-1 phase does possess a fine and fibrous structure and is resistant to crack propagation. This leads to higher physical strength and higher reducibility than microstructures composed predominantly of SFCA.[23]

The mineralogical composition of sinter in the basicity range, 0.8–1.8, shifts toward a higher proportion of SFCA and a glassy phase as alumina content in the sinter blend increases:[24]

- Standard sinter blend (1.5 wt.% Al_2O; 5.5 wt.% SiO_2): SFCA = 3.24 wt.%; secondary hematite = 0.75–1.5 wt.%; secondary magnetite = 0.63–1.61 wt.%; and glass = 7.95 wt.%.
- Sinter blend of gibbsite ore (4.6 wt. % Al_2O; 5.5 wt. % SiO_2): SFCA: 4–12 wt. % and glass: 11 wt. %.
- The mineralogical composition (in wt.%) of typical fluxed sinter (CaO/SiO_2 approximately 2-2.2) from India[25] is SFCA: 4.9 wt.%; secondary hematite: 0.71 wt.%; secondary magnetite: 1.13 wt.%; and glass: 5.0 wt.%.

8.7.4 Sinter MgO

The influence of MgO addition on sinter mineralogy vis-à-vis the properties of sinter has been studied in detail by Panigrahy et al.[3]

- The most significant effect of MgO addition was the suppression of hematite and Ca-ferrite phases and the increase in the magnetite phase.
- The magnetite phase formed during the sintering is more stable with the addition of MgO and oxidizes less during the cooling stage of sinter. This reduces the proportion of reoxidized hematite. Most of the MgO is picked by the magnetite phase to form mixed spinels of type (Fe, Mg)O.Fe_2O_3. The magnetite may, however, convert into calcium ferrite on cooling under oxidizing conditions. The kinetics of this reaction depend on whether the magnesia is added as dolomite or as serpentine. When MgO is introduced in the form of dolomite, the amount of calcium ferrite increases marginally, and when MgO is added in the form of serpentine, the calcium ferrite increases considerably.

The mineralogy of sinter is crucially decided by the process of diffusion of Mg^{2+} into magnetite lattice to form mixed spinels during the sintering.[3]

8.8 SINTER MINERALOGY AND QUALITY PARAMETERS OF SINTER

The types of phases present in sinter, their composition, amount/volume fraction, as well as morphologies have a very strong influence on the quality parameters of sinter, and its size distribution (see **Sec. 9.2**).

8.8.1 Sinter Mineralogy

Sinter is composed of various mineral phases, each having its own strength and reducibility.

The minerals, monocalcium ferrites, formed in the self-fluxed sinter have reasonable high reducibility and compressive resistance. The olivines ($Ca_xFe_{2-x}SiO_4$) with $x = 0.5$ have the highest compressive strength, though the lowest reducibility. The hematite mineral phase has the highest reducibility followed by calcium ferrite (see **Sec. 9.2.2**).

8.8.2 Sinter Porosity

The sinter consists of several pores of varying sizes and shapes in addition to the mineral phases.[23] These can be broadly categorized as macropores and micropores. The microporosity denotes the pore radius of sinter samples, say, below 10 micron and up to 0.001 micron, whereas the pore radius of samples, typically in the range 10–100 micron corresponds to macroporosity. The macro- and microporosities of the sinter samples can be measured by a mercury porosity meter at lower and higher pressures: typically, these are up to 400 K Pa and 400 m Pa, respectively.

Reduction Properties The reduction properties of sinter are decided by the microporosity within the structure (see **Sec. 9.2.2**). Jeulin et al.[26] have observed that reducibility of the sinter samples increased with their pore specific area and decreased with the range of pore co-variance. Coalesced fine pores can increase the surface area of sinter to achieve a substantial reducibility. The proportion and structure of the micropores are important for the reaction by the reducing gases and, hence, in determining the sinter reducibility.[1] The sinter macropore structure, on the other hand, influences its strength. It is, therefore, important to optimize the pore structure in the sinter to improve its reducibility while maintaining the sinter strength. The SFCA phase stabilizes the fine pores produced during the reduction process, leading to increased porosity and higher reducibility.[27]

Thermal Effect The sinter mineralogy as well as its porosity are significantly influenced by the manner in which the mix material undergoes sintering and the sinter is subsequently cooled. The bed temperature profile in terms of peak temperature and the period of sintering (above 1,100°C) matter in this respect. If expansion of the heat pattern is suppressed, fine pores could be prevented from coalescing into coarser pores. This could be achieved by the addition of sized coke and limestone in sinter mix.

Size Parameters The sintering of ore fines having a reduced size produces the sinter with increased microporosity and decreased size of pores.[28,29] Besides, the pore distribution of sinter also gets unified when the ore of reduced size is sintered.[30]

The number of fine pores in the sinter also increases when smaller-sized fractions of limestone or/and coke breeze are used in the sinter mix.[26] Typical bench scale experiments[28,29] at 2.1 sinter basicity (CaO/SiO_2) show that when the size of coke breeze in the sinter mix was reduced, microporosity of the sinter increased from 5.2% to 10.4%. It is important to note that while the microporosity of the sinter changed quite significantly, its total porosity (consisting of micro- and macropores) remained more or less the same: 11% and 13%, respectively, with normal- and reduced-sized coke in the sinter mix. Hosotani et al.[31] have also observed that number of fine pores increased in the sinter when smaller sized fractions of limestone or/and coke breeze were used in the sinter mix. The elimination/reduction of (very) finer size fractions from coke and limestone also helps in this respect.

8.8.3 Sinter Morphology

The effect of CaO and SiO_2 and other parameters on the morphology of precipitated iron ore sinter during reduction is schematically shown in **Figure 8.6**. The morphological characteristic of sinter is influenced by:

a. Chemical composition: The characteristic of sinter changes from fibrous to porous to dense with increasing the silica content and/or decreasing the CaO content of sinter.
b. Sintering conditions: With increasing the sintering temperature and/or sintering time, the sinter becomes more dense.

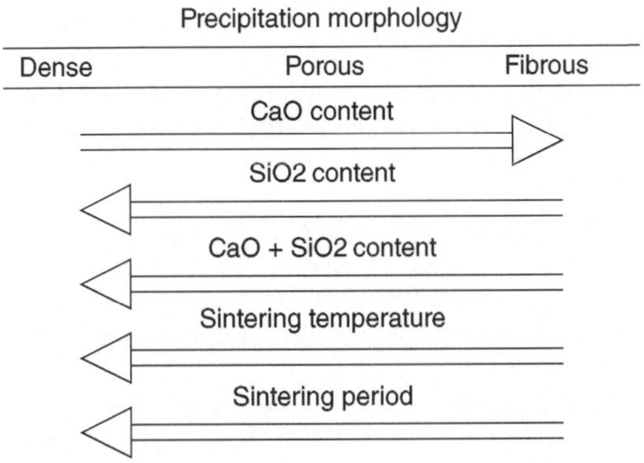

FIGURE 8.6 Influence of different factors on precipitation morphology (after Ref. [32])

REFERENCES

1. Loo, C. E. and W. Leung, Factors influencing the bonding phase structure of iron ore sinters, *ISIJ Inter.*, 43, (2003), 1393–1402.
2. Cohen, E., Radiological studies of the process of sintering iron ores, *JISI*, 175(3), (1953), 160.
3. Panigrahy, S. C., P. Verstraeten, and J. Dilewijins, Influence of MgO addition on mineralogy of iron ore sinter, *Metallurgical Transactions B*, 15B, (March 1984), 23–32.
4. Nicola, V. Y., Y. Scarlett, M. I. Pownceby, I. C. Madsen, and A. N. Christensen, Reaction sequences in the formation of silico-ferrites of calcium and aluminum in iron ore sinter, *Metallurgical and Materials Trans.*, 35B, (2004), 929–936.
5. Patrick, R. C. and M. I. Pownceby, Stability of silco-ferrite of calcium and aluminum (SFCA) in air-solid solution limits between 1240°C and 1390°C and phase relationships within the Fe_2O_3-CaO-Al_2O_3-SiO_2 (FCAS) system, *Metallurgical and Materials. Trans. B*, 33B, (2002), 79–89.
6. Patrick, R. C. and R. R. Lovel, Leaching dicalcium silicates from iron ore sinter to remove phosphorus and other contaminants, *ISIJ Int.*, 41(2), (2001), 128–135.
7. Scarlett, N. V. Y., I. C. Madsen, M. I. Pownceby, and A. N. Christensen, In situ x-ray diffraction analysis of iron ore sinter phases, *J. Appl. Crystallogr.*, 37, (2004), 262–268.
8. Van Den Berg, T. and J. P. R. de Villiers, An assessment of the production of fine material in iron ore sinter, *Mineral Processing and Extractive Metallurgy (Trans. Inst. Min. Metall. C)*, 118, (2009), 214–221.
9. Ostwald, J. Mineralogy and microtexture of Australian iron ore sinters, *BHP Tech. Bull.*, 25(1), (1981), 13–20.
10. Raju K. S., Petrographic studies of ten samples of iron ore sinters, Report by Indian Bureau of Mines (IBM), Nagpur, Submitted to the Research & Development Centre, *Steel Authority of India Limited*, (RDCIS–SAIL), Ranchi, August 1981.
11. Ishikawa, Y., Y. Shimomura, M. Sasaki, Y. Hida, and H. Toda, Proc. 42nd Iron making conf., Atlanta meeting, April 17-20, 1983, Vol. 42, *Iron and Steel Society of AIME*, Ironmaking Division, Warrendale, PA, pub. (1983), p.17.
12. Zhou, H., L. Zihao, M. Cheng, R. Liu, and K. Cen, Effect of flame front speed on the pisolite–ore sintering process, *Applied Thermal Engineering*, 75, (2015), 307–314.
13. Bhagat, R. P., Effect of bed height on strength of sinter and other sintering indices, *Steel India*, 33(2), (2010), 184–191.
14. Yadav, U., S. A. Ranjan, and B. K. Das, *Tata Search*, (2006), 109–115.

15. Ball, D. F., J. Dartnell, J. Davidson, A. Grieve, and R. Wild, *Agglomeration of Iron Ore*, Heinermann Education Books Ltd., London, (1973), Sec. 2, Chap. 11.
16. Bagnail, E. J., Influence of feed material properties on sinter for blast furnaces, in: *Agglomeration–77*, Proc. *Inter. Symp. on Agglomeration*, Atlanta, GA, March 6–10, 1977, Ed. K. V. S. Sastry, Amer. Inst. of Mining, Metallurgical & Petroleum Engineers, New York, 2, (1977), 587.
17. Ravat, Y. F. and A. Chatterjee, *Tisco. Tech. Jaur.*, (April 1979), 51–61.
18. Bogandy, L. V. and H. J. Engell, *The Reduction of Iron Ores* (in English), Springer-Verlag-Berlin Heidelborg, New York, (1974), Chap. 1.
19. Matsuno, F., Changes of mineral phases during the sintering of Fe_2O_3–CaO-SiO_2 *Trans. ISIJ*, (19), 1979, 595–604.
20. Mazaneck, E. and S. Jasienska, *JISI*, 202, (April 1964), 319–324.
21. Kissin, D. A. and T. I. Litvinova, *Stal.* (in English), 5, (1960), 318–323.
22. Kawaguchi, T., K. Kuriyama, S. Sato, K. Takata, and T. Miyake, in: Proc. 6th *International Iron and Steel Cong.*, Nagoe, Japan, 1990, Iron and Steel Institute of Japan, 40–47.
23. Lu, L., R. J. Holmes, and J. R. Manuel, Effects of alumina on sintering performance of hematite iron ores, *ISIJ Inter.*, 47(3), (2007), 349–358.
24. Yamaoka, Y., S. Nagaoka, Y. Yamada, and R. Ando, *Trans.* Iron Steel Inst. Jpn., 14, (1974), 185.
25. Mukherjee, T. and J. A. Whiteman, Structure of fluxed sinter, *Ironmaking & Steelmaking*, 12, (1985), 151–156.
26. Jeulin, D., J. L. Latailleur, and A. M. Schneider, A new technique for the quantification of iron ore sinter morphology, in: *Agglomeration–77*, Proc. *Inter. Symp. on Agglomeration*, Atlanta, GA, March 6–10, 1977, Ed. K. V. S. Sastry, Amer. Inst. of Mining, Metallurgical & Petroleum Engineers, New York, Chap., 29, 526–537.
27. Bristow N. J. and C. E. Loo, Sintering properties of iron ore mixes containing titanium, *ISIJ Inter.*, 32, (1992), 819–828.
28. Bhagat, R. P., U. S. Chattoraj, M. C. Goswami, D. P. Singh, and S. K. Sil, Effect of size parameters of mix ingredients on the porosity and reduction characteristics of sinter, *Steel Research International*, 78(6), (2007), 451–454.
29. Bhagat, R. P., U. S. Chattoraj, and S. K. Sil, Porosity of sinter and its relation with sintering indices, *ISIJ Inter.*, 46(11), 2006, 1728–1730.
30. Sato, E. S., M. Ichidate, K. Kato, and T. Kawaguchi, *Tetsu-to-Hagame*, 69, (1983), S744.
31. Hosotani, N. Konno, K. Yamaguchi, T. Orimoto, and T. Inazumi, Reduction properties of sinter with fine dispersed pores at high temperature of 1273 K and above *ISIJ Inter.*, 36, (1996), 1439–1447.
32. Pandey, B. D., U. S. Yadav, D. N. Jena, and B. K. Das, Reduction and reduction degradation characteristics of sinter, *Tata Search*, 2000, 17–22.

9 Sinter Quality: Theoretical Consideration and Plant Practice

ABBREVIATIONS

AI *Abrasion index*
LOI *Loss on ignition*
RDI *Reduction-degradation index*
RI *Reducibility index*
TI *Tumbler index*

9.1 SINTER QUALITY AND CONTRIBUTING FACTORS

9.1.1 Quality Parameters

Important quality parameters of sinter that do influence the iron-making process are:

Tumbler Index (TI): The TI of sinter denotes strength of sinter under cold conditions. The index is a measure of sinter's resistance to breakdown during handling and transportation, whereas the abrasion index **(AI)** is a measure of dust-generating tendency.

Reducibility of Sinter (RI): Sinter reducibility is defined as the ease with which the oxygen combined with it can be removed. This is also measured and expressed as kinetics of reduction of sinter with ascending reducing gases. A higher reducibility of the burden materials contributes to the enhanced gas efficiency in the blast furnace (BF), affecting the coke rate and productivity of the furnace positively. The sinter reducibility is influenced by the intracrystalline structure of minerals, lattice diffusion of sinter, and its porosity. The reducibility index is an inherent property of the sinter that is influenced by its chemical composition and the sintering (and cooling) conditions.

Reduction–Degradation Index (RDI) of Sinter: The index denotes disintegration of sinter during reduction at low temperature after being charged into the furnace. The sinter of higher RDI degrades more during its reduction in the upper part (stack zone) of the furnace and hence is not preferred. Sinter RDI is of greater significance in blast furnace iron making. Understandably, the fines generated inside the stack zone of a blast furnace during reduction affect the permeability of this zone adversely, increase the pressure drop, and disturb the gas distribution.

These factors result in a decrease in driving rate and adversely affect CO utilization with the ultimate consequences of higher coke rate and lower productivity.[1,2]

Softening and Meltdown Characteristics of Sinter: These characteristics of sinter show the manner in which the sinter undergoes softening and melts at the high temperature (bosh/hearth region) of the blast furnace. The permeability for reducing gas in the cohesive zone of the blast furnace is greatly reduced once the burden material starts softening. This results in substantial pressure drop (ΔP) across this zone, affecting the furnace productivity adversely. The pressure drop is guided by the softening and meltdown characteristics of the burden materials.

Annexure 2 lists the test procedures of these quality parameters and their relevance in iron making.

9.1.2 Contributing Factors in Sinter Quality

The quality parameters of sinter are guided by its mineralogical and morphological characteristics, which are influenced by the temperature profile across the sintering bed. The width of the combustion zone and the speed of the transversing flame front are important in this respect. For improved sintering indices, the heat transfer front and flame front should match so that heat generated through the combustion of coke and other chemical reactions is effectively utilized while the gases transverse through the sintering bed. That is, the bed temperature profile (see **Figures 6.14a** and **6.14b**) should not be broadened.

The combustion of coke along with the velocity of the descending gas (through the bed) decide:

a. The maximum temperature reached
b. The holding time above 1,100°C (1,373 K)

The temperature of 1,373 K is important (see Sec. **6.6.4**) as the majority of the sintering reactions take place above this temperature. The holding period above 1,373 K is also important for stability of these reactions and assimilation of the reactant minerals.

The contributing factors affecting the quality parameters of sinter (**Figure 9.1**) are:

a. Raw Material Characteristics: The characteristics of raw materials that affect sinter quality include (a) nature of ores, (b) size of ore fines and its distribution, and (c) content of ultrafines (say, below 0.25 mm) in fluxes and coke.

   Nature of Ores The granulation of sinter mix is affected by the chemical purity of the ore: the presence of clay minerals (alumina), content of loss on ignition (LOI), etc. (see **Sec. 6.8.2**). High LOI ores make a positive contribution to granulation in view of their high moisture retention capacity. Besides, the granulation of mix improves when the ore with

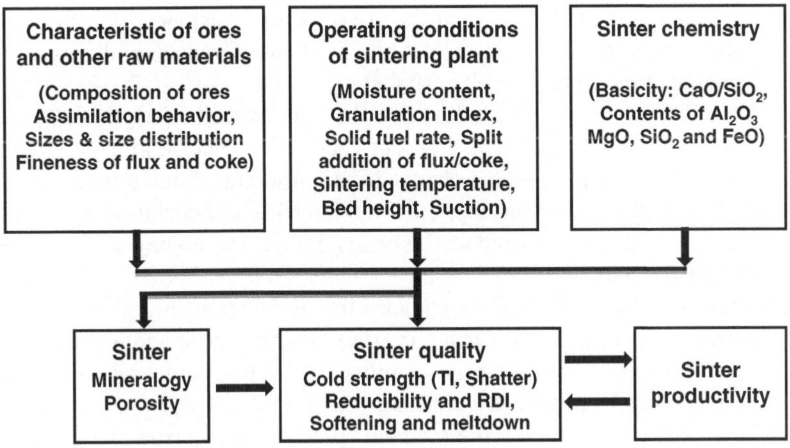

FIGURE 9.1 Schematic representation of the factors that influence the sinter qualities (cause and effect with reference to the sinter qualities and productivity vis-à-vis its mineralogy)

high Fe and low LOI content is blended with those of high LOI. Such type of blending may be essential to incorporate a higher amount of ultrafines, also to improve the bulk density of ores.

The amount of moisture available for the granulation depends on the internal porosity of ores, which varies from one source to another (see Sec. **6.5.7**). The bonding by diffusion decreases in case the mix constituents are poorly assimilated.

Particle Size The particle sizes of ores, solid fuel, fluxes, and their distribution patterns in the mix influence the granulation during balling and transfer of heat from the hot gases to solids and vice versa (see **Sec. 7.4.1**). The heat transfer coefficient between solid particles and combustion gases increases with the decrease in particle diameter, which results in an increase of the average mass temperature of the particle.[3]

On the other hand, if the proportion of coarse grains is high, the heat content of the waste gas cannot be sufficiently absorbed by the solid; besides, the bonding by diffusion decreases, which affects assimilation of the mix constituents on sintering. The coarser grains act as a heat sink as the heat content of the waste gas cannot be sufficiently absorbed by the solid.

Content of Fines (below 0.25 mm) in Fluxes and Coke The pore network of the sinter cake is improved when the finer-sized fractions from flux are eliminated while it is charged in the sinter mix. This facilitates a uniform gas flow in the sintering bed. Besides, this is effective in shortening the high temperature holding time with improvement in the high temperature reducing properties and an increase in the sintering speed. Consequently, the reduction properties (RDI and RI) of sinter get improved.

Sawada et al.[4] report that the use of sized limestone particles in the sinter mix increases the formation of calcium ferrite(s) and magnetite phases in the sinter (see **Sec. 10.6.1**).

b. Sinter Chemistry: This includes sinter basicity as well as MgO, FeO, and Al_2O_3 contents of sinter. The effect of chemical composition on the sinter quality is mainly through the change in the structural characteristics of the sinter. At higher sinter basicity, the sinter strength is contributed by the formation of diffusion-initiated ferrite bonds. Hence, the sinter possesses high strength and reducibility.
c. Operational Parameters: This includes the granulation (index) of the sinter mix, temperature of sintering: maximum temperature and holding time of the maximum temperature as reflected by bed temperature profile (see **Sec. 6.6.4**, **Figure 6.14a**), consumption of solid fuel, and sintering speed. The combustion and sintering reactions proceed in a narrow but effective zone only when the sinter mix is properly granulated. This is more important when the ores have difficult sintering characteristics.

   The process of (selective) granulation neutralizes the adverse effect of alumina in iron ore and improves the positioning of coke particles in the sinter mix. Consequently, the combustion and heat transfer is affected (see **Sec. 6.5.6, 7.4.3**).

   Diffusion-initiated bonds can be initiated by adopting the low-temperature sintering process with a decrease in the consumption of coke breeze. This is facilitated at higher sinter basicity with the addition of flux.[5,6]

These variables are interinfluenced and together decide the textural parameters of sinter: the mineralogy and porosity. These are detailed in **Sections 6.5.2, 6.5.7, 6.6.5**, **6.7**, **8.6**, and **8.7**.

9.2 SINTER MINERALOGY AND ITS QUALITY PARAMETERS

The types of phases present in sinter, their composition, amount/volume fraction, as well as morphologies have a very strong influence on the quality parameters of sinter, also the yield of sinter from strand and its size distribution (see **Sec. 8.8**). It is believed that these sinter quality parameters are decided by the inherent strength of the bonding phases present and the structure of pores.

9.2.1 Sinter Mineralogy

Sinter is composed of various mineral phases, each having its own strength and reducibility. Theoretically, there should be a relationship between the quality parameters of sinter and its mineralogy as well as its microtexture.

Mazenek and Jaisinka[7] studied the reducibility and compressive resistance of different minerals in self-fluxing sinter and observed that olivines ($Ca_xFe_{2-x}SiO_4$) with x = 0.5 had the highest compressive strength (above 56) followed by monocalcium ferrites which are formed in fluxed sinters. The hematite mineral phase has the highest reducibility, followed by calcium ferrite. **Table 9.1** shows the compression

TABLE 9.1
Compression Resistance and Reducibility of Mineral Constituents in Self-Fluxing Sinter[7]

Mineral	Compression resistance, kg/mm^2	Percent reduction
Hematite	26.7	49.9
Magnetite	36.9	26.7
Monocal. ferrite	37.0	40.1
Dical. ferrite	14.2	28.5
Olivines ($Ca_xFe_{2-x}SiO_4$)		
X = 0 (fayalite)	20.0	1.00
X = 0.5	56.6	2.7
X = 1 crystalline	23.3	6.6
X = 1 (glassy)	4.6	3.1
X = 1	30.3	10.5
X = 1.5	16.2	4.2

resistance and reducibility of various minerals present in sinter. The reducibility has been expressed in terms of percentage after the mineral is reduced for 40 minutes at 850°C. The reducibility of mineral phases present in sinter decreases in the order: hematite > calcium ferrite > magnetite > olivines (also see **Sec. 9.4.2** for detail).

Besides the content of hematite, its distribution is important for improved sinter reducibility. The reducibility decreases significantly in case the hematite phase in sinter is embedded by the glassy phase as the pores within the sinter that provide path for reducing gas are not open. In addition, the glassy phases in sinter have the lowest reducibility (see **Table 9.1**).

The sinter mass formed during the sintering also undergoes mineralogical changes during the cooling stage; the multicomponent magnetite is converted to multicomponent hematite, which has lost CaO and MgO but retained Al_2O_3 as a precipitate under suitable oxidizing conditions during the cooling stage. It is claimed that this form of secondary hematite, also known as skeletal rhombohedral hematite, thus formed in sinter with volume expansion is responsible for a reduction-degradation effect in the blast furnace. The minerals, calcium ferrites, in sinter retain a compact structure without forming pores during the entire reduction process.[4,8] The ferrite minerals arrest the propagation of cracks while the sinter is degraded during reduction at low temperature.

SFCA With the presence of Al_2O_3 and SiO_2 in iron ore, the Al and Si species diffuse though the calcium ferrite minerals to form silico-ferrites of calcium and aluminum, SFCA (SiO_2-Fe_2O-CaO-Al_2O_3) minerals of two different types depending on the purity of the iron ore (see **Secs. 6.10.3** and **8.5.2**). Low-grade ores (Fe less than 62 wt.%) typically form SFCA as a part of final structure; whereas high-grade ores (Fe more than 65 wt. %) will largely form SFCA-1. The SFCA-1 phase in fluxed sinter possesses a fine and fibrous structure and is effective in arresting the crack propagation. This leads to higher physical strength and higher reducibility than microstructures composed predominantly of SFCA. This also works in the case of goethitic ore.

Content of MgO and Impurities The most significant effect of MgO addition happens to be the suppression of hematite and Ca-ferrite phases and the increase in the magnetite phase. Most of the MgO is picked by the magnetite phase to form mixed spinels of the type (Fe, Mg)O.Fe_2O_3.[9] The presence of a controlled amount of impurities or additives (like CaO and SiO_2) in sinter can alter its intracrystalline structure. Besides, the lattice diffusion, porosity, and surface orientation within the sinter get changed. In this regard, the elimination of the first step of the reduction reaction (Fe_2O_3 to Fe_3O_4), which is the principle cause of degradation, is important. The crystallization of silica minerals present in sinter has a favorable effect on the reduction sequence. The porous morphology of precipitated iron also facilitates the reduction reactions.

Dicalcium Silicate The presence of dicalcium silicate in fluxed sinter affects the sinter strength. Polymorphism of dicalcium silicate could lead to localized internal stresses and disintegration of the sinter. Hence, prevention of polymorphism of dicalcium silicate is very important in improving the strength of fluxed sinter. The ferrites of calcium can stabilize dicalcium silicates if silicate content in the ferrite bonds does not exceed 25 wt. %.

9.2.2 Sinter Porosity

The sinter consists of several pores of varying sizes and shapes in addition to the mineral phases. These can be broadly categorized as micropores and macropores (see **Sec. 8.8.2**). The porosity of sinter, including size, shape, and volume of pores, is important in determining the structural strength of sinter.[10] Sinter can be treated as that of polycrystalline brittle materials with reference to porosity,[11] which—along with pore size distribution—affects the sinter strength.[12,13]

Reducing the size of pores in sinter unifies the pore distribution of sinter; consequently, its reducibility gets improved. This is owing to the increased specific area of sinter pores.[14]

The pores in sinter provide a diffusion path for reactant gases and surface area (pore wall surface) for reactions. Hence, the reducibility of sinter is facilitated with the pores as it provides increased contact surface area for the reducing gases to react with iron-bearing minerals. In the absence of pores, reaction can only occur through the exterior surface and will result in an unreacted sinter core, and the overall reduction rate will be low.

The sinter with high porosity undergoes volume expansion and stresses during reduction. The presence of cracks and irregular pores as well as poor fracture toughness of phases present in the matrix promotes degradation of sinter during its reduction.[9] The extent of size degradation of sinter depends upon the ability of cracks to easily propagate within the sinter matrix.[15] Cracks propagate within the sinter matrix and get arrested at the pores. New cracks are initiated from the sharp edges/corners of irregularly shaped pores and they further propagate and weaken the sinter. In comparison, finer and spherical pores having smooth and round edges suppress initiation of new cracks and thereby improve the strength. The propagation of cracks while the sinter is degraded during reduction gets arrested by the ferrite minerals present in sinter.

9.2.3 Pore Size

In addition to the sinter porosity (presence of pores), the shape and size of pores within the sinter matrix do influence reduction parameters of sinter at low temperature and reducibility and RDI significantly. (See **Secs. 10.7.2** and **8.8.2**.) Maeda et al.[16] reported that the reduction of sinter starts in its portions in contact with micropores and that the proportion and structure of the micropores supplied by the reducing gas play an important role in determining the reducibility of the entire sinter. Matsuo et al.[17] performed the reduction experiments with the synthesized pellets of different micropores. They report that while the overall porosity was equal, the pellets with micropores alone were reduced at a higher rate than the pellets constituting both macropores and micropores.[17].

Fine pores, initially formed, coalesce into coarser pores during the sintering process. However, these pores could be prevented from coalescing into coarser ones if expansion of the heat pattern is suppressed. The number of fine pores in sinter can be increased by decreasing the small fraction of the limestone and coke used.[18,19]

9.2.4 Porous Morphology

Besides the pore size, the orientation of pores also affects the sinter reducibility. The sinter reducibility decreases if the pores in sinter are closed. Conversely, the pores in sinter increase the reducibility if these are open (see **Sec. 8.8.3**).

A porous surrounding iron layer with interconnected ores is formed on reduction of the fluxed sinter, which facilitates the reduction reaction. The creation of new pores as well as the destruction of existing pores happens to be a common phenomenon during the reduction of sinter by reactant gases. In addition to that, the phases that are present around the pores affect the sinter reducibility. The reduction of magnetite to wustite markedly increases the number of micropores (0.1–10 micron) in size and, when the melt starts to form at 1,150 to 1,200°C, the fine pores are sequentially closed.[20]

Pore Stability The stability of pores during reduction is important. The pores, if they coalesce during the reduction, accelerate the low temperature breakdown of sinter during reduction (leading to higher RDI). This can be prevented/reduced by:

a. *Stabilization by Ferrites:* Bristow et al.[21] have studied the effect of porosity on reducibility and found that SFCA stabilized the fine pores produced during the reduction process, leading to increased porosity and higher reducibility.

b. *Increasing the Number of Fine Pores:* Matsuo et al.[17] suggested that if the coalescence of fine pores is suppressed and the number of small pores is increased in the sintering process, the surface area of sinter can be increased to achieve a substantial improvement in its reducibility. This is because the reduction area of sinter increases with an increasing number of fine pores.[22] This can be achieved when sized fluxes and coke are incorporated in the sinter mix (see **Sec. 8.8.2**).

9.3 COLD STRENGTH

The cold strength of sinter depends on the strength of the individual mineral components. Loo et al.[10] studied the fracture toughness of various mineral phases and found the following order of decreasing strength: Primary hematite (1.00) > secondary hematite (0.88) > magnetite (0.55) > SFCA (0.63) > glass (0.45).[22] The figure in () denotes the fracture toughness in $MNm^{3/2}$.

In addition, the strength of the bonding phase as a result of slagging does affect the overall strength of sinter. Strength could be related to the fluidity of the melt formed by the adhering layer of the particles.[4]

The major factors affecting sinter strength are:

- Void fraction and specific surface area of sinter[10]
- Chemical composition of the adhering layer
- Pore size and loss of ignition of the core particles
- Flame front speed (FFS)[23]

9.3.1 Strength of Sinter vis-à-vis Sintering Speed

The sinter strength increases with decreasing sintering speed on account of:

- The maximum temperature in the sinter bed increases with decreasing sintering speed. This leads the sintering process to continue at higher temperatures so that the sinter product undergoes more fusion. An increase in melt amount and a decrease in melt viscosity would result in the improvement of sinter strength.
- The residence time above 1,100°C becomes longer with decreasing FFS, as a consequence the assimilation of minerals accelerates during the sintering and the content of SFCA increases.

Figure 9.2 shows the effect of increasing the FFS on sinter strength (TI) and other sintering indices.[23] It is apparent that TI can be maintained up to a point (with increasing FFS) and then it decreases.

There is a need to optimize the FFS as lowering the sintering speed has a negative impact on sinter productivity while TI of sinter increases. In other words, if the productivity is maximized by increasing the sintering speed, the sinter TI would decrease. Typical pot grate experiments show that sinter strength was inversely related to the sintering speed. However, the correlation was not so strong.[24] This

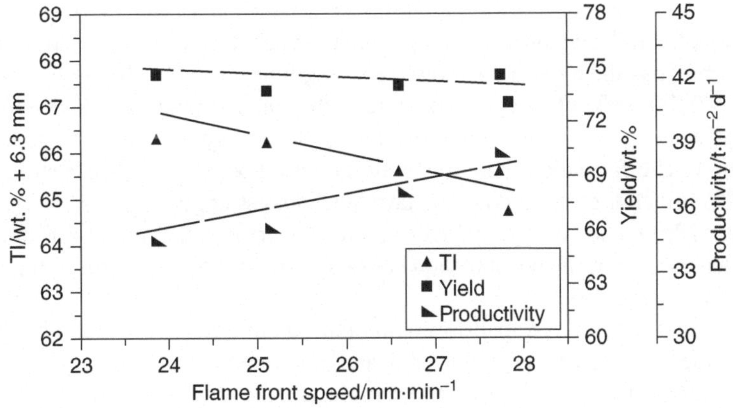

FIGURE 9.2 Influence of flame front speed on the sintering indices (after Ref. [23])

suggests that the improvement in cold strength of sinter was not solely due to the lowering of sintering rate and that other factors do influence the strength. For instance, there are certain types of ores, such as pisolite ore, that assimilate very quickly, and as a result, the level of unassimilated material is very low. In this case, the sinter quality could be maintained while undergoing a higher flame front speed.

9.3.2 Sinter Chemistry and Strength

Important parameters affecting the tumbler strength are content of gangues (alumina and silica) and sinter basicity (contents of CaO and MgO).

Sinter Alumina Alumina in sinter promotes the formation of SFCA, which should be beneficial for sinter strength. However, the presence of alumina in the lattice of iron ore weakens the sinter. Alumina increases the viscosity of the primary melt that forms during the sintering process, which leads to a weaker sinter structure with more interconnected irregular pores.[25] (See **Sec. 10.2.1.**)

Sinter Silica Silica combines with FeO and CaO to form compounds with low melting points: $FeO.SiO_2$, $2FeO.SiO_2$, $CaO.SiO_2$, etc. Temperature of primary melt formation drops with the increase in silica content and basicity of adhered fines. This situation is favorable for the subsequent assimilation reaction at the solid-liquid interface between the fines and the nucleus particles.[26]

Basicity Effect An increase in sinter basicity decreases the iron content of the slag phase by replacing part or all of the FeO by CaO and MgO. Lime combines with iron oxide to form low melting compounds—a minimum of which are required in order to produce strong sinter. These are monocalcium ferrite ($Fe_2O_3.CaO$) and FeO.CaO having the melting points 1,205°C and 1,120°C, respectively. The properties of the melts formed during sintering decide the structure of the bonding phases originated in the sinter.

The ferrites crystallize as interlocking lathes and form a strong framework to bind neighboring grains of ore together. The mineral, monocalcium ferrite, has the highest compressive strength next to olivines ($Ca_xFe_{2-x}SiO_4$, x = 0.5) (see **Table 9.1**). An increase in the sinter strength is also due to the lesser formation of the glass, fayalite, and possibly iron silicate phases as the ferrite formation increases with increased basicity.[27] (See **Sec. 8.7.1.**) The mineral, glass, and faylite have much less strength.

Sinter MgO The content of MgO in sinter decreases the amount of molten phase in sinter; the phenomenon has two opposing effects on sinter strength, which are as follows.

- MgO restricts the melt formation during sintering. The binding among the sinter particles is lowered and the sinter porosity increases. The strength decreases due to the combined effect of previous two.[12]
- As the molten phase in sinter is lowered, the shrinkage decreases. Hence, cracking during cooling of sinter is reduced and the sinter strength increases.

In view of the aforementioned phenomena, an optimum value of the MgO content can be worked out. The sinter strength increases with increasing its MgO content up to 1.5 wt. %. The addition of dolomite beyond 4 wt.% in the mix reduces the strength of sinter[28] on account of a higher formation of the glassy phase in sinter, which leads to a high stress concentration in its structure.

9.3.3 Coke Addition

The strength of sinter increases with an increase in coke content but flattens at high coke levels.[29] The strength (TI) of nonfluxed sinter is contributed mainly by slag bonding with the formation of FeO in sinter as the fuel consumption is increased.[30] Since the FeO content in sinter has an adverse effect on sinter reducibility, the rate of reduction also decreases with increasing coke consumption.

A much higher amount of fuel in sinter blends broadens the bed temperature profile. and the carbon value of coke is not fully utilized. This is reflected by a higher amount of CO gas in exhaust. (see Secs. **6.6.4, 6.14.3 and 6.14.6.**) A decrease in temperature maximum and widening of the combustion zone is reported with an increase in the solid fuel content in relation to the sintering of Rajhara ore from India.[31]

The effect of coke addition on sinter strength is interinfluenced by its basicity (CaO/SiO_2). The tumbler index (TI) increases significantly with an increase in coke breeze in the sinter mix at lower sinter basicity, ($CaO/SiO_2 < 2$), whereas, at higher sinter basicity, the effect becomes marginal (see **Sec. 10.5.3**).

9.3.4 Bed Height

Sinter strength is markedly influenced by suction pressure and bed height.[28] An increase in bed height contributes to the strengthening of sinter at equivalent coke consumption. This is due to the fact that the sinter emanating from the top layers of the strand remains weak. In the case of a deeper sinter bed, this (sinter from the top layer) will represent a smaller proportion of the total sinter produced compared to a shallower one. In view of this, the amount of recycled sinter will decrease.

The positive effect of increasing bed height in imparting strength to sinter is more than increasing the solid fuel in the mix. This is due to the fact that at higher fuel consumption, the strength of sinter is contributed by its FeO component, whereas favored mineralogical constituents are predominant at a higher bed height (and having a higher reducibility also) (see **Sec. 8.6.3**).

The gaseous fuel requirement for ignition, the solid fuel consumption, decreases[34] with increasing the bed height.

9.4 SINTER REDUCIBILITY

The sinter reducibility is important in affecting the furnace performance. The sinter while descending through the furnace need to be quickly reduced with the ascending reducing gas at lower temperature. The reducibility is influenced by the bonding and types of mineral phase in sinter and its morphological characteristics. The process and operational variables affecting the sintering reducibility are described below.

9.4.1 Nature of Ore and its Size

The ores from different sources differ in their assimilation characteristics (see **Sec. 6.8.3**). The bonding by diffusion decreases in case the mix constituents are poorly assimilated. Hence, the sinter reducibility decreases.

The sinter reducibility is also affected by the granulation during balling on account of the size of ores and their distribution. The profile of bed temperature in terms of the maximum temperature attained and its holding time is modified favorably in case the sinter mix is properly granulated. This is more important in case the ore constitutes a higher amount of ultrafines. Besides, the adverse effect of higher alumina in ore could be neutralized through adopting selective granulation in the balling stage (see **Sec. 6.5.6 and 7.4.3, Figure 7.4; Sec. 10.3.3, Figure 10.2**).

9.4.2 Sinter Chemistry

FeO Content Reducibility increases with increased (Fe^{3+})/(Fe^{2+}) ratio in sinter.[32] That is, the hematite content (Fe^{3+}) increases and the magnetite or wustite content (Fe^{2+}) decreases the sinter reducibility. Besides the level of hematite phase in sinter, its distribution is important for improved sinter reducibility. The reducibility decreases significantly in case the hematite phase in sinter is embedded by the glassy phase as the pores within the sinter, which provides path for reducing gas, are not open. In addition, the glassy phases in sinter have the lowest reducibility (see **Table 9.1**).

Figure 9.3 shows the degree of reduction (in percentages) of different minerals in fluxed sinter, also hematite and magnetite minerals as a function of reduction time.[33] The mineral, CaO. FeO.SiO_2, has the lowest reducibility. In comparison, calcium

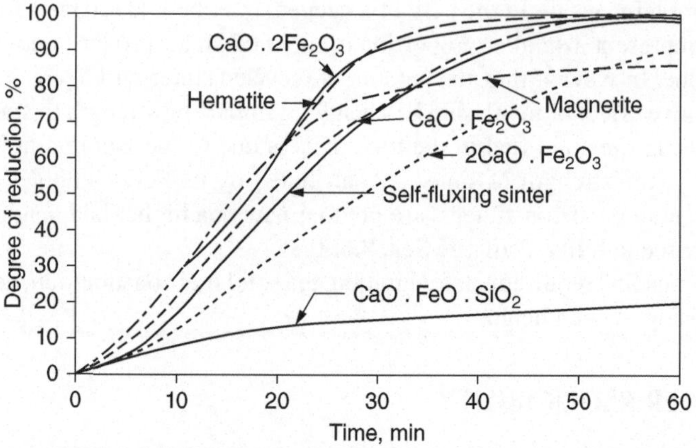

FIGURE 9.3 Degree of reduction of various minerals constituents in the fluxed sinter vis-à-vis hematite and magnetite minerals as function of reduction time (after Ref. [33])

diferrite ($CaO.2Fe_2O_3$) has the highest reducibility at increased reduction time (above 30 min. of reduction). Below 20 min. of reduction, hematite has approximately the same degree of reduction as that of $CaO.2Fe_2O_3$ or CaO, Fe_2O_3.

Sinter Basicity Low melting point mineralogical constituents, such as calcium ferrites, are formed with increasing sinter basicity. Formation of these constituents starts at a lower temperature, even at solid state through diffusion. The minerals, calcium ferrites, retain a compact structure without forming pores (cracks) during the entire reduction process, thus leading to improved RDI. The viscosity and liquidus temperature of slag bonding decreases at increased sinter basicity so that sintering can be achieved at a lower temperature. Lesser coke is required in the sintering, which increases the degree of oxidation and sinter reducibility.

The regimes of sinter basicity when its reducibility is higher are:

a. Most of the iron oxide (hematite) can be made free by selecting its basicity such that whole silica is combined with lime to form calcium silicate. Such a type of sinter is formed within the basicity range of 1.3–1.4 and has the highest reducibility.
b. Another regime of higher sinter reducibility prevails at higher basicity with the formation of a sufficient amount of monocalcium ferrite (CF), besides hematite. The mineral, CF, has the highest reducibility next to hematite (see **Table 9.1**). This has been corroborated with the sintering plant data.[34]

Content of MgO and Al_2O_3 The addition of dolomite in the sinter mix, in general, lowers the sinter reducibility, which is attributed to the formation of the minerals magnesium ferrite and magnesiowustite, both having poor reducibility.[6,35]

The alumina has refractoriness and hence requires a higher temperature in getting assimilated into the iron-bearing minerals and ferrites (to form SFCA). This increases the fusion affecting the sinter reducibility adversely.

9.4.3 Operating Practices

Granulation Index The process of granulation improves the positioning of coke particles in the sinter mix; consequently, the combustion and heat transfer are improved (**Sec. 6.14.6, 7.4.3**).

Low-Temperature Sintering The sintering at low temperature (predominantly through diffusion) helps to improve the reducibility as the amount of glassy phase is reduced in the sinter and a higher proportion of calcium ferrites is formed. The condition is achieved when the solid fuel rate and sintering temperature are decreased. The sinter produced in this case is mainly through diffusion bonding. A larger proportion of calcium ferrite and free magnetite are produced, mainly through diffusion in this case.

9.5 SINTER REDUCTION–DEGRADATION INDEX (RDI)

The degradation characteristic of sinter is important in affecting the furnace performance in terms of its productivity and coke rate. In case of poorer sinter RDI, a higher amount of fines are generated during its reduction which affect the permeability of the stack zone. The physical and chemical phenomena responsible for sinter RDI during its reduction as well as the process- and operation-related variables affecting the sinter RDI are mentioned below.

9.5.1 Mechanism of Sinter RDI

Although the fundamental cause of reduction degradation is well known, the mechanism that leads to severe breakdown of the sinter structure is only partially understood. It is still not possible to explain why degradation of sinter increases markedly when the sintering process is subjected to a minor change, besides on a marginal change in the sinter reducibility index.[22] The principal cause of degradation of sinter lies in the first step of its reduction reaction: Fe_2O_3 is converted to Fe_3O_4. The reduction reaction involves structural change with hexagonal structure of α-hematite transforming to cubic structure of magnetite. This is associated with volume expansion,[36] which could be around 4%.[37] The phenomenon of volume expansion during reduction results in the generation of internal stresses within the sinter matrix that are relaxed by crack propagation. The cracks are formed both within hematite grains and at the grain boundaries, embrittling the material.[38]

The sinter, as burden material, must reduce to iron, preferably in the upper part (stack zone) of the furnace; at the same time, this should degrade to minimum. Hence, the fundamental cause of RDI is the reduction of hematite to Fe. The factors

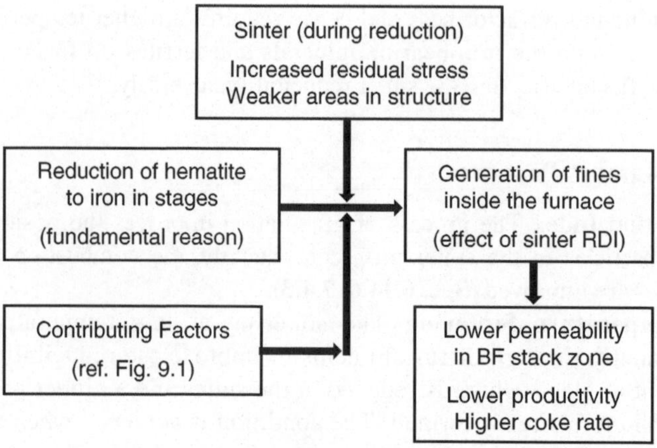

FIGURE 9.4 Schematic showing mechanism of low-temperature reduction of sinter in the furnace

contributing to it include the sinter chemical composition, sinter mineral/phase composition, nature of porosity, and process conditions.[39] **Figure 9.4** schematically shows the factors contributing to the RDI.

The glassy phase formed in sinter (see Sec. **6.7**) is amorphous and brittle, and its (excessive) presence is detrimental to sinter RDI. Whereas the diffusion-imitated tough bonding phase in the sinter matrix provides resistance to the crack propagation during its reduction as schematically shown in **Figure 9.5**.[22]

9.5.2 Sinter Chemistry

Sinter Basicity The structure of sinter in the superfluxed ranges {[(**CaO+ MgO**)/(**SiO_2+ Al_2O_3**)] ~ 1.6.} is marked by a continuous ferrite frame,[40] which retains a compact structure without forming pores during the entire reduction process. The mineral is also able to arrest the propagation of cracks initiated at the hematite crystals while the sinter is degraded during reduction at low temperature.[6] In this respect, the increase in magnetite (along with the decrease in secondary hematite), also the increase in fine acicular calcium ferrite phases, helps to improve the sinter RDI. Several researchers have observed that the volume of secondary hematite decreases when pseudo-particles with different CaO content—one with high CaO (limestone, mill scale, and iron ores) and another with low CaO (return fines, coke, and iron ores)—are processed. This results in a decrease in sinter RDI.[4,8]

Sinter MgO The increased replacement of CaO with MgO in the form of dolomite at basicity (CaO+ MgO)/(SiO_2+ Al_2O_3) = 1.6–1.9 improves (lowers) the sinter RDI, also its softening and melting properties. However, the strength of sinter, its reducibility, and sinter productivity decrease marginally.

The beneficial effect of MgO on sinter RDI may be attributed to fact that MgO stabilizes the magnetite phase in sinter. Therefore, the formation of

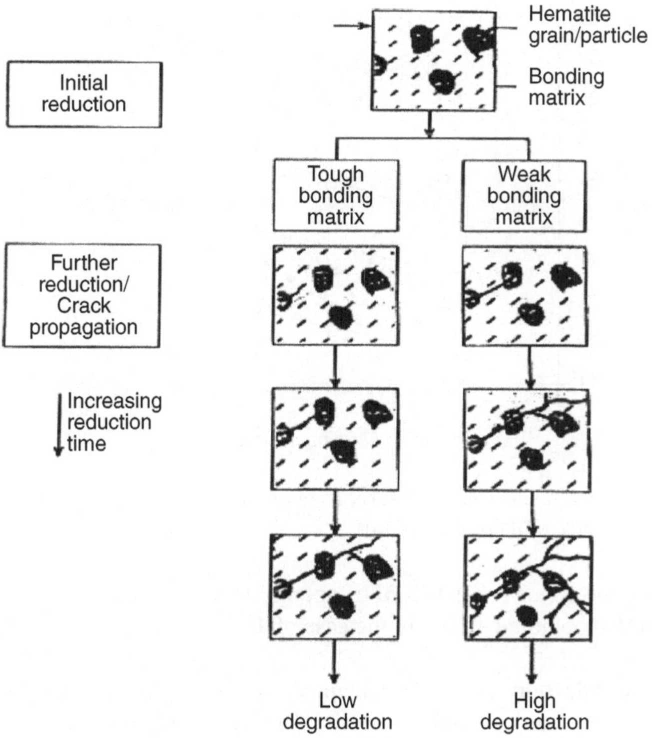

FIGURE 9.5 Schematic representation of the reduction-degradation of sinter (after Ref. [22])

secondary (reoxidized) hematite from magnetite during the cooling stage of sinter decreases.[41] The magnetite may, however, be converted into calcium ferrite on cooling under oxidizing conditions. The kinetics of this reaction depend on whether the magnesia is added as dolomite or as serpentine; in the case of dolomite, the increase in calcium ferrite in sinter is marginal. When MgO is added in the form of serpentine, the content of calcium ferrite in sinter increases considerably having a beneficial effect on sinter RDI. However, the addition of MgO through magnesium silicates (in the form of pyroxenite/dunite) has limitations. The magnesium silicates should be incorporated in the mix only when the silica content of iron ore is low; else, the silica load in sinter will increase.

Sinter FeO The sinter RDI is partially related to the amount and type of hematite in its structure. A possible reduction of secondary hematite in the sinter structure leads to an improved (lower) sinter RDI. A possible approach in this regard is to increase the FeO content of sinter. Sinter FeO partially stabilizes the magnetite against its oxidation to (secondary) hematite during the cooling stage.

Higher FeO sinters display excellent sinter RDI—such a correlation has been observed in the case of Tata Steel (India) ores.[42] A 2% increase in the FeO content in

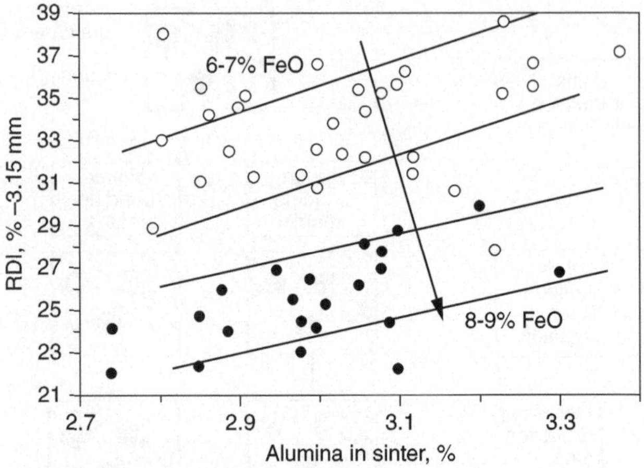

FIGURE 9.6 Effect of alumina and FeO content on sinter low-temperature reduction degradation characteristics, RDI (after Ref. [44])

sinter has improved (lowered) its RDI by approximately eight basis points. The result is similar to that obtained earlier in the case of Tata steel ores.[43] (See **Figure 9.6**.)

Magnetite Addition The FeO content of sinter can be adjusted by changing the coke rate in sinter blends. However, it may be cheaper to do so through addition of an appropriate type and amount of magnetite to sinter blends. Yamaoka and coworkers[45] found that both sinter strength and RDI improved as the addition of magnetite in sinter blends increased. This trend continues until the magnetite content in the blends reaches about 50%. However, the FeO content of sinter affects the reducibility of sinter adversely, and the addition of magnetite concentrate to hematite sinter blends will also have negative effects on sintering productivity due to low reactivity of magnetite. It is, therefore, important to find an optimum FeO level to improve RDI without compromising the other sinter properties.

Sinter Alumina Alumina content in sinter adversely affects the low temperature reduction degradation (RDI) characteristics of sinter. Industrial experience shows that within a range of 10–10.5% CaO in sinter, an increase of 0.1 wt. % alumina in sinter raises the sinter RDI by two points[44] (see **Figure 9.6**).

The effect of alumina on RDI of sinter is interinfluenced by its lime (CaO) content (see **Figure 10.1**). The adverse effect of alumina on sinter RDI could be neutralized to some extent by incorporating more lime. Another way to decrease the RDI of sinter is to increase its FeO content (**Figure 9.6**).

The mechanism responsible for the deterioration of sinter RDI with increasing alumina is partially understood. According to Matsuno et al.,[36] the alumina in sinter weakens the strength of glassy silicates and promotes large columnar SFCA (see **Sec. 6.10.3, Figures 6.20, 8.4**) both of which are detrimental to crack propagation resistance. This was further supported by Bristow and Loo's work.[17]

The addition of magnetite, MgO, and CaO in the mix is reported to be effective in neutralizing the adverse effects of alumina and improving the RDI characteristics of sinter.

The MgO, when added in the sinter mix, reacts with α Al_2O_3 and forms a new phase, $MgO.Al_2O_3$, by solid state reaction:

$$MgO(s) + Al_2O_3(s) = MgO.Al_2O_3(s)\ \Delta H = -35600\,(J/mol)$$

Gibb's free energy (ΔG^0) of the reaction is related to temperature T (K):

$$(\Delta G^0)\ (J/mol) = -35600\ \ -2.09T$$

At T = 1373 K, ΔG^0 = −38469.57 J/mol. Thus, at 1,100°C, the thermodynamic requirement for producing $MgO.Al_2O_3$ is met.

Sinter Silica In contrast to alumina, a higher proportion of silica in the ore fines/mix appears to have a beneficial effect on sinter RDI. The microstructure of sinter is modified with the silica and the formation of the optimum form of SFCA (silico-ferrites of calcium and aluminum), SFCA -1 (see **Sec. 8.7.3**). This form of SFCA is fine, fibrous, and hence has resistance to crack propagation with the consequence of reducing the RDI value.

Content of Gangue Minerals While a higher percentage of CaO+SiO_2 in sinter is beneficial for the precipitation of the SFCA phase, a minimum level of alumina in sinter could also be essential to stabilize these phases. This means that the presence of constituting gangue minerals (CaO + MgO + SiO_2 + Al_2O_3) in sinter mix is required to some extent in order to improve sinter RDI.

The presence of a controlled amount of impurities or additives (like CaO and SiO_2) in sinter can alter its intracrystalline structure and hence can lower the sinter RDI. The first step of the reduction reaction (Fe_2O_3 to Fe_3O_4), which is the principal cause of degradation, could be eliminated to some extent by the presence of a controlled amount of impurities in sinter since it changes the sequence of its reduction reaction.

The relationship between percent of gangue content and RDI of sinter is shown in **Figure 9.7**.[22]

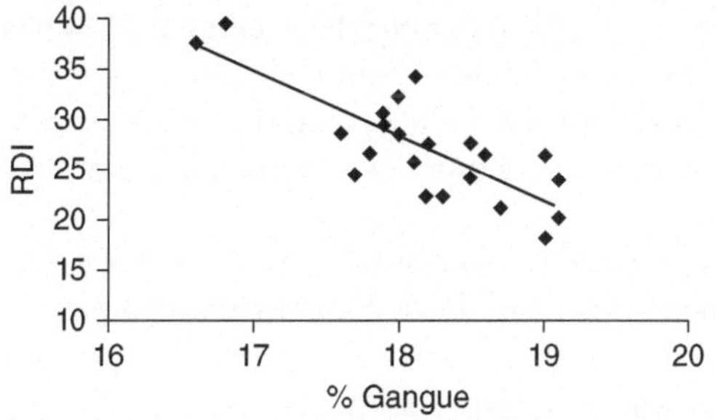

FIGURE 9.7 The relationship between percent of gangue content and sinter RDI (after Ref. [22])

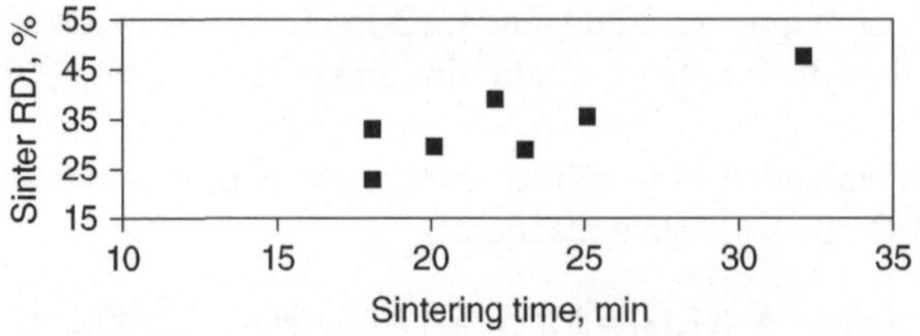

FIGURE 9.8 Relationship between sintering time and sinter RDI (BH 550 mm) (after Ref. [42])

9.5.3 Vertical Speed of Sintering

Sinter RDI decreases with the decrease in time of sintering (i.e., with an increase in sintering speed). This means that the permeability of the bed that undergoes the process of sintering has significant influence on sinter RDI.[42] **Figure 9.8** shows the results from pot sintering experiments (bed height 550 mm). The sinter RDI increases with increasing sintering time. That is, faster sintering helps in improving (lowering) the sinter RDI. The presence of secondary hematite in the sinter matrix is one of the most significant reason for increasing its RDI. This is formed through the reoxidation of magnetite (also from ferrites after the mineralogical changes) in the cooling stage of sintering and, at higher sintering speed, the reoxidation of magnetite.

9.6 SALIENT WAYS TO IMPROVE SINTER RDI

9.6.1 Sinter Alumina: Its Reduction and Neutralization

a. Reduction of alumina content in sinter
   This can be possibly achieved through the enrichment of ore fines, substituting the mix materials of high alumina content with the lower ones, also partial replacement of coke by petroleum coke, anthracite, etc. Coke contains high ash, in some cases more than 28%.
b. Neutralizing the adverse effect of alumina in sinter
   - The effect of alumina on sinter RDI is interinfluenced by its CaO and FeO content, which can be possibly exploited to reduce the RDI by optimally selecting the CaO and FeO regime of sinter (see **Figure 9.6**). A higher FeO in sinter can be achieved by providing a reducing environment when coke in the sinter blend is increased. However, the incorporation of magnetite in the blend is more helpful.
   - The sinter mix consisting of high alumina ore, when blended with highly reactive components followed by sintering, decreases the sinter RDI (see **Sec. 10.3.3**).
   - Pregranulation of the sinter blend to selectively localize the alumina in sinter also reduces its RDI (see **Figures 10.2, 10.3**).

9.6.2 Improving the Flux Size

The RDI of fluxed sinter is significantly improved (lowered) when below 0.5 mm-sized fractions are eliminated from the flux.[46,47] The calcination of limestone and its assimilation into hematite and other mineral phases gets adversely affected when limestone contains excess fines (size below 0.5 mm) in the mix. The use of coarse (sized) limestone in the sinter mix improves (lowers) the sinter RDI. This is due to the creation of the favored texture in sinter: the formation of acicular calcium ferrites; also, the shelter hematite increases in this case.

The addition of limestone having a coarser size increases the volume of free melt in the sinter. Consequently, the sinter porosity is modified; the pores become more spherical and sinter RDI is improved[48] (see Sec. **10.6.1**).

9.6.3 Split Addition of MgO-Bearing Fluxes[49]

The sinter quality parameters, tumbler index, as well as RDI are greatly influenced by not only the quantity of MgO-bearing fluxes (wt.% MgO) in the sinter but also the manner how these are distributed in mixing and balling stages during the granulation of sinter mix. MgO-bearing fluxes can be late added into the sinter mix (during granulation stage) so that a major portion of MgO remains on the surface of quasiparticles and lower content of MgO in its core. The split addition of MgO-bearing fluxes in such a manner produces sinter with its varied chemistry: lesser MgO is put inside and more MgO remains in its outside. Refer to **Figure 10.10** and **Figure 10.12** for the structure of quasiparticles; also **Figure 10.11** for the schematic of the arrangement. Such distribution of MgO in the structure of sinter improves its strength, also its RDI. The rationale is based on the following facts:

- In general, the tumbler Index (TI) decreases (and the abrasion index increases) with the increase in MgO content of sinter. Whereas, the addition of MgO in sinter blend on RDI has an opposing effect: the RDI improves (decreases) with increasing MgO content.
- The sinter TI is basically a property of core. Hence, with lesser MgO inside (in the core) of sinter, the TI should increase; whereas the AI should decrease with such a distribution.
- The RDI, being a surface property, should decrease with more MgO outside of the sinter.

Exploratory studies to reduce the disintegration of sinter during its reduction at low temperature have been made by impregnating the mix material with diluted solutions of alkaline earth halides, $CaCl_2$, $MgCl_2$, etc. Upon heating to 600°C, the added halides decompose through complex solid-state reactions. The reaction product mainly is confined to the sinter surface and reacts with iron oxide to form spinel, like $MgO.Fe_2O_3$ compounds having two functions, as follows. (These result toward the improvement of sinter RDI.)

a. The compound is able to reduce the transformation of hematite to magnetite on sintering. Therefore, more of the hematite mineral remains on the sinter surface in its primary form, which adds to the sinter strength on reduction.

b. The species mostly is confined to the surface of Fe_2O_3 particles and is able to diffuse into the subsurface layer at thermal treatment above 500°C.[49] Thus, the disintegration of sinter during its low-temperature reduction is inhibited/reduced.

9.6.4 Addition of Polymeric Additives

It is reported that a group of polymers could be incorporated in the mix as a binder for green pelletization that (partially) substitute bentonite (Sec. 12.4.4).[50] A typical study shows that the addition of similar additives in the sinter mix in the proportion of 1 kg/T results in significant improvement in the tumbler index (TI) and RDI of the sinter. However, the strand productivity decreases marginally compared to the base figure.[51] An improvement in the granulometry of the sinter blend has been observed with the addition of polymers; the medium-sized mix fractions increased by 15% compared to the base level.

9.6.5 Inert Gas Injection[52]

The presence of secondary hematite in sinter contributes toward its RDI. The secondary hematite is formed following the oxidation of the magnetite phase in sinter during the cooling stage. Inhibiting reoxidation of magnetite at this stage could be useful in improving (reducing) the sinter RDI, which can be executed through injecting an inert gas (say N_2) after the sintering process has attained the burn-through point. The hypothesis is based on the fact that during the cooling stage, heat is transferred from (hot) sinter to the incoming gas in the sinter bed from the top, and the composition of the incoming gas is immaterial from the heat transfer point. The injection of N_2 in the gaseous stream, while it does not affect the heat transfer adversely, reduces the availability of oxygen and thereby inhibits the reoxidation of magnetite.

REFERENCES

1. Tiwari, G., Y. Venkateshwarlu, R. Mohanty, and U. N. Behra, in Proc. *Trends in Beneficiation and Agglomeration of Iron Ore (CORAS 2001)*, September 25-26, 2001, SAIL-Ranchi, 191-201.
2. Bhagat, R. P., H. S. Ray, and S. K. Gupta, A relationship between strength of sinter and the ratio of return fines output to return fines input, *NML Technical Journal*, 34(3–4), (1992), 20–27.
3. Kasai, E., E. S. Wu, and Y. Omori, Factors governing the agglomeration granules after sintering, *ISIJ Int.*, 31(1), (1991), 17–23.
4. Sawada, I. K., K. Yashiko, and T. Takashaki, *Tetsu-to-Hagane*, 71 (1985), 1880.
5. Kramer, C. M. and R. M. German, J. *American Ceramic Soc.*, 61(7–8), (1978), 340–342.
6. Wynnyckyj, J. R. and T. J. Fahidy, *Metall. Trans.*, 5 (5), (1974), 991–1000.
7. Mazaneck, E. and S. Jaisehnka, *ISIJ* (London), 201, (1963), 60–63.
8. Hosotani, Y, N. Konno, K. Yamaguchi, T. Orimoto, and T. Inazumi, Reduction properties of sinter with fine dispersed pores at high temperatures of 1273 K and above, *ISIJ Inter.*, 36, (1996), 1439–1447.

9. Panigrahy, S.C., P. Verstraeten, and J. Dilewinjns, Influence of MgO addition on mineralogy of iron ore sinter, *Metallurgical Trans.*, 15B, (1984), 23–32.
10. Loo, C. E., C. E. Wan, and V. R. Howes, Mechanical properties of natural and synthetic mineral phases in sinters having varying reduction degradation indices, *Ironmaking & Steelmaking*, 15(6), (1988), 279–285.
11. Kasai, E., W. J. Rankin, R. R. Lovel, and Y. Omori, An analysis of the structure of iron ore sinter cake, *ISIJ Int.*, 29(8), (1989), 635–641.
12. Kasai, E., W. J. Rankin, J. F. Gannon, The effect of raw materials properties on bed permeability during sintering, *ISIJ Int.*, 29, (1989), 33–42.
13. Koranne, V. M. and R. V. Ramna, Effect of basicity on Pore characteristics of sinter, *Tata Search*, (2009), 81–84.
14. Jeulin, D., J. L. Latailleur, and A. M. Schneider, A new technique for the quantification of iron ore sinter morphology, in: *Agglomeration–77 (*Proc. *Inter. Symp. on Agglomeration)*, Atlanta, GA, March 6–10, 1977), Ed. K. V. S. Sastry, Amer. Inst. of Mining, Metallurgical & Petroleum Engineers, New York, (1977), Chap. 29, 526–537.
15. Bristow, N. J. and C. E. Loo, Sintering properties of iron ore mixes containing titanium, *ISIJ Inter.*, 32(7), (1992), 819.
16. Maeda, T. and Y. Ono, *Tetsu-to-Hagane*, 72, (1986), 775.
17. Matsuo, N., T. Maeda, and Y. Ono, *CAMP-ISIJ*, 4, (1991), 1081.
18. Bhagat, R., P. Uday, S. Chattoraj, M. C. Goswami, D. P. Singh, and S. K. Sil, Effect of size parameters of mix ingredients on the porosity and reduction characteristics of sinter, *Steel Research Int.*, 78(6), (2007).
19. Bhagat, R., P. U. S. Chattoraj and S. K. Sil, Porosity of sinter and its relation with sintering indices, *ISIJ Inter.*, 46(11), (2006), 1728–1730.
20. Naito, M., A. Okamoto, K. Ono, and Y. Hayashi, *Tetsu-to-Hagane*, 71 (1985), S49.
21. Bristow N. J. and A. G. Waters, Role of SFCA in promoting high temperature reduction properties of iron ore sinters, *Mineral Processing & Extractive Metallurgy*, Section C, 100 (January–April 1991), C1–C10.
22. Pandey, B. D., U. S. Yadav, D. N. Jena, and B. K. Das, Reduction degradation characteristics of sinter, *Tata Search*, (2000), 17–22.
23. Zhou, H., Z. Liu, M. Cheng, R. Liu, and K. Cen, Effect of flame front speed on the pisolite–ore sintering process, *Applied Thermal Engineering*, 75, (2015), Fig. 4, 307–314.
24. Chakraborty, D., M. C. Goswami, and R. P. Bhagat, Effect of process variables on sintering indices in relation to high LOI hematite ores fines of Gua mines, in: Proc. *International Seminar on Mineral Processing Technology (MPT 2010)*, org. by Indian Institute of Mineral Engineers (IIME) and National Metallurgical Laboratory (CSIR), Jamshedpur, December 15–17, 2010, 772–778.
25. Loo, C. E., and W. Leung, Factors influencing the bonding phase structure of iron ore sinter, *ISIJ Inter.*, 43, (2003), 1393–1402.
26. Debericant, D., C. E. Loo, and M. F. Hutchens, Effect of iron ore particle assimilation on sinter structure, *ISIJ Inter.*, 44, (2004), 1308–2004.
27. Ostwald, J., Mineralogy and microstructure of Australian iron ore sinter, *BHP Technical Bulletin*, 25, (1981), 13–20.
28. Higuchi, T. K., M. Naito, M. Nakano, and Y. Takamoto, Optimization of chemical composition and microstructure of iron ore sinter for low temperature drip of molten iron with high permeability, *ISIJ Inter.*, 44, (2004), 2057–2066.
29. Yadav, U. S., B. D. Pandey, B. K. Das, and D. N. Jena, Influence of magnesia on sintering characteristics of iron ore, *Ironmaking & Steelmaking*, 29, (2002), 91–95.
30. Cappel, F., E. Pfaff, and H. Kortmann, *Preprint Iron Making Conf.* Ranchi, org. by BF & SPp Operating Committee, R&D Center, July 10–11, 1989.

31. Gupta, S. K. and F. M. Bazanov, The influence of the physic-chemical characteristics of the fluxed sinters from Rajhara iron ore on their mineralogical analysis –Part I, *Trans.* Indian Institute of Metals (IIM), (June 1966), 75–81.
32. Gupta S. K. and E. F. Vegman, The influence of the physico-chemical characteristics of the fluxed sinters from Rajhara iron ore on their mineralogical analysis- Part II *Trans.* Indian Institute of Metals (IIM), (Dec. 1966), 227–229.
33. Watanabe, S., Fundamental studies of self-fluxing sinter, in: *Agglomeration–62*, Ed. W. A. Knepper, Interscience Publisher, New York, (1962), 865–894.
34. Bhagat, R. P., H. S. Ray, and S. K. Gupta, Effect of some process parameters on the performance of a sintering plant at Bhilai Steel Plant—a statistical analysis, *Trans.* IIM, 41, (1988), 465–474.
35. Panigrahy, S. C., and R. Verstraetan, and J. Dilewijins, Effect of MgO addition on strength characteristics of iron ore sinter, *Ironmaking & Steelmaking*, 11, (1984), 17–22.
36. Matsuno, F., S. Nishikida, and H. Ikesaki, Strength deterioration of samples of iron ore ~ 5–10% Cao systems during the reduction at 550°C in 30% CO-N2 gas, *Trans.* Iron & Steel Inst. Jpn, 24, (1984), 275
37. Inazumi, T., K. Shinada, and M. Kawabe, *Tetsu-to-Hagane* (J. Iron Steel Inst. Jpn), 68(15), (1982), 2207–2214.
38. Edwards, H. B., B. L. Daniell, and R. L. Samuel, Structural changes accompanying the reduction of polycrystalline hematite, *JISI*, (April 1965), 361.
39. Bhagat, R. P., S. K. Sil, U. S. Chattoraj, M. C. Goswami, D. P. Singh, and B. K. Das, Pot sintering studies for improvement in reducibility and RDI values of sinter, *Tata Search*, (2006), 127–134.
40. Pimenta, H. P. and V. Seshadri, Characterization of structure of iron ore sinter and its behavior during reduction at low temperature, *Ironmaking & Steelmaking*, 29(3), (June 2002), 169–174.
41. Hsieh, L. H., and J. A. Whiteman, Effect of raw material composition on the ineral phases in lime-fluxed iron ore sinter, *ISIJ Inter.*, 33, (1993), 462–473.
42. Bhagat, R. P., S. K. Pradhan, M. C. Goswami, B. Nandi, and T. Venugopalan, Effect of LOI content of iron ores and other process variables on the reduction properties of sinter, in: Proc. XXVI *International Mineral Processing Congress 2012*, New Delhi, India, September 24–28, 2012, org. by Indian Institute of Mineral Engineers, Jamshedpur, and Indian Institute of Metals, Kolkata.
43. Choudhary, M. K. and B. Nandy, Effect of variation in sinter FeO on sinter characteristics, *Tata Search*, (2007), 119–123.
44. Kumar, C. U., R. V. Ramana, S. Ali, A. K. Das, A. Kumar, A. K. De, and T. Mukherjee, *Tata Search*, (1995), 20–25.
45. Yamoaka, Y., H. Hotta, Y. Takasaki, and S. Ohzeki, in: *Australia/Japan Extractive Metallurgy Symp.*, the Australasian Institute of Mining and Metallurgy, Victoria, Australia, (1980), 245.
46. Bhagat, R. P., Studies on calcination of limestone and its assimilation by hematite during the sintering using thermo-gravimetric and X-ray diffraction analysis, *Scandinavian Journal of Metallurgy*, 21, (1992), 246–250.
47. Bhagat, R. P., S. K. Gupta, and H. S. Ray, Heat transfer consideration for improvement in sintering indices, in: Proc. 48th *Iron Making Conference*, ISS-AMIE, Chicago, (1989), 481–490.
48. Nakano, M., S. Kasama, and Y. Hosotani, *Tetsu-to-Hagne*, 85(10), (Oct. 1999), 711–716.
49. Larrea, M. T., A. Cores, A. Gutierrez, and A. Formoso, Surface behaviour of magnesium chloride-coated iron ore sinter, *Journal of Materials Science*, 23, (1988), 1151–1158.

50. Belenko, E. V., L. P. Vakhrushev, L. I. Voevodin, V. A. Gorbachev, and D. Y. Usol`Tsev, *Steel in Translation*, 35(2), (2005), 12–14.
51. Bhagat, R. P., Improvement in quality parameters of super-fluxed sinter from Indian iron ores: NML's experience, in: Proc. Xii *Inter. Conf. on Mineral Processing Technology (MPT–2011)*, org. by Hindustan Zinc Limited and Indian Institute of Mineral Engineers, Udaipur, India, October 20–22, 2011, Paper Fe 33, Technical Session Ii, Fe-1 (Iron Ore Beneficiation), (2011), 1–9.
52. Sinha, M., H. Nistala, and A. Biswas, Effect of inert gas injection on sinter RDI, J. *Metallurgy and Material Science*, 57(3), (2015), 173–178.

10 Process and Operational Variables with Respect to Sintering

ABBREVIATIONS

EPMA	*Electron probe micro analysis*
LOI	*Loss on ignition*
RDI	*Reduction-degradation index*
RI	*Reducibility index*
SFCA	*Silico-ferrites of calcium and aluminum*
TI	*Tumbler index*

10.1 PREAMBLE

The process and operational variables that do influence the sintering indices—namely, productivity of sinter and its quality—are interinfluencing and guided by their "limiting conditions" with respect to the sintering process. These variables have a simultaneous effect on the sinter productivity and its quality. The effect of increasing or modifying a particular variable on a typical index, say sinter RDI, could have a negative effect on another index, say sinter reducibility or sinter productivity. In view of this, the variables need to be seen in totality and not in isolation as it is difficult to modify a particular variable solely on the fact that this (should) have a positive effect on a typical sintering index.

The variables related to process, product, and operation are as follows.

a. **Process-related:**
    - The characteristics of ores and their size parameters
    - The sources of fluxes, their characteristics, and size parameters
    - Granulation parameters

b. **Product-related:**
    - Chemical constituents of sinter in terms of sinter basicity, content of alumina, and MgO largely decide the textural properties of produced sinter and, hence, the sinter quality. Stoichiometrically, the chemical composition of sinter is the result of the constituents of the sinter mix, which are important in affecting the sinter yield and flame front speed. Hence, the sinter productivity has an indirect correlation with the sinter chemistry.
    - Targeted mean size of sinter and its size range decide the recirculation of return fines and strand (machine) productivity.

c. **Operational-related:**
   - The amount and type of solid fuel added and the mode of its addition during granulation
   - The amount and type of fluxes added and the mode of their addition during granulation
   - The sintering operation, such as ignition condition, bed height, suction pressure, any disturbances in the sinter machine operation (frequency of machine stoppages), utilization of sensible heat, recycling of waste gases, etc.

10.2 IRON ORE: CHARACTERISTICS AND SIZE PARAMETERS

The chemical composition of iron ore, together with the thermal condition of the sinter blends, determine the primary melt formed during sintering and, consequently, the sinter structure and quality. With the ore being a major constituent of the sinter mix, its mean size and its range affect the mix granulometry, which decides the sintering efficiency and works better in its optimum but in a limited range.[1–3]

The sinter-ability of ores varies, which depends on their:

a. Chemical composition: the presence of gangue constituents, content of loss on ignition (LOI) and alumina
b. Mineralogical and morphological characteristics
c. Properties at higher temperature: melt formation and assimilation behavior
d. Size parameters: mean size, variance, minimum, and top sizes
e. Balling characteristics: depends on types of ores in conventional balling

10.2.1 Chemical Composition

Low-grade ores (Fe less than 62 wt. %) form SFCA (SiO_2-Fe_2O-CaO-Al_2O_3) as a part of the final structure; whereas, high-grade ores (Fe more than 65 wt. %) will largely form SFCA-1. The SFCA-1 phase does possess a fine and fibrous structure and is resistant to crack propagation. This leads to higher physical strength and higher reducibility than microstructures composed predominantly of SFCA formed from low-grade and high-alumina ore (see **Secs. 6.10.3** and **8.7.3**).

Higher SiO_2 in the sinter will induce the formation of glassy phases in the sinter and will reduce the strength of the sinter.

Higher alumina content in iron ore demands an elevated sintering temperature and a longer sintering time to promote melt formation and hence requires a higher amount of solid fuel. Higher sintering temperature has a consequence on the reducibility and strength parameters of sinter due to the formation of unfavorable sinter structure. Alumina increases the viscosity of the primary melt formed during the sintering process, leading to a weaker sinter structure with more irregular interconnected pores. The sinter product is usually characterized by a high FeO content and a large volume of bonding phases.

Alumina content of ore plays an important role in the sintering process as it adversely affects the reduction behavior of sinter in the stack zone of the blast furnace. Besides, the sinter productivity gets affected. In view of this, the role of alumina in sinter and its management has been separately mentioned in **Sec. 10.3**.

10.2.2 Mineralogical and Morphological Characteristics of Ores

High LOI of iron ores has a detrimental effect on the quality of sinter and its productivity. These ores (typically from India) are a combination of hematite-goethite and hematite-limonite having difficult sinter-ability. The sintering performance is adversely affected due to high fusibility and high porosity of iron ore fines (see **Sec. 6.8.2**). The sintering process has to be developed to tackle these adversity by adopting the techniques of preheating the iron ore fines, granulation of sinter mix, etc.

10.2.3 Properties of Ores at Higher Temperature

The characteristic of iron ores at high temperature is important in affecting the sintering process.[4] A faster tendency of melt formation is generally observed in porous ore and/or ore with high gangue content. Whereas, the massive to laminated type of ores containing higher Fe[5] has a higher assimilation temperature.[6–7] An increase in assimilation temperature (for this type of ore) results in a lower amount of melt, which could produce sinter of higher reducibility and consequently higher RDI.[8]

10.2.4 Size Parameters of Ores

The grain size of iron ore has a major role in deciding sinter mineralogy as it is easier for fine particles to be assimilated in the melt. By limiting the particle size of the feed materials, the reaction between the lime and other constituents is greatly facilitated. The period of fusion need not be prolonged; and hence, the profile of bed temperature is conducive when sized ore is used. The effects are:

- An increase in the shatter index of the sinter.[9]
- Less coke is required in sintering.
- An increase in the sinter reducibility index.

When the ore fines with reduced size are sintered, the microporosity (volume percentage of pores having radius below 10 μm) of sinter increases and its pore size decreases.[10–11] Reducing the size of pores in the sinter unifies the pore distribution of sinter; consequently, its reducibility gets improved.[12] This is owing to the increased specific area of sinter pore.[13] In addition, the density of sinter also increases.

10.2.5 Balling Characteristics of Ores

The balling characteristics of ores vary according to their textural characteristics, which affect the bed permeability. The bed permeability is lower when the mix consists of a higher proportion of massive- to hard-laminated hematite ore than that which consists of a higher proportion of soft-laminated to goethite mineral.[14] This results in poorer sinter productivity.

The massive to laminated ore, when blended in higher proportions with laminated to soft-laminated ore, had increased green permeability and, subsequently, an improved productivity.[14]

Case Study The study is in reference to typical Indian iron ores from the Noamundi and Joda mines of Tata Steel. Noamundi ore is massive- to hard-laminated and has higher Fe and lower LOI contents compared to Joda ore. The productivity of sinter, also its reduction–degradation index (RDI), with the Noamundi ore were inferior to that when Joda ore was used.

The bed permeability of the sinter mix with Noamundi ore was lower in view of its poorer balling characteristics than Joda ore. A lower amount of melt in the case of Noamundi ore could produce sinter of higher reducibility and consequently higher RDI.[8] Noamundi ore and Joda ore, when blended in some proportion[6] (say in the ratio 70:30), have resulted in improved sintering indices.[10,11]

10.3 ALUMINA CONTENT OF ORE

10.3.1 Sinter Mineralogy

Alumina promotes the formation of SFCA, which should be beneficial to the strength. Instead, the presence of alumina in sinter lowers the strength of (its) mineral constituents. This is due to the presence of high alumina in their lattices. In contrast, low alumina ores (Fe more than 65 wt. %) will largely form the SFCA-1 phase, which possesses a fine and fibrous structure and is resistant to crack propagation during reduction.

10.3.2 Effect of Alumina on Sinter Quality

Case Study: The results of an industrial trial at the Kimistu No. 3 sintering machine show that lowering of Al_2O_3 content in sinter to 1.50% from the base level (Al_2O_3 1.85%) improves its cold strength. Keeping the sinter cold strength that has been raised by lowering Al_2O_3 content, at the same level has enabled the sintering plant for low-energy (low FeO content) operation. As a result, the sinter reducibility increased from 60% to 68%. Consequently, the shaft efficiency at the stack zone of the blast furnace improved. Besides, FeO containing melt form has decreased in the cohesive zone (above 1,150°C), and the chance of pores being stopped up by the melt has diminished.[15] This has led to better permeability at the cohesive zone of the blast furnace (see **Sec. 3.6.6**).

10.3.3 Tackling the Adverse Effect of Alumina

Quite often, the adverse effect of alumina in iron ore is perceived only from the furnace point of view. The fact that higher alumina does affect the performance of a sinter plant in terms of quality of the product (sinter) and its productivity is inadvertently missed. Therefore, we need to evolve sintering methodology such that the adverse effect of Al_2O_3 in sinter is neutralized. Obviously, this is more important in the steel plants where sinter constitutes a significant proportion of the blast furnace burden.

Reduction of Alumina Content in Sinter In addition to iron ore fines, coke breeze and other raw materials contribute about 30% of the alumina in sinter. Their reduction in sinter could be approached by adopting the following:

a. Beneficiation of ores, also of lean grade and slimes (see **Sec. 2.11**).
b. Substitution by raw materials containing lower alumina in sinter blends should reduce the overall alumina level in sinter.
c. Partial replacement of coke breeze by petroleum coke, anthracite coal, and other fuels including biomass has been tried and remains a distinct opportunity to lower the total alumina content of sinter. Since the coke breeze contains a significant amount of ash (in some cases above 22 wt. %) any reduction in its consumption should decrease the gangue in sinter and increase its Fe content.

Neutralizing the Adverse Effect of Sinter Alumina The adverse effect of high alumina content in iron ore could be neutralized by following appropriate sintering process:

a. **CaO/FeO Content in Sinter: Figure 10.1** shows the effect of alumina on RDI of sinter is interinfluenced by its lime (CaO) content. The adverse effect of alumina on sinter RDI could be neutralized to some extent by incorporating more lime. Another way to decrease the RDI of sinter is

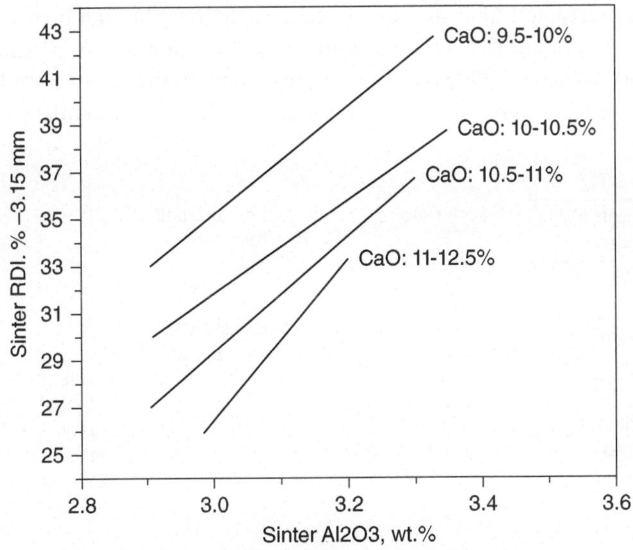

FIGURE 10.1 Effect of alumina on sinter RDI at various CaO levels (after Ref. [16])

to increase its FeO content (see **Figure 9.6**). The operative data of sinter making at Tata Steel show that RDI index (within the range of 25–35%) could be reduced by about eight points for a 2 wt.% increase in FeO (range: 6–8 wt.%).[16] Besides, the sensitivity (dependence) of RDI on sinter alumina (the slope) also decreases at its higher FeO. However, FeO content of sinter affects its reducibility adversely; hence, the increase in sinter FeO has limitations.

Magnetite Ore Sinter FeO can be increased through incorporating higher coke in the sinter blends. However, it is more appropriate to incorporate magnetite in the blends as both sinter strength and RDI improves when magnetite is added,[17] and the trend continues till the magnetite content reaches about 50% of the ore mix.

The magnetite mineral has low reactivity and, hence, has negative effects on sinter productivity. Besides, an increase in FeO content of sinter (on account of magnetite addition) affects its reducibility adversely; hence, the blend should be optimized in terms of the ore constituents.

b. **Blending with More Reactive Ore:** Blending the sinter mix consisting of high alumina ore with highly reactive components followed by sintering decreases the sinter RDI. High alumina iron ore, if sintered alone, has limited melt in the sinter matrix; whereas the sintering of highly reactive ore yields excess melt but lacks sufficient nuclei within the sintering structure. If these are blended together, this will dilute the adverse effect of alumina and limit the complete assimilation of reactive ore. This is schematically shown in **Figure 10.2**.[18]

c. **Localization of the Alumina Effects:** Unreacted alumina forms weak centers in sinter. In the case of fine alumina-bearing minerals, it may be possible to dissolve them completely in the primary melt to avoid any unreacted alumina. In the case of coarse alumina-bearing minerals, it is better to keep them unreacted. Therefore, their effect could be made localized. One of the most promising measures to control melting reactions during the sintering process is selective granulation. Unlike

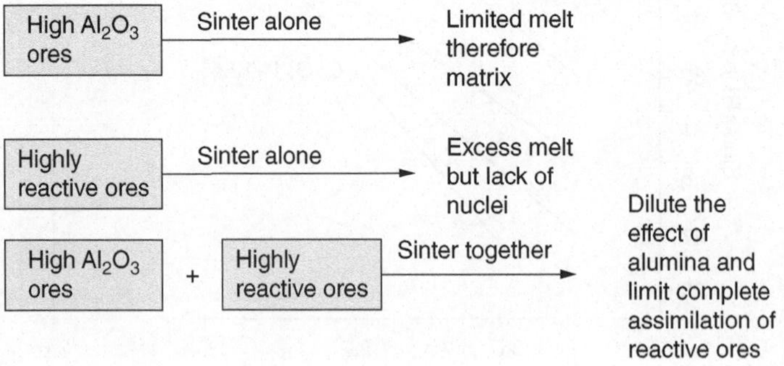

FIGURE 10.2 Schematic arrangement showing the dilution of alumina effect with the addition of highly reactive low alumina ores (after Ref. [18])

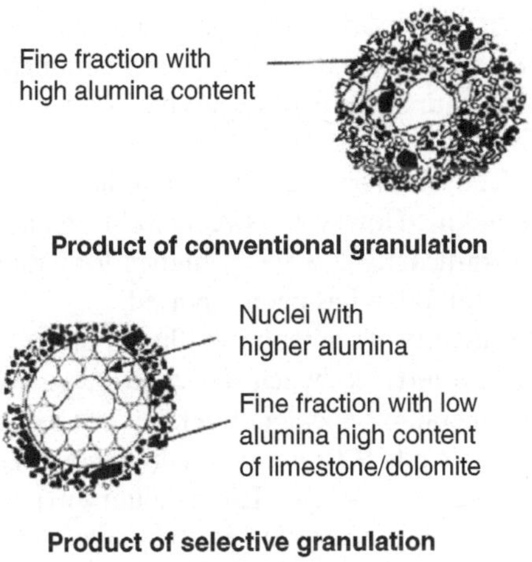

FIGURE 10.3 Conceptual view of conventional and selective granulation (after Ref. [19,20])

conventional ones, high alumina ores are pregranulated to form numerous small alumina-rich granules that act as nuclei in the subsequent granulation stage[19,20] (see **Sec. 7.4.3**).

d. **Selective Granulation:** In the conventional granulation process, the composition of the adhering layer, which mainly participates in the melt forming reaction, is uniform in the mixing drum. Hence, the alumina (with the ore fines) goes to the granule surface, which results in higher fusion temperature and poor bond formation during sintering.

Haga et al.[19,20] have developed a selective granulation process in which the high alumina fraction of iron ore fines is separated out and pelletized to a size where they act as nuclei particles rather than going into the finer part of the pseudo-particles. These high alumina nuclei particles are then fed into the remaining base mix for further granulation (**Figure 10.3**). Following this selective granulation technique, high alumina particles are forced/subjected into the nuclei of the sinter feed, and the adverse effect of high alumina in ore fines is neutralized.[19,20] It was found to effectively promote melt formation in sinter products, with an increase in open pore ratio and pore size. (Also see **Sec. 10.8.1, Figures 10.10**, **10.11**, **Sec. 7.4.3,** and **Figure 7.4**.)

Figure 10.3 shows a schematic view of pseudo-particle formation of base mix in conventional and selective granulation methods.

10.4 SINTER BASICITY

The basicity, CaO/SiO_2, of sinter is well correlated with its mineralogical and morphological characteristics.[21] Calcium-rich ferrites are formed with increasing basicity.[22] Because of the fact that these minerals impart beneficial properties of both strength and reducibility to fluxed sinter,[23] one can safely infer that under certain conditions it is possible to combine both strength and reducibility in the same product by adding lime. The belief for a long time that quality parameters of sinter—namely, strength and reducibility—are incompatible parameters[24] is no longer in vogue.

Nonfluxed Sinter: The tumbler strength of nonfluxed sinter is contributed mainly by slag bonding. However, its reducibility on account of its higher FeO content is significantly low.[24] A higher formation of fayalite with increased coke consumption has been reported.

Fluxed Sinter: The quality parameters of fluxed sinter are guided by the formation of calcium ferrites, which are contributed with increasing sinter basicity. The mineral, monocalcium ferrite, has the highest reducibility next to hematite (see **Table 9.1**); it also has high compressive strength next to olivines ($Ca_xFe_{2-x}SiO_4$, $x = 0.5$). The calcium ferrites retain a compact structure without forming pores during the entire reduction process and improve the resistance of sinter to degrading during reduction.[25] A greater formation of SFCA-1 in fluxed sinter is effective in arresting the crack propagation. This also works in the case of goethitic ore.

A major portion of the secondary hematite is formed during oxidation of hematite in the cooling stage. This form of hematite is the most disadvantageous to the RDI of sinter (see **Sec. 9.2.1**). With the increase in sinter basicity, the ferrites are formed, which may not participate to decompose to hematite in a larger extent, and the amount of reoxidized hematite in sinter decreases.[26]

Fluxed sinter loses strength during hydration[27] of the unassimilated lime owing to which there is an increase in volume by 10%. To prevent this, limestone should be crushed to 90–98% below 3 mm size.

Limiting Condition of Basicity The basicity, CaO/SiO_2 around 2, is possibly not conducive with regards to the degradation property of sinter in reducing medium at low temperature.[28] Structure of the sinter could be heterogeneous around this basicity level with a mixture of some basic and some acidic phases in it.[21] This possibly causes short cracks in the sinter that adversely influence the RDI of the sinter.

10.5 MGO CONTENT OF SINTER

10.5.1 Mineralogy and Quality of Sinter

The replacement of CaO in sinter by MgO leads to an increase in the liquidus temperature of the melt phase.[29] Consequently, the sintering reaction undergoes differently from that with a CaO addition alone. Hence, a sinter with different microtexture is produced.[21] The addition of MgO increases the magnetite phase, which picks most of the MgO to form mixed spinels of the type, $FeO.MgO.Fe_2O_3$.[30]

An increase in magnesia in sinter blends can retard the formation of reoxidized hematite during the cooling stage of the sinter,[26] which results in improved (lowered) sinter RDI. However, under oxidizing conditions, the magnetite is converted into calcium ferrite on cooling. The kinetics of this reaction depend on whether the magnesia is added as dolomite or as serpentine. When MgO is introduced in the form of dolomite, the amount of calcium ferrite decreases slightly, and when MgO is added in the form of serpentine, the calcium ferrite increases considerably.

The sinter porosity under some conditions increases with an increase in MgO content.[30] A higher MgO in the mix restricts the melt formation during sintering, resulting in increased porosity. As a result, the sinter reducibility and its RDI do increase. The tumbler strength decreases with an increase in porosity.[25] MgO also improves the softening-melting characteristics.[31]

10.5.2 High MgO Sintering Practice

Broadly, the productivity of sinter increases and its tumbler index (TI) improves with the decrease in sinter MgO. However, some blast furnaces could undergo high MgO slag practice, essentially to increase the slag fluidity (see **Sec. 9.3.2**) in the case of high alumina slag practice in the furnace. In view of this, the sinter plant is required to produce high MgO sinter.

10.5.3 Interinfluence of Sinter Basicity and Its MgO Content vis-à-vis Coke Content

The effect of MgO content of sinter and its basicity is interinfluenced. **Table 10.1** shows typical results of the pot (size 300 × 300 mm cross section) sintering experiments (bed height 500 mm) using hematite ore. A vacuum of 14,700 Pa could be maintained and heat input for ignition was 100 × 10^6 J/(m^2-min). The sinter productivity has been computed following shatter test (3 drops × 2 m height). The fraction above 10 mm size has been considered as sinter in this case. In order to establish the interactive effect, solid fuel vis-à-vis MgO content of sinter and its basicity, the coke (ash = 34 wt.%) in sinter mix has been varied widely: from 8 wt.% to 4 wt.%.

TABLE 10.1
Effect of Sinter Basicity and Coke in Sinter Mix on the Sintering Indices at Different MgO Contents Showing the Interinfluence of These Parameters (Bed Height 500 mm)

Input Parameter			Output Parameter				
Sinter Basicity	Sinter MgO %	Coke in mix, %	Yield % +10	Sintering Time, min	Cooling Time, min.	Prod T/m^2/h	Tumb. Index
2.0	3.0	8	44.28	20.5	14.5	1.14	68.8
2.0	2.0	8	35.02	13.5	8.5	1.21	63.6
2.5	1.5	8	39.71	11.5	10.5	1.54	70.1
2.5	1.5	4	46.34	12.5	7.5	1.87	73.9
2.5	2.5	4	44.65	13.0	7.0	1.64	72.2

The sinter mass at 8 wt. % coke in the mix was over-fused. At sinter basicity (CaO/SiO_2): 2.0 and MgO: 3.0 wt. % the sintering and cooling were much slower and the productivity was mainly contributed by increased yield. With decrease in MgO content to 2 wt. % from 3 wt. %; the TI of sinter decreased significantly from 68.8% to 63.6%. When sinter basicity was raised to 2.5 and its MgO content reduced to 1.5%, the TI increased from 63.6% to 70.1%. Besides, the productivity increased significantly from 1.21 to 1.54 T/(m^2-h), which was primarily due to a decrease in the sintering time (**Table 10.1**). These results were, however, at a much higher coke content, at 8 wt. % of the mix. Keeping the sinter basicity at 2.5 and its MgO content at 1.5 wt. %, when the amount of coke in mix was reduced to 4 wt. %, a substantial increase in strand productivity (1.87 T/m^2-h) could be achieved. With a further increase of sinter MgO to 2.5 wt. %, strand productivity of 1.64 T/m^2-h was obtained; the TI of sinter was also reasonably high (72.2%).[32]

10.5.4 Alternate Resources of MgO

Though the calcination temperature of dolomite is lower than limestone, the crystal size of dolomite is approximately 20 times larger than that of limestone. Dolomite, in view of its larger crystal size, takes a longer time (compared to limestone) to calcine (and get assimilated).[29] This is disadvantageous considering the temperature condition prevailing during the sintering process and the short reaction time.[5] In view of this, alternate resources of MgO, such as dunite and pyroxenite, have been attempted in sintering and currently are in practice in some sintering plants.[33]

Dunite has a typical advantage of substituting dolomite for the MgO requirement. Dunite, being a silicate of magnesium, does not require calcination. The average crystal size of dunite is about one-fourth that of dolomite. Therefore, dunite would be more easily assimilated into the sinter melt than dolomite and would require less heat (i.e., less coke breeze during sintering).

Dunite introduces SiO_2 along with MgO in the mix. Individually, these are high melting compounds. Their presence in the chemical combination with each other provides better assimilation during sintering, which results sinter of higher strength. In some cases, quartzite is directly charged to the blast furnace to lower the alumina/silica ratio in the slag (for its better fluidity). The usage of dunite in the sinter mix should be preferred rather than to directly charge quartzite in the blast furnace in view of the aforementioned reason.

Dunite has excellent potential to replace dolomite in the sintering process. However, it is costly. Pyroxenite is another alternative. Pyroxene is a group of important rocks forming silicate minerals found in many igneous and metamorphic rocks.

10.6 FLUXES: SIZE PARAMETERS, CALCINATION, AND ASSIMILATION

The manner how the flux is assimilated by hematite during sintering and to what extent decides the sinter mineralogy. A higher formation of calcium ferrite(s) and magnetite phases in sinter, besides, its pore network improved when the flux in size range 3–0.25 mm was used.[34,35]

10.6.1 Particle Size of Flux

A sinter plant receives the fluxes, limestone, and dolomite, in the lump form from mines, and they are subsequently sized after being crushed. Normally, fluxes of below 3 mm-sized fractions are used in sintering. The fluxes, if these have undergone higher crushing, will contain excess fines with size below 0.5 mm. The net effect is that the sinter mix at higher basicity (CaO/SiO_2 = 2 or more) will contain a higher amount of "ultra-fines" with size below 0.25/0.5 mm. In this situation, the calcination of limestone and its assimilation into hematite gets adversely affected, which results in poorer sintering indices (see **Secs. 6.8.3** and **9.6.2**).

Laboratory studies have shown that the sintering speed increased when limestone in the size range 3–0.5 mm[27]/3–1 mm[34] was charged compared to the base level with a total of below 3 mm-sized fractions in the sinter mix (see **Sec. 7.7.2** and **Table 10.2**). More importantly, the elimination of undersized fraction (below 0.5 mm) from the flux improved the sinter RDI while the RI value was marginally affected. Besides, the TI also improved (**Table 10.2**).

Sawada et al.[35] report that the use of sized limestone particles in the sinter mix increases the formation of calcium ferrite(s) and magnetite phases in sinter. The elimination of finer-sized fractions from flux is reported to improve the pore network of the sinter cake to ensure a uniform gas flow in the sintering bed. Besides, it is effective in shortening the high temperature holding time with improvement in the high temperature reducing properties and an increase in the sintering speed.

10.6.2 Calcination and Assimilation of Fluxes

Flux has poor balling characteristics compared to iron ore. The finer fraction of flux does not pelletize properly during the balling of the sinter mix. As a result, the sinter mix gets compacted at higher sinter basicity when the fluxes with excess ultrafines are blended (in the mix). This contributes to the worsening of the bed permeability, and the flow of gas is channelized through the surrounding of larger

TABLE 10.2
Effect of Flux Size on the Sintering Indices (Ore Size: –10 mm, Bed Height 550 mm)

Indices	Values					
Flux size, mm		below 3			3 to 0.5	
Vertical speed of sintering, m/min	23.6	23.6	23.8	25.1	25.1	25.0
Tumbler index, % +6.3 mm	67.6	67.8	68.2	69.3	69.0	69.1
RDI, % –3.1 mm	27.3	28.2	26.8	26.1	26.4	25.6
RI, %	ND	74.2	75.4	81.1	79.9	82.3

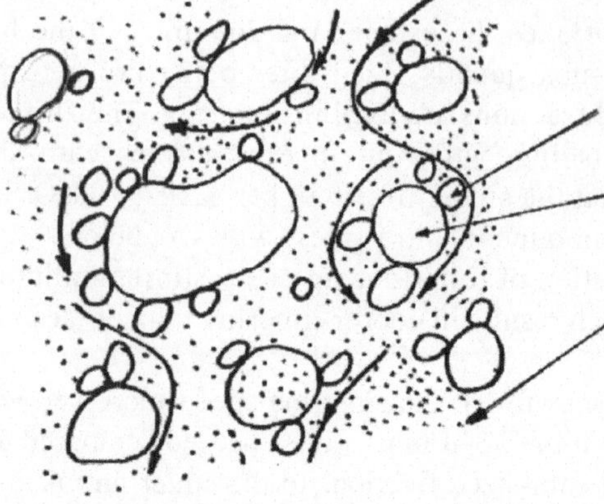

FIGURE 10.4 Schematic showing the poor aerodynamics when the flux contains a higher amount below 0.5 mm-sized fractions at sinter basicity 3.25 (after Ref. [36])

particles leaving the segregated fine fraction of flux uncalcined as schematically shown in **Figure 10.4**. This situation prevails at higher sinter basicity, especially when the fluxes contain too much finer (below 0.25/0.5 mm) fraction.

Bed Temperature Profile Typical pot sintering experiments at 3.25 sinter basicity, carried out using factorial design (see **Table 10.3** for treatment combinations), establish how the bed temperature profile varies when

TABLE 10.3
Comparison of the Sintering Indices at 3.25 Basicity Ratio at the Factorial Treatment Combinations, ae and ac. The Sintering Indices at the Modified Parameters from ae Are also Shown in the Last Two Columns

	Treatment Combination	ae	ac	ae (mod.)[1]	ae (mod.)[2]
A	Basicity ratio (CaO/SiO_2)	3.25	3.25	3.25	3.25
B	Coke in sinter mix, wt.%	5.5	5.5	5.5	5.5
C	Flux size (Dw), mm	0.67	1.47	0.67	
	+3/–3+1/–0.5, (mm) wt.% of flux	0/27/55	20/32/34	0/27/55	0/40/34
D	Coke size (Dw), mm	0.67	0.67	1.07	1.07
E	Bed height, mm	315	265	290	290
Response					
Sintering speed, mm/min		9.6	23.9	20.7	22.6
Yield (+10 mm/5 drops), wt.%		13.1	40.3	25.0	47.5
Shatter index, % +10 mm (5 drops)		70.7	77.0	69.0	77.7
Productivity(+10 mm/5 drops), T/(m²-h)		–	0.71	0.76	0.82

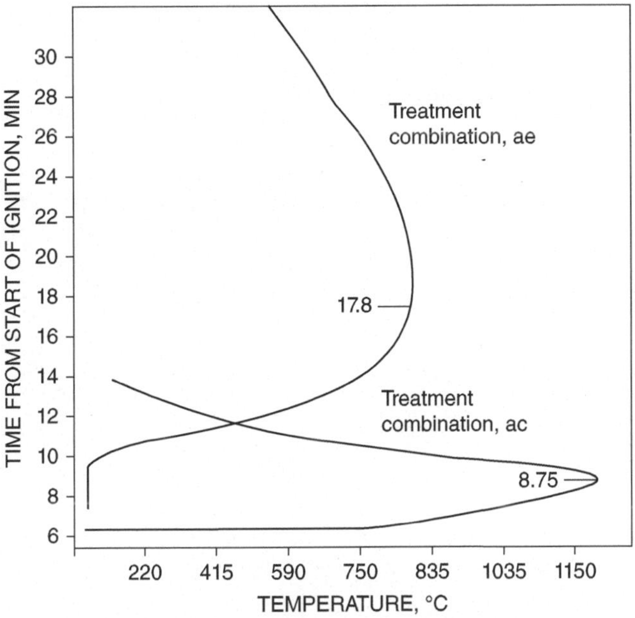

FIGURE 10.5 Bed temperature profile on sintering of hematite ore: Effect of flux size at sinter basicity 3.25 (refer to Table 10.3 for treatment combination of the factorial design) (after Ref. [36])

limestone of different size (range) is used in the sinter mix. At the treatment combination, **ae,** limestone had lower weight average diameter (Dw) at 0.67 mm having 55 wt. % below 0.5 mm size. A much broadened bed temperature profile[36] was observed (**Figure 10.5**). This is in view of poorer thermal efficiency discussed in previous paragraph. In case of treatment combination, **ac,** the flux size (Dw) being at 1.47 mm, the temperature profile was normalized.

Modification in Experimental Conditions The modification in experimental conditions leads to the improvement in sintering indices (**Table 10.3**).[37] The factorial combination, **ae (mod.)**^{1} represents the experiments were carried out at the base level of bed height (290 mm) while keeping other parameters the same as that of **ae**. The factorial combination, **ae (mod.)**^{2} represents the experiments were carried out at the base level of bed height (290 mm), while the coarse size (above 3 mm) of flux in the sinter mix was completely eliminated (kept at zero), and the finer sized (below 0.5 mm) fraction in flux was kept at 34 wt.% (lower level). The sinter productivity was higher at the treatment combination, ae (mod.)^2 which was contributed by increase in sintering speed and sinter yield.

Thermal and X-Ray Diffraction Analyses The thermal and X-ray diffraction analyses of the sinter and return fines samples are important tools to show whether or not the flux is fully calcined (and assimilated by hematite)

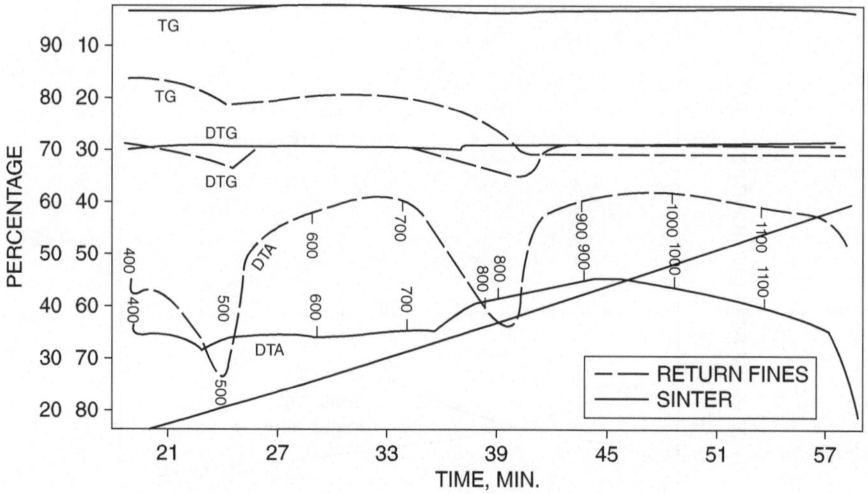

FIGURE 10.6 Curves of thermal analysis of the return fines and sinter samples on sintering of hematite ore with flux size skewed toward −0.5 mm (after Ref. [36])

during the sintering process or not[36] **Figure 10.6** typically shows TG/DTG plots of sinter and return fines samples with regards to the treatment combination **ae.** Sharp endothermic peaks of weight loss at around 500°C and 825°C in the case of return fines show that it contained a substantial amount of limestone that has not undergone calcination in the process of sintering. These peaks, in the case of sinter, were marginal.

The presence of limestone and CaO, which remained unassimilated, in the samples of sinter (to a much lesser extent) and return fines were observed through the X-ray diffraction analysis also.

10.6.3 Optimization of Crushing Scheme

The production of limestone in a narrow size (3–0.5 mm) range could be achievable to some extent through an optimally designing crushing and classification scheme. The sieve analysis of the crushed limestone, produced from the hammer mill in a typical sinter plant over a period, shows that the size (range) of crushed product varies widely and depends on the hammer conditions.[36]

Therefore, it is possible to produce flux in a narrow size range by adopting these measures:

- A proper design of higher erosion-resistant hammers and their regular replacement in the hammer mill may be required in this regard.
- Multistage crushing of limestone and size classification of the crushed product may be adopted to avoid the overcrushing of −0.5/−1 mm-sized fractions: the −0.5 mm fractions are screened out before the crushed product is fed to another crusher (**Figure 10.7**).

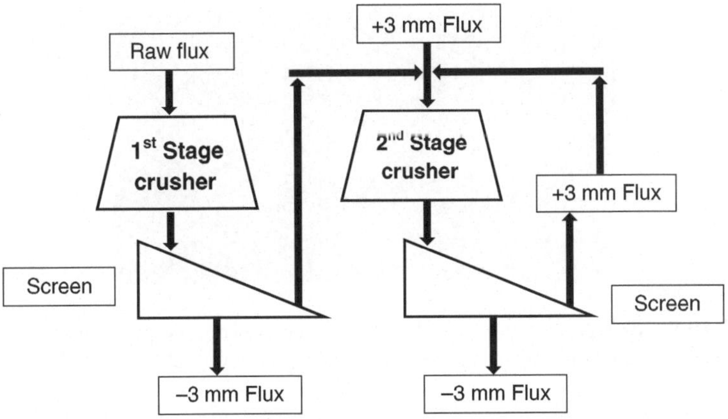

FIGURE 10.7 Schematic of the multistaged crushing of limestone

10.7 SOLID FUEL

10.7.1 Thermal Effect

The amount of solid fuel and its characteristics in terms of its size, reactivity, and chemical constituents do affect the bed temperature profile during the sintering process. A higher maximum bed temperature and longer holding time are reported with increasing the amount of coke breeze. The presence of excessive coke in the sinter mix, however, results in a broadened temperature profile though there is an increase in the peak temperature.

An increase in the peak temperature of the sinter bed with the addition of coke improves the slag bonding. This results in greater formation of lower oxides of iron (FeO), which contributes to the sinter strength. The quantity of secondary hematite in sinter increases with a decrease in fuel rate. In this case, the sinter porosity increases and consequently more pore surface is available where the oxidation of magnetite to hematite takes place during the cooling of sinter (**Sec. 8.8.2**).

10.7.2 Size of Coke and its Distribution

The mean size of coke and its distribution pattern affect the sintering indices. The equilibrium temperature at which combustion of coke takes place increases as its initial particle diameter is increased.[38] The excessively fine, also the excessively coarse, coke particles in the mix require higher fuel consumption to attain the same sinter characteristics while the sintering rate is simultaneously decreased. The sintering zone gets broadened in either of the cases with lower peak bed temperature; this also results in higher levels of CO in exhaust gases.[39]

The chemical consistency factor, especially with respect to FeO, improves with proper sizing of coke particles.[40]

The coarse-grained coke particles remain free, that is, these do not take part in the granulation process and remain unattached to other materials. Besides, larger-sized

FIGURE 10.8 EPMA (carbon scan) of fluxed sinter showing the presence of fine carbon (white dots) throughout the sinter matrix (magnification 1,000×) (after Ref. [42])

coke particles cause local superheating and longer combustion time. Whereas, fine coke remains well-embedded into granules of sinter mix and their combustion rate is decreased[41] (see **Sec. 6.14.4**). Very fine carbon particles in the mix granules may also remain unburned in the sintering process. This is evident from typical electron probe microanalysis (EPMA) scan of the fluxed sinter (CaO/SiO_2: 2.2) (**Figure 10.8**) showing the presence of fine carbon particles.[42]

The optimum coke particle sizes for better combustion efficiency and productivity are 0.25–3 mm, which are neither so fine nor so coarse.[43,44] By increasing the fraction of medium-size particles (3–0.25 mm), the heat pattern could be tailored in such a manner that heating time is reduced; whereas cooling time is lengthened at high temperatures. This region results in lower FeO formation.

Effect on Sinter Mineralogy Fine pores, initially formed during the sintering process, coalesce into coarser pores. These pores could be prevented from coalescing into coarser ones if expansion of the heat pattern is suppressed (see **Sec. 6.8.2**). The number of fine pores in the sinter could be increased by decreasing the smaller fraction of coke. At the same time, too much coarser-sized coke in the sinter blend decreases the sinter microporosity.

Sinter microporosity is important with respect to its reducibility (see **Sec. 9.2.3**). In the case of pellets, those with micropores alone reduce at a higher rate than the pellets constituting both macropores and micropores.[45]

Laboratory investigation with typical Indian hematite ores has shown that when the size of coke in the sinter mix increased from the base level (100 wt.% below 3.3 mm) to all below 6.3 mm, the sinter microporosity decreased from 10.4 vol.% to 5.2 vol.%. While the microporosity of these samples changed quite significantly, the total porosity (consisting of micro- and macropores) of the sinter samples remained more or less the same, 13% and 11%, respectively[27] (see **Secs. 8.8.3** and **9.2.3**).

Staged Crushing Crushing of coke in roller mills and size classification in stages, similar to that adopted for the flux preparation (see **Figure 10.7**), could be adopted for the production of sized coke. The coke crushing facility in Schwelgern sinter plant uses two-stage roller mills equipped with hydraulic gap adjustment mechanisms for improving the coke grain size (see **Figure 2.8**).[46]

10.7.3 Mode of Coke Addition

The sintering process and the quality parameters of sinter produced are also dependent on the manner the coke is added in the mix at the granulation stage (see **Sec. 6.14.4**). The following schemes are adopted:

a. Delayed introduction of coke into the sinter mix. This involves when coke is fed into the raw mix after all the other materials (iron ore, fluxes, return fines, etc.) had been mixed and granulated. The scheme affects the positioning of the finer coke particles in the granules and finally produces the granules with coke coating. The coating may contain some portion of ore and fluxes but to a lesser amount than with normal mixing. Because the coke is readily accessible to the incoming gas constituting oxygen, the combustion will proceed at a faster rate.
b. The entire coke is split into two fractions: the first one is put into the mixing drum along with the other mix ingredients while the rest (second) one is added in the second (balling) drum. This facilitates positioning the coke particles in the outer adhering fine layer (AFL) and hence better combustion.

   Hou et al.[47] have observed that the effects of 50% coke late addition (LA) are better than those for all late addition. When all the coke particles are added outside, permeability would decrease. The effects when coke with size fraction above 1 mm is added late are better than those of all LA and 50% LA. The results show it is not suitable to add coke particles of smaller size (below 1 mm) outside.
c. The technique of divided coke addition according to sizes has been attempted for the efficient combustion of coke. The entire coke is fractionated into two: the coarser and the finer one and added separately for granulation. The rationale for such a scheme lies in the fact that large coke particles remain free (i.e., unattached to other materials), whereas the fine coke particles are well-embedded into granules that decrease their accessibility for the combustion.[41] According to Ishikawa et al.,[48] the large coke particles should only have a thin layer of adhering fines to facilitate combustion, and fine coke particles should be located on the outermost part of the adhering fines layer of a granule for their better accessibility by oxygen to facilitate combustion.

Teo et al.[41] studied the positioning of coke in the granule during the sintering process. **Figure 10.9a** represents the micrograph when the coke is added into the mix along with ore fines. Larger coke particles are on the surface of AFL. **Figure 10.9b** represents the micrograph when the coke addition to the sinter mix is delayed. Fine coke particles (as indicated by arrows) concentrated on the surface of the AFL. This positioning of coke enhances its combustion rate (with delayed addition).

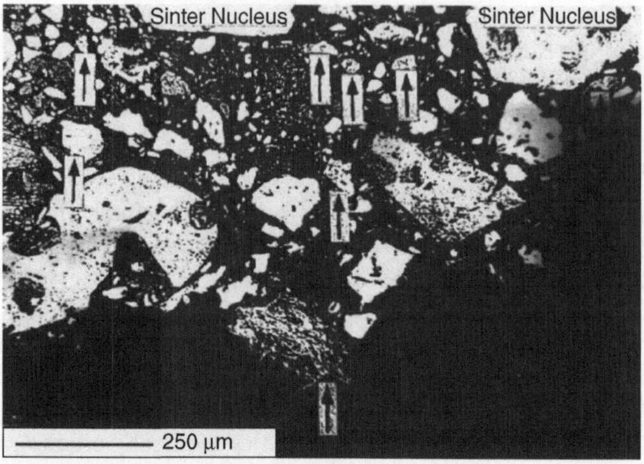

(a)

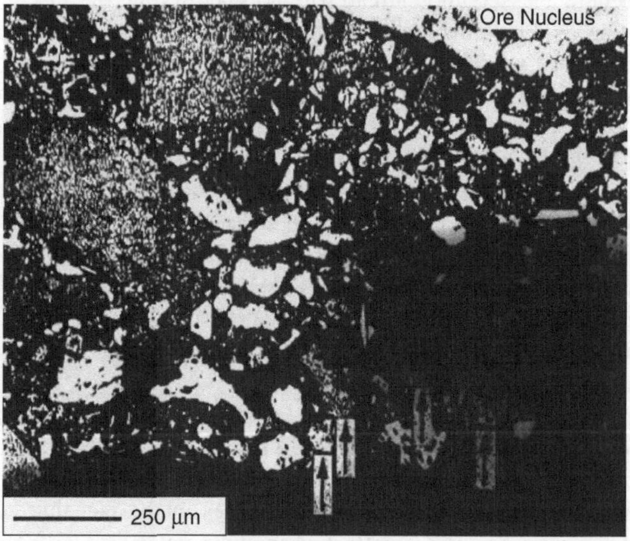

(b)

FIGURE 10.9 (a) Micrograph showing the distribution of coke within the adhering fines layer of a return sinter nucleus; coke particles are indicated by arrows (after Ref. [41]) (b) Micrograph of a granule formed using the late addition technique; coke particles are indicated by arrows (after Ref. [41])

10.7.4 Ash Content of Coke

Stoichiometrically, an increase in coke ash content decreases the Fe content of sinter with a simultaneous increase in silica and alumina content. This adversely affects the furnace performance when the sinter (of lower Fe content) is charged. A rough estimate indicates that with a 1% increase of iron concentration in the burden, the blast furnace productivity increases by 1.5–2.5% and coke rate decreases by 0.8–1.2%.[49,50]

The adverse effects (of higher gangue content in sinter) are:

a. Higher content of ash in coke and/or higher specific consumption of coke add to extra alumina in sinter blends, which decreases the quality of sinter by virtue of the formation of unflavored sinter mineralogy (see **Sec. 8.7.3**).
b. A higher ash content of coke decreases the sinter productivity, though marginally.[51] The coke breeze requirement (for sintering) also gets adversely affected.

The adverse effect can be neutralized by:

a. Reducing the specific consumption of solid fuel
b. Partial replacement of coke with alternate solid fuel(s) that have lower ash content

10.7.5 Reduction in Specific Coke Consumption

The benefits of sintering with a reduced amount of coke is:

- Decrease in flux requirement at the same basicity level and, consequently, the Fe content of sinter increases. This leads to the conservation of coke and fluxes, which ultimately reduces the cost of the production.
- Decreasing fuel consumption results in a lower FeO content in sinter with lesser amounts of glassy phase. Consequently, the sinter reducibility improves.
- Reduction in environment pollution—the combustion of coke in the sintering process generates the gaseous pollutants, NO_X and CO. The emission will be less when the coke combustion is reduced.

Coke Consumption and Sinter Productivity An analysis of performance of 116 sintering plants throughout the world shows interesting observations.[52] Indices were in terms of:

a. Area of sinter machine (50–600 m^2)
b. Specific productivity (0.8–2.10 T/m^2-hr)
c. Consumption of solid fuel (30–100 kg/T)
d. Consumption of gaseous fuel (40–280 MJ/T)

Findings were:

a. Both the solid fuel and gaseous fuel consumptions were less, but productivity was higher in 44 sintering plants out of the aforementioned 116. The indices were:

   Solid fuel: 40 kg/T, gaseous fuel: 81-120 kg/T, productivity: 1.9–2.0 T/m^2-hr.

b. Both the solid fuel and gaseous fuel consumptions were higher, but productivity was lower in 65 sintering plants out of the aforementioned 116. The indices were:

   Solid fuel more than 70 kg/T, gaseous fuel: 201–240 kg/T, productivity 0.91–1.0 T/m^2-hr.

The above findings show that no correlation existed between specific coke consumption and strand productivity. Hence, the coke rate could be decreased without sacrificing the strand productivity (also while maintaining/improving the sinter quality). This is more relevant in the case of India and elsewhere where the solid fuel requirement is higher than that of their peers. Hence, there exists a greater scope for reduction of coke consumption in the process.

Possible approaches toward the reduction in the coke requirement in the sintering process are:

a. **Decreasing the Liquidus Temperature** The requirement of solid fuel decreases with a decrease in the liquidus temperature of the melt phase. The possible ways to achieve this are:
   - Through the addition of a lower amount of dolomite
   - By reducing the alumina content from ore/other mix materials

   Cappel et al.[53] have researched 91 pot grate sintering tests with 10 different ore mixes between the years 1985 and 1988. They observed that the solid fuel consumption in sintering increases as the Al_2O_3 content of sinter mix rises and/or the bed height is reduced. The fuel consumption rises disproportionately to the increase in the Al_2O_3 content of the sinter.

b. **Promoting Oxidation during Sintering** Coke consumption could be decreased if the sintering conditions favor the oxidation process, which is exothermic. Consequently, the oxidation condition leads to sinter with lower FeO content and possibly lower NOx and SOx levels in the exhaust.

c. **Diffusion Controlled Low Temperature Sintering** It is possible, in principle, to force the sintering process to adopt a greater amount of diffusion at relatively lower temperatures.[32,54] Japanese studies have shown that additional savings in solid fuel consumption and further improvements in reducibility without impairing the mechanical strength of the sinter can be attained by lowering the sintering temperature. This can be worked out by optimally selecting the process parameters, including the control of the sinter chemistry.[26,28] This approach may yield a product (sinter) having more favorable mineralogical and morphological characteristics.

   Section **10.5.3** mentions that the parameters—namely, basicity of sinter and its MgO content and coke rate—are interinfluenced with respect to the sintering indices—namely, sinter productivity and TI[32] (see **Table 10.1**). This can be exploited to reduce the specific consumption of solid fuel.

d. **Disproportionate Coke across the Sintering Bed** In the normal sintering process, the coke (in the sinter blend) is distributed uniformly. The result is that while the top portion of the sinter bed remains weak due to the chilling effect from incoming air from top, the bottom portion (of the bed) could be overheated due to the downward transfer of heat from the sintering zones. More return fines are generated from the upper bed region compared to lower bed region. In view of this, an appropriate segregation of coke is desirable to enhance the sintering temperature of the upper part. That is, a higher amount of coke is required in the top layer of the bed (to compensate for the chilling effect), while the bottom portion requires a lesser amount of coke (to avoid overheating).

The charging of the mix in two layers—the top portion with higher content of coke and the bottom portion with lesser coke—improves the tumbler strength of the sinter and its yield at the same coke and/or decreases the coke consumption while maintaining the same sinter quality (see **Sec. 5.6.5** and **Figure 5.10**).

e. **Improved Granulation** Improved granulation of the sinter mix with divided coke addition during the mixing process[47] helps to save coke in the sintering process. The amount of coke saved was in the range of 0.2 wt.% of the sinter machine feed using modified mixing.[55]

The coke rate can also be reduced while following the scheme with modification of the scheme mentioned previously in **(d)**. The sintering with charging the sinter mix in two layers of the sintering bed is as follows:

Upper sintering-mix layer: The mix having improved granulation process with rolling of fuel on the granules
Lower sintering- mix layer: Normal sinter mix

The scheme promotes the effective combustion of fuel, which has a positive effect on sinter quality as shown in **Table 10.4** (upper and lower figures denote the indices of the upper and lower layers). The sintering-machine productivity improves by 1.5–2%.

10.7.6 Use of Alternate Fuels

The usage of alternative fuels in sintering in the USSR as early as the seventies has been reported.[56,57] Raw petroleum coke, calcined petroleum coke, and anthracite coal are some of the alternatives tried by various sinter plants.[58,59] The usage of alternate fuels in the sintering process becomes attractive on the basis of the following:

a. When the amount of coke breeze produced internally (by the coke oven) in the steel plant is less and cannot meet the requirement of the sinter plants.
b. The alternate fuels, mentioned previously, have a lesser amount of ash content. Hence, the acid insoluble gangue in the final sinter becomes less, and consequently, the Fe content increases in the sinter.

TABLE 10.4
Indices on Account of Disproportionate Addition of Coke in the Sintering Bed (after Ref [55])

Parameter	Base Period	Experimental Period
Residual carbon content, %	0.19/0.16	0.11/0.09
FeO %	15.68/15.81	18.53/18.72
Mechanical strength % (+5 mm)	80.4/87.6	89.3/93.7
Mechanical strength % (+13 mm)	38.7/71.3	65.6/84.7

TABLE 10.5
Sintering Indices at 4.6% Carbon in the Mix (Moisture: 8%) Using LECO Fines/Coke

Indices	Coke Breeze	LECO Fines	% Improvement
Sinter Fe, %	45.60	46.20	1.30
CaO/SiO_2	2.09	2.33	11.50

However, care has to be adopted in selecting the (alternate) fuel and optimizing the sintering conditions. Different solid fuels have different reactivity, and the bed temperature profile changes accordingly. The coke reactivity increases with the increasing total amount of catalytic minerals in crystalline phases, such as metallic iron and iron sulfides.[60]

One typical study is with respect to the usage of lignite char in sintering as a substitute of coke. The char fines generated, called LECO fines, in the Neyveli plant of India has less ash content, around 7–8%, as compared to conventional coke breeze, which has 25–28%. This fact makes the replacement of coke breeze by lignite char attractive. **Table 10.5** shows difference in the chemical composition of sinter using coke breeze and LECO fines at their respective dosage equivalent to 4.6 wt. % carbon in sinter mix. The improvement in sinter Fe was 1.30 wt. % compared to that when coke breeze was used. However, LECO fines are more reactive than coke breeze, which may result in broadened temperature profile during sintering when they are used.[51]

Flue dust has about 30 wt. % carbon content (see **Sec. 5.5.3**). Its addition though, could substitute a part of the carbon requirement, thereby decreasing the vertical speed of sintering and, hence, the strand productivity.

10.8 NEWER COATING AND GRANULATION TECHNIQUES

The productivity of sinter and its quality are dependent on the combustion efficiency of coke added to the sinter mix. Besides, the manner of coke combustion during the sintering process is also important. The coating of fine coke on the surface of quasiparticles (of the sinter mix) and segregation of coke across the vertical depth in the sintering bed reduces the coke consumption considerably[61] (see **Figure 5.10**). Besides, the improved granulation adds to maximizing the sinter productivity through the enhancement of bed permeability.

10.8.1 Coating of Flux and Coke

In contrast to the conventional granulation, the granulation of the sinter mix when the fluxes and coke are coated on the surface of the quasiparticles results in improvement of the textural characteristics of the sinter product. The minerals, calcium ferrites, having high strength, form the outer part of the sinter. The secondary hematite, which is formed subsequent to the oxidation of primary hematite, appears when the

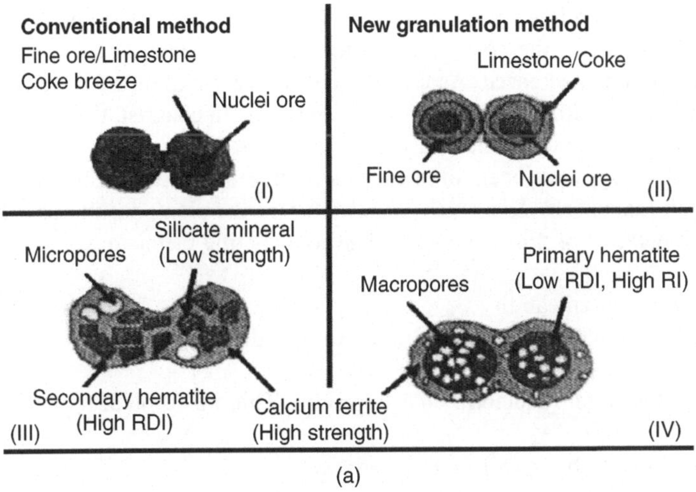

FIGURE 10.10 Schematic of the quasiparticles (I) and (II), also the sinter structure (III) and (IV), with conventional and newer methods of granulation, respectively (after Ref. [62])

conventional granulation is followed (**Figure 10.10**). The mineral, secondary hematite, does possess a lower resistance characteristic against breakdown during low temperature reduction (higher RDI). The TI of sinter increases, while its RDI decreases, when more MgO is pushed on its surface (see **Sec. 9.6.3**).

The control of the melting reactions in the sintering process could be achieved through improved granulation with coating of limestone and coke on the surface of quasiparticles. While the sinter mix containing iron ore, return fines, metallurgical wastes, and other ferruginous materials is charged from the receiving end of the drum mixer, as per conventional practice, limestone and coke are added through the discharge end of the drum. **Figure 10.11** shows the schematic of such an arrangement that is practised in Kurashiki No. 2 plant.

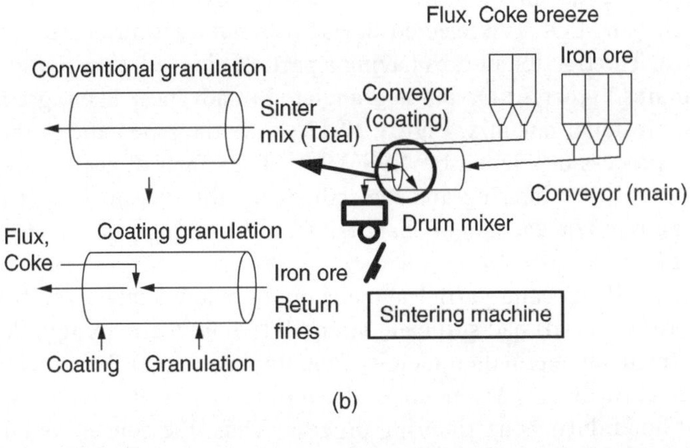

FIGURE 10.11 Schematic of granulation: quasiparticles coated with flux and coke particles (after Ref. [63])

The fine coke particles remain embedded into the granules in the normal granulation process when the coke is added to the sinter blends in bulk. Hence, their combustion rate is decreased. As a consequence, the width of the coke combustion zone broadens, resulting in lower sinter productivity. In contrast, the late addition of coke improves the sinter productivity (see **Sec. 7.7.3**).

The technology has been implemented in the JFE Kurashiki (No. 2) sinter machine, in Japan.[62–64] The sinter productivity (in the plant) has increased through the implementation of this technology. The contributing factors are:

a. Enhanced permeability
b. Increased sinter yield by forming a strong calcium-ferrite matrix

Besides, the sinter reducibility has improved. This was/is on account of:

a. The sinter matrix retains a fine porosity
b. The structure of sinter consists of a greater amount of primary hematite relict particles, which are highly reducible.

The TI of sinter and its RDI depend not only on the MgO content of sinter but also on the manner how MgO is distributed in the sinter matrix. The coating of MgO-bearing fluxes, dolomite, and/or magnesium silicate (pyroxenite/dunite) on the outer surface of the sinter mix during the granulation helps in the distribution of MgO-bearing fluxes to produce sinter with a varied sinter chemistry: less MgO inside (core) and more MgO outside (crust) This improves the sinter RDI and its reducibility (see **Sec. 9.6.3**).

10.8.2 Selective Granulation

The process of "selective granulation" has been developed by Nippon Steel Corporation (now Nippon Steel & Sumitomoto Metal Corporation, **NSSMC**), in Japan, to improve the sintering indices in case the iron ore fines contain a higher amount of clay and LOI. The selected clayish (limonitic/pisolitic) ores are screened at 3 mm size. The fraction of ores having a particle size less than 3 mm, thus separated, contains higher Al_2O_3 and is granulated in advance. Then, granulates are added to the residual raw mix. **Figure 10.12** shows the schematic of the selective granulation process developed by **NSSMC**.[19] The effect of such granulation on the efficiency of the sintering and the indices (quality parameters of sinter and sinter productivity) on account of changes in bond-forming part are also shown in **Figure 10.12**.

The green balls (pseudo-particles) made out of below 3 mm-sized fractions (of higher clayish material) had sufficient strength to hold their strength during handling and effectively retain their identity (remain segregated). This enables the rest of the material (with a lower amount of deleterious chemical components) to form a diffusion bond during the sintering process. Thus, the concentration of Al_2O_3 in the bond-forming material decreased, while the calcium ferrites ($CaO.Fe_2O_3$) increased. This led to an increase in the melt formation of the sinter products. The

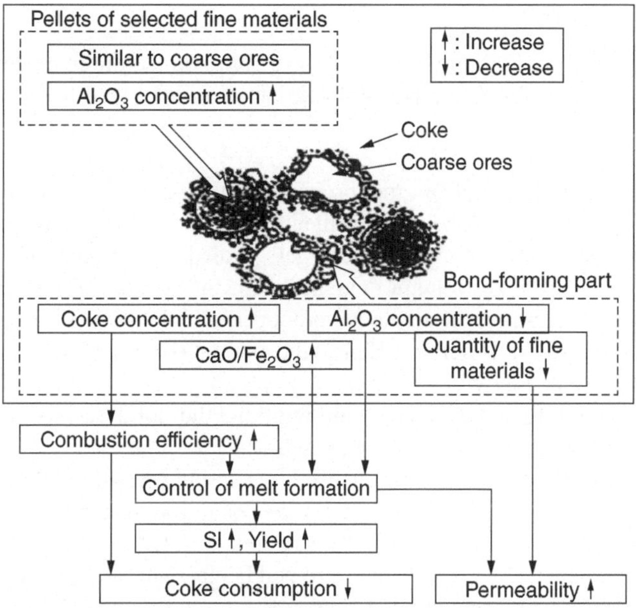

FIGURE 10.12 Concept of the selective granulation method with respect to limonitic ore developed by NSSMC, Japan; the effect on the sintering effectiveness (resulting in increased productivity and improved reducibility) is also shown (after Ref. [19])

fact that ferrites have higher strength and reducibility adds to the improvement of these indices in produced sinter. In addition to that, the sinter (produced with modified granulation) has more open pores, which improves its reducibility. The combustion efficiency also gets improved (resulting in increased sinter productivity) as the concentration of coke in bond forming part increases (see **Figures 7.3, 7.4, Sec. 7.4.3, Figures 10.3, 10.10** and **Sec. 10.3.3**).

10.9 OPERATING PRACTICE

The following operating practices of the sinter machine and sinter-making process have a significant influence on the sintering indices–namely, cold strength, reducibility, and RDI.

10.9.1 Moisture Content

The moisture content of the sinter mix does influence the reduction properties of sinter and its strength.[25,65] Also, there is optimum granulating moisture value of mixed iron ore fines in relation to RDI.[66] The optimization of the moisture content of the sinter blend is essential for stability of the sinter bed.[67] However, the content of optimum moisture for granulation depends on the size and nature of ores and is material specific (see **Secs. 6.5.7** and **7.5.1**).

10.9.2 Bed Height

A bed height at/above 700 mm happens to be a common feature in modern sintering plants. The rationale for the sintering operation at an increased bed height has the following benefits:

a. The sinter emanating from the top layers of the strand remains weak. The assimilation of the mineral constituents is not complete in this portion as the sinter mass (of the top portion) gets cooled by the incoming air, and some calcined fluxes remain unassimilated, In the case of a deeper sinter bed, this will represent a smaller proportion of the total sinter produced compared to a shallower one. Hence, the amount of return fines that is recirculated will decrease.
b. The gaseous fuel requirement for ignition, the solid fuel consumption, decreases with increasing the bed height.
c. Sinter strength is markedly influenced by suction pressure and bed height. An increase in bed height increases the yield and strength of the sinter[l] at the same coke rate; conversely, a decrease in the consumption of fuel could be achieved with an increase in bed height. The sinter mean size also improves by increasing the bed height.[68]

   Maintenance of high bed height operation by ensuring stable permeability in the sinter bed results in the improvement of sinter yield.[69]
d. The positive effect of increasing bed height in imparting strength to sinter is more than increasing the solid fuel in the mix. At higher fuel consumption, the strength of sinter is contributed by its FeO component; whereas favored mineralogical constituents in sinter are predominant at a higher bed height having higher reducibility also. Cappel et al.[52b] have observed that FeO in sinter is directly correlated with the consumption of coke in the sintering process. Since the FeO content in sinter has an adverse effect on sinter reducibility, the rate of reduction also decreases with increasing coke consumption.

   While considering the rationale for going to a higher bed height, one needs to consider the fact that a higher bed height demands a high-capacity fan (with higher suction); consequently, a higher infiltration of air (leakage) across the sintering machine is expected unless the strand is properly sealed.

10.9.3 Suction Pressure

An empirical relationship between the suction pressure under the grate and sintering time has been established. The logarithm value of sintering time decreases with the increase in suction pressure. A linear correlation has been established between the suction pressure at the wind main (X1) and the input of raw mix to the strand (I):[70]

$$I = 1.42 + 0.00146 \times X1 (\text{corr. coeff.} = 54\%)$$

The range of I is between 2.03 and 2.52 (T/m^2-h), and of X1 (Pa) is between 5,100 and 7,000.

However, increasing the suction pressure has a restriction. An increase in suction pressure gives more option of leakage through the pallet sides in an operating sintering plant.

10.9.4 Frequency of Machine Stoppages

Many times, a sinter plant has to stop the sintering machine for some time due to on-the-spot rectification of disturbances contributed by a variation in process or operation. This is besides the planned shutdown for maintenance. The sintering materials on the bed become weak and fragile on account of uneven heat transfer and thermal treatment. The weakening of sinter continues for some time; also, the amount of return fines that is generated and recirculated subsequently becomes higher and is partly fused.

In view of this, the quality of the sinter is badly affected (due to the frequent stoppages of the machine). An analysis of sintering plant operational data show that the sinter yield (Y) was directly correlated with the efficiency of utilization of sinter machines (X) according to the following empirical equation:[71]

$$Y = 30.4 + 0.096\ X, \text{corr. coeff.} = 60.7\%.$$

Range of variation of X (in percent) was 71.4–89.5.

10.9.5 Stockpiling of Sinter

Sinter when stockpiled, showed exposed pieces of calcined limestone and deteriorates in its sizes when handled. The exposed calcined limestone (CaO) reacts with humidity in the air, forming calcium hydroxide. In general, the sinter reclaimed from stock is reported to have a higher RDI. This is caused by the mechanical stress imparted to the sinter by the extra handling of stock sinter. Attempts have been made to control the limestone size distribution in order to have more stability of sinter during handling and storage.[54]

The strength parameter of the sinter deteriorates with time on storage of the sinter.[27]

REFERENCES

1. Yang Y. H. and N. Standis, *Fundamental mechanism of pore formation in iron ore sinters and pellets, ISIJ Inter.*, 31, (1991), 468-477.
2. Kasai, E., W. J. Rankin, and J. F. Gannos, The effect of raw materials properties on bed permeability during sintering, *ISIJ Inter.*, 1, (1989), 33–42.
3. Kawaguchi, T., K. Kuriyama, S. Sato, K. Takata, and T. Miyake, in: *Proc. 6th International Iron and Steel Cong.*, Nagoe, Japan, Iron and Steel Institute of Japan, (1990), 40–47.
4. Yang L. X. and L. Davis, Assimilation and mineral formation during sintering for blends containing magnetite concentrate and hematite/pisolite sintering fines, *ISIJ Inter.*, 39, (1999), 239–245.

5. Nandy, B., V. M. Korane, M. K. Chowdhury, T. Chakraborty, and A. Kumar, Effect of ore quality on sinter strength, *Tata Search*, (2011), 34–43.
6. Yadav, U. S., V. M. Korane, and B. K. Das, Role of different iron ore fines on sintering, *Tata Search*, (2005), 101–109.
7. Upadhyay, R. K., S. K. Sit, T. K. Chakraborty, M. V. S. Rao, and S. Verma, Role of geology in improving quality of iron ores at Joda: Six Sigma Approach, *Tata Search*, (2008), 101–109.
8. Bhagat, R. P., S. K. Pradhan, M. C. Goswami, B. Nandi, and T. Venugopalan, Effect of ores granulometry and related parameters on the sintering indices, *Tata Search*, 1, (2013), 89–94.
9. Windeborn, H. and F. Cappel, Limits of the physical aspects of the sinter process, in: *Agglomeration*, Ed. W. A. Knepper, Intersciences, New York, (1962), 1041–1065.
10. Bhagat, R. P., U. S. Chattoraj, M. C. Goswami, D. P. Singh, and S. K. Sil, Effect of size parameters of mix ingredients on the porosity and reduction characteristics of sinter, *Steel Research Int.*, 78(6), (2007), 451–454.
11. Bhagat, R. P., U. S. Chattoraj, and S. K. Sil, Porosity of sinter and its relation with sintering indices, *ISIJ Inter.*, 46(11), (2006), 1728–1730.
12. Sato, S., M. Ichidate, K. Kato, and T. Kawaguchi, *Tetsu-to-Hagame*, 69 (1983), S744.
13. Jeulin, D., J. L. Latailleur, and A. M. Schneider, A new technique for the quantification of iron ore sinter morphology, in: *Agglomeration–77 (Proc. Inter. Symp. on Agglomeration)*, Atlanta, GA, March 6–10, 1977, Ed. K. V. S. Sastry, Inst. of Mining, Metallurgical & Petroleum Engineers, New York, (1977), Chap. 29, 526–537.
14. Boranbaev, B. M., T. M. Srinivasan, S. S. Dataa, N. K. Kakkar, R. P. Bhagat, S. Dharnipalan, B. R. Choudhary, D. S. Thakur, V. N. Tripathy, and N. Dass, Sinterability of various ore fines used in Bhilai Steel Plant, *Steel India*, 5 (1982), 115–123.
15. Kajuyoshi, Y., U. Hiramidsu, K. Takuya, M. Shinichi, O. Hiroshi, and A. Shigeru, Influence of Al_2O_3 on reduction-meltdown behavior of sinter in blast furnace, *ISIJ Inter.*, 34, (1994), 664.
16. Kumar, C. U., R. V. Ramana, S. Ali, A. K. Das, A. Kumar, A. K. De, and T. Mukherjee, *Tata Search*, (1995), 20.
17. Yamaoka, Y., S. Nagaoka, Y. Yamada, and R. Ando, *Trans. ISIJ*, 14, (1974), 185.
18. Lu, L., R. J. Holmes, and J. R. Manuel, Effect of alumina on sintering performance of hematite iron ores, *ISIJ Inter.*, 47, (2007), 349–368.
19. Haga, T., A. Ohishio, T. Nzono, and K. Uekawa, Control techniques of the melting reaction in sintering process by the fine part granulation of clayish iron ores, *Tetsu-to-Hagane*, 83(2), (1997), 103–108.
20. Haga, T., A. Ohishio, Y. Hida, H. Fukuda, and N. Ogata, Improvement of the melting reaction in sintering process by the fine part selective granulation of clayish iron ores, *Tetsu-to-Hagane*, 83(4), (1997), 233–238.
21. Ostwald, J. and J. Ostwald, Mineralogy and micro-texture of Australian iron ore sinters, *BHP Technical Bulletin*, 25 (1), (1981), 13–20.
22. Scarlet, N.V.Y., M. I. Pownceby, I. C. Madsen, and A. N. Christensen, Reaction sequences in the formation of silico-ferrites of calcium and aluminum in iron ore sinter, *Metallurgical and Materials Trans.*, 35B, (2004), 929–936.
23. Ahsan, S. N., T. Mukherjee, and J. A. Whiteman, *Structure of fluxed sinter, Ironmaking & Steelmaking*, 10, (1983), 54–64.
24. Deghe, M. S. and P. M. Mehta, *Trans. IIM*, 28, (1975), 154–163.
25. Koranne, V. M., A. S. Reddy, T. Chakraborty, and A. Kumar, Effect of basicity on pore characteristics of sinter, *Tata Search*, (2009), 145–148.
26. Hsieh L. H. and J. A. Whiteman, Effect of raw material composition on the mineral phases in lime-fluxed iron ore sinter, *ISIJ Inter.*, 33, (1993), 462–475.

27. Bhagat, R. P., S. K. Sil, U. S. Chattoraj, M. C. Goswami, D. P. Singh, and B. K. Das, Pot sintering studies for improvement in reducibility and RDI values of sinter, *Tata Search*, (2006), 127–134.
28. Zhou, Q., Y. Ren, and X. Yang, *Acta Metall.* Sin. (China), 18(6), (1982), 635–644.
29. Pangrahy, S. C., M. Rigaud, and P. Dilewijns, Effects of replacement of limestone by dolomite and other MgO bearing materials on production rate of iron ore sinter, *Ironmaking & Steelmaking*, 11 (5), (1984), 246–252.
30. Panigrahy, S. C., P. Verstraeten, and J. Dilewinjns, Influence of MgO addition on mineralogy of iron ore sinter, *Metallurgical Trans.*, 15B, (1984), 23–32.
31. Yadav, U.S., B. D. Pandey, B. K. Das, and D. N. Jena, Influence of magnesia on sintering characteristics of iron ore, *Ironmaking & Steelmaking*, 29, (2002), 91–95.
32. Bhagat, R. P., Reduction of solid fuel in sintering of Indian iron ore, *Science of Sintering*, Belgrade, Serbia, 47, (2015), 205–213.
33. Mukherjee, T., A. Chatterjee, L. M. Chatterjee, S. S. Gupta, and A. De, Tata Steel's experiences with dunite sinter, *Tata Search*, (1996), 16–23.
34. Hosotani, N. Konno, K. Yamaguchi, T. Orimoto, and T. Inazumi, Reduction properties of sinter with fine dispersed pores at high temperatures of 1273 K and above, *ISIJ Inter.*, 36, (1996), 1439–1447.
35. Sawada, I. K., K. Yashiko, T. Takashashi, *Tetsu-to-Hagane*, 71, (1985), 1880.
36. Bhagat, R. P., S. K. Gupta, and H. S. Ray, Heat transfer consideration for improvement in sintering indices, in: Proc. 48th Iron Making Conference, *ISS-AMIE*, Chicago, (1989), 481–490.
37. Bhagat, R. P., Studies on calcination of limestone and its assimilation by hematite during the sintering using thermo-gravimetric and X-ray diffraction analysis, *Scandinavian Journal of Metallurgy*, 21, (1992), 246–250.
38. Nehleton, M. A., *Combustion & Flame*, 3, (Sept. 1965), 311–316.
39. Loo, C. E., R. P. Williams, L. T. Matthews, Influence of material properties on high temperature zone reactions in sintering of iron ore, *Mineral Processing & Extractive Metallurgy (Trans.IMM)*, Section C, 101 (1992), C7–C15.
40. Toda H. and K. Kato, *Trans. ISIJ*, 24, (1984), 178–186.
41. Teo, C. S., R. A. Mikka, and C. E. Loo, Positioning of coke particles in iron ore sintering, *ISIJ Inter.*, 32, (1992), 1047–1057.
42. Bhagat, R. P., *Studies to increase reducibility and decrease RDI of sinter*, Project completion report (CLP 0029), CSIR-NML Jamshedpur, April 2004.
43. Karabasov, Y. S., A. N. Pokhvishev, E. F. Shkurko, and V. S. Valvin, *Steel in the USSR*, 5, (1975), 583–584.
44. Kornilova, N. K., E. F. Vegman, and S. E. Lazutkin, *Steel in the USSR*, 3, (1973), 1–2.
45. Matsuo, N., T. Maeda, and Y. Ono, *CAMP-ISIJ*, 4, (1991), 1081.
46. Peters, K.-H., K. Kersting, and H. Beer, Ways and means of reducing the energy consumption in iron ore sintering, Proc. 4th *Inter.* Symp. On Agglomeration, Toronto, Canada, 1985, C. E. Capes, Ed., Ontario, Iron & Steel Society of AMIE CIM, 1985, 157–167.
47. Hou, P., S. Choi, E. Choi, and H. Kang, Improved distribution of fuel particles in iron ore sintering process, *Ironmaking & Steelmaking*, 38, (2011), 379–385.
48. Ishikawa, Y., K. Sugawara, and Y. Umezu, in: Proc. 2nd *Inter. Symp. Agglomeration*, Vol. 2, Ed. K. V. S. Sastry, Am. Inst. Mining, Metallurgical & Petroleum Engrs., New York, (1977), 503.
49. Wegmann, E. F., *A Reference Book for Blast Furnace Operators*, Mir Publ., Moscow, (1984), 202.
50. Bhagat, R. P., H. S. Ray, and S. K. Gupta, Improvement in performance of the blast furnace at Bhilai through statistical analysis and improvement in burden preparation, *ISIJ Inter.*, 31, (1991), 669–676.

51. Bhagat R. P., Replacement of coke breeze by leco-fines in iron ore sintering, *Trans. IIM*, 52, (1999), 299–304.
52. Gebenkin, G. A., *Soviet Expert*, RDCIS-Bhilai Sub Centre, 1982.
53. Cappel, F., E. Pfaff, and K. Heinrich, Reduction of coke consumption during sintering of iron ores, in: *Preprint*, Iron Making Conf., Ranchi, January 10–11, 1989, org. by BF and SP Operating Committee, R & D Center, SAIL, Ranchi.
54. Bhagat, R. P., S. K. Pradhan, M. C. Goswami, B. Nandi, and T. Venugopalan, Effect of LOI content of iron ores and other process variables on the reduction properties of sinter, in: *Proc. XXVI International Mineral Processing Congress*, New Delhi, (2012).
55. Colquhoun, L. W., Production of sinter for large blast furnaces, *BHP Tech. Bull.*, 25(1), (1981), 68–73.
56. Ten, S. B., A. G. Mikhalevich, B. M. Boranbaev, B. I. Kolokol'tsev, and V. V. Dyabin, Making fuel utilization in sintering more effective, *Stal*, 10, (1980), 868–870.
57. Globin, A. G., The use of anthracite fines for sintering iron ores, *Metallurgia*, 2, (1971), 14.
58. Kowalczyk, T. F. and P. N. Peronis, *Petroleum coke as the main fuel in the sintering process*, Ironmaking Conference Proceedings, 47, (1988) 617.
59. Chakraborty, D. S., B. K. Das, and A. Kumar, Use of alternate solid fuels in iron ore sintering, *Tata Search*,(2003), 77.
60. Grigore, M., R. Sakurovs, D. French, and V. Sahajwalla, Influence of mineral matter on coke reactivity with carbon di-oxide, *ISIJ Inter.*, 46, (2006), 503–512.
61. Lu, L. and O. Ishiyama, Recent advances in iron ore sintering, Mineral Processing and Extractive Metallurgy *(Trans. IMM, C)*, 125(3), (2016), 132–139.
62. Oyama, N., K. Taketa, and N. Fuji, Development of new coating granulation technology of limestone and coke breeze, *JFE Technical Report*, 22, (2008), 32–37.
63. Oyama, N., H. Sato, K. Taketa, T. Ariyama, S. Matsumoto, T. Jinno, and N. Fuji, Development of coating granulation process at commercial sintering plant for improving productivity and reducibility, *ISIJ Inter.*, 45, (2005), 817–826.
64. Oyama, N., Y. Iwami, et al., Development of secondary fuel injection technology for energy reduction in the iron ore sintering process, *ISIJ Inter.*, 51, (2011), 913–921.
65. Xuewei, L. V., C. Bai, G. Qiu, S. Zhang, and M. Hu, Moisture capacity: definition, measurement, and application in determining the optimum water content in granulating, *ISIJ Inter.*, 50(5), (2010), 695–701.
66. Yanaka, H. H. and Y. Noda, *Yamaoke Nippon Kokan Tech.* Report, 115, (Nov. 12–17, 1986).
67. Inazumi, T., M. Fujimoto, S. Sato, and K. Sato, Effect of sinter-cake load reduction by magnetic force on iron ore sintering, *ISIJ Inter.*, 35(4), (1995), 372–379.
68. Magedanz, N., Untersuchung Der Finflubgroben Auf Die. Kornungsverteilung Von Eisenerz Sinter/*Investigation of the Factors Influencing the Grain Size Distribution of Iron Ore Sinter, doctoral thesis*, TU Berlin, Germany, (1983).
69. Obata, H., H. Takahashi, M. Nakamura, T. Natsumi, and K. Komamutra, High productivity operation at Chiba No. 4 sinter plant, *ISIJ Inter.*, 31(5), (1991), 478–486.
70. Boucraut, M. and Michard, in: *Blast Furnace, Coke Oven and Raw Material Proceedings*, AIME Detroit Meetings, (1962), 345–369.
71. Bhagat, R. P., H. S. Ray and S. K. Gupta, Effect of some process parameters on the performance of a sintering plant at Bhilai Steel Plant—a statistical analysis, *Trans. IIM*, 41, (1988), 465–474.

11 Pelletization Process

ABBREVIATIONS

CCS *Cold crushing strength*
CPR *Composite pre-reduced*
DR *Direct reduction*

11.1 PELLETIZATION PROCESS

Figure 11.1 shows the schematic of the pelletization process. The process involves four stages:

a. Raw material preparation
b. Formation of green balls or pellets
c. Induration of the green pellets
   - Drying of green pellets
   - Preheating
   - Firing
d. Cooling, storage, and transport of pellets

11.2 RAW MATERIALS AND THEIR PREPARATION

The characteristics of raw materials, in terms of their chemical constituents and size analysis, are important in affecting the pelletizing process and the quality of pellets.

Raw materials used in producing pellets can be grouped into:

a. Ferrous: Iron ore concentrate in its finer form (size: below 100/76 micron), after elaborate beneficiation process (Fe in product should be around/above 64 wt.%), is used in the pelletization process.
b. Binders: This type of material facilitates the green pelletization and ensures sufficient strength during the heat hardening of pellets and thereafter. Bentonite—a type of clay mineral—is commonly used. Organic binders can be/are also used to substitute bentonite partially.
c. Additives: Additives are limestone and/or dolomite, and coal/coke breeze, all ground to – below 76 micron. These are incorporated in the pellet mix in limited quantities in order to:
   - Modify the metallurgical properties of pellets. Fluxes (limestone, dolomite, and magnesium silicate) are used to incorporate lime in pellets so as to control the basicity as needed in the downstream iron-making processes.
   - Partially fulfill the heat requirement in induration. For example, coal/coke breeze is used as a fuel supplement.

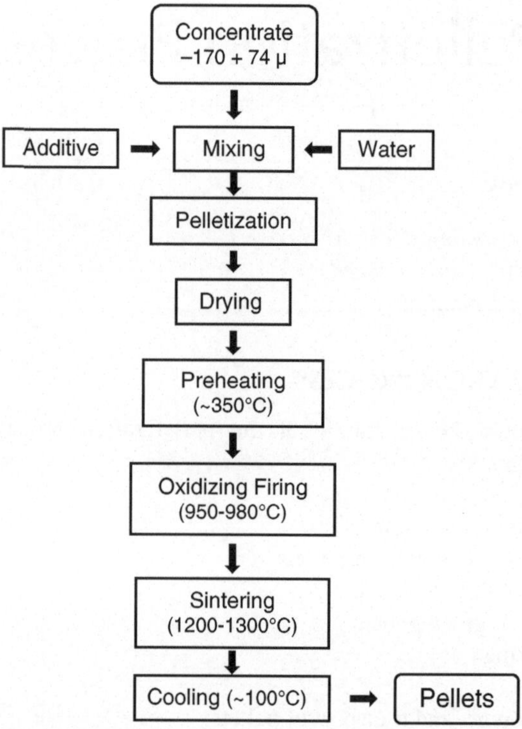

FIGURE 11.1 Schematic of the pelletization process

11.2.1 Iron Ore

Pellet feed is the iron ore concentrate with a minimum of 64 wt.% Fe and (SiO_2 + Al_2O_3) 3.5 wt,% maximum (see **Sec. 2.1**). In case the ore is of a high grade (Fe more than 65 wt.%), it needs dry processing: communition and sizing at the mine end before the ground ore is dispatched to the pellet plant (iron-making unit). Low-grade iron ore and slimes are upgraded through beneficiation in order to enrich Fe content and reduce gangues containing silica and alumina in the product. The topic is detailed in **Chapter 2**. As these are treated in a wet circuit, removal of water from the product is essential.

The grain size of concentrate and its size distribution are important in the pelletization process as these parameters are well correlated with the drying rate during induration and the abrasion index of indurated pellets. Normally, the ore concentrate having a particle size of below 150 micron (100 percent passing with blain index more than 1800) is used.

11.2.2 Fluxes

Fluxes are used to incorporate lime and/or MgO in pellets in order to tailor their chemistry and to improve their metallurgical performance. Limestone and/or dolomite are used as fluxes. Major impurities in limestone (and dolomite) are the contents of silica and alumina. Normally, these materials do not require beneficiation. These are excavated from mines, dry crushed, and ground to below 150 micron size for its use in pellet making.

Limestone, in its original form, does not have any effect in improving the green ball formation. It is liable to the same conditions for green ball formation as the ore itself. It should have adequate fineness to ensure that after the induration of green pellets carbonate present is fully dissociated, and CaO should completely react with gangue and hematite. Free CaO should not be present as it reacts with moisture (atmospheric) to form hydrate, which causes weakening of pellet structure.

As an alternative to dolomite, dunite and/or pyroxenite are also used as an Mg source to improve the metallurgical performance of fired pellets. Dunite is comprised of pure olivine ($2MgO. SiO_2$). This rock is preferred because of its high MgO content and favorable $MgO:SiO_2$ ratio. Theoretical values of MgO and silica are 57 wt.% and 43 wt.%, respectively. Pyroxenite comprises an enstatite variety of pyroxene ($MgO. SiO_2$) and is the second-best alternative as an MgO source. Theoretical values of MgO and silica are 40 wt.% and 60 wt.%, respectively.

11.2.3 Solid Fuel

Solid fuel in a limited quantity is added into the pellet mix in order to partially substitute the gaseous (or liquid) fuel requirement during the induration process of green pellet. Coal/coke breeze is generally a less-expensive source of heat. These solid fuels reduce the hematite to magnetite during the induration process; at the same time, they accelerate the pelletizing process. Consequently, these fuels bring the inherent fuel value of a hematite green ball equal to that of a magnetite pelletizing feed. Besides, the fuels on combustion provide heat and lower the consumption of the more expensive burner fuel.

11.2.4 Binders

Binders are incorporated in pelletization to provide sufficient strength of the green balls following pelletization, also to impart adequate strength in the pellets after induration. These include mainly bentonite, lime, and organic binders.

Bentonite is a good binder and widely used in the pelletization process. The green balls, with bentonite as a binder, have superior properties in terms of drop strength and compressive strength compared to other binders. Additionally, the indurated pellets (with bentonite) have better mechanical and metallurgical properties.

Hydrate lime [$Ca(OH)_2$)] acts as both a binder and a basic additive in the pelletization process. It is used with ores that already have some porosity for self-agglomeration. Hydrate lime contributes almost no silica to the final product. It helps to improve the Fe grade in pellets. It acts as flux when the pellet is reduced in the blast furnace and improves the metallurgical performances of the pellet. The use of hydrate lime, as binder, is attractive when fluxed pellets are being produced.

Organic polymers are also used as a binder to partially substitute bentonite. Their advantage lies in the fact that they do not add to impurities in the pellet mix. However, these binders alone cannot provide adequate strength to the pellets after firing. Besides, these are costlier.

The specifications of raw materials in the pelletization are given in Section **2.1**.

11.3 PELLETIZATION STEPS

11.3.1 Mixing

Ground iron ore concentrate and additives are conveyed into individual storage bins. The raw materials drawn from the respective bins at the desired rate are fed to the mixer for homogenization. The binder, bentonite, constitutes only 0.6–0.9 wt.% of the ore concentrate but has major role in the whole pelletization process. Hence, it is important that the binder is evenly distributed in the ore mix so that no portion of the concentrate is left without a binder. In view of this, bentonite and other additives are required to be carefully proportioned to the ore concentrate and intimately mixed so that a uniform and homogeneous pelletizing mix is produced. Mixing is carried out using horizontal paddle mixers. This is a horizontal tube through which a rotating shaft is passed that is equipped with paddles. The paddles knead and blend the components of the mix. Water is introduced in the mixer to provide the required moisture content in the pellet mix.

11.3.2 Pelletization (Balling)

The formation of green pellets is referred as balling (see **Chapter 12**). On an industrial scale, the operation is carried out in a drum or disc (see **Figures 12.1a** and **12.1b**). Iron ore, mixed with the additives, is fed to balling unit. Initially, small nuclei of pellets are formed. On the addition of water, the nuclei grow bigger into balls and then into pellets as they pick up loose grain particles during their travels through the drum or disc. The green pellets, discharged from the balling discs, are conveyed to a double deck roller screen (see **Figure 12.4**). The green pellets are screened to separate the product size fraction (9 to 16 mm in diameter) for induration. The undersize fractions (below 9 mm) are returned to the balling unit as seeds. The oversize fractions are usually crushed and returned (to the balling unit). The balling units usually are connected to roller screens used for separating undersized pellets.

Disturbances in the balling operation increase:

a. The amount of materials being recycled to the pelletizing operation.
b. The variation in the production rate of the sized green pellet fractions (surging); excessive surging causes problems not only in the balling circuits but also in the induration machine.

Mechanical interlocking also plays an important role in binding. The maximum strength of the green balls will be attained by compacting material with minimum porosity and sufficient water to saturate the voids. More rolling ensures more effective collisions among the particles and improves green strength.

11.3.3 Induration

On-size green pellets are fed to the indurating machine for heat hardening. The induration process (see **Chapter 14**) consists of the following stages. Each is carried out at a different temperature range.

a. Drying of green pellets
b. Preheating of dried (green) pellets
c. Firing of preheated pellets for (heat) hardening
d. Cooling of fired pellets

The induration machine is of two types: (a) straight-grate machine or (b) grate-kiln-cooler (see **Secs. 14.4** and **14.5**, respectively). In the former process, the wet pellets are dried, preheated, indurated, and cooled on a continuously moving grate without intermediate transfers. The process air, that is introduced for pellet cooling, is circulated from the cooling zone of the grate in a multipass manner to the other process zones to obtain maximum possible thermal efficiency. Only the relatively cool and moisture-laden gases are discharged to process gas cleaners and then back into atmosphere. In the later process, while the process of drying and preheating is carried out in a straight grate machine, similar to the Dwight-Lloyd type sintering machine, the firing is done in a rotary kiln. Subsequently, the cooling of fired pellets is carried out in the rotary cooler.

11.4 POLLUTION CONTROL AND ENERGY CONSERVATION

Pollution prevention is defined as the use of processes, practices, materials, products, and/or energy that avoids or minimizes the creation of pollutants and waste and that reduces overall risk to human health or the environment.

11.4.1 Emission of Pollutants

Like the sintering process, a pelletizing plant is also a polluting unit, and the pollution is from the particulate as well as gaseous emissions (see **Sec. 5.4.1**). The emission of these pollutants from a pelletizing unit is, however, less than that from a sinter plant (see **Table 5.1**). The emission of harmful gases (SOx and NOx) is less when the pelletizing feed is magnetite concentrate compared to hematite concentrate. However, the feed characteristics, process of induration, and local conditions decide the pollutant emissions in both cases. For instance, solid fuel containing 1% S alone, if completely burned, would produce 400 mg of SO_2 Nm^3 off-gas.[1]

Emissions from the operation of a pelletizing unit arise from:

a. Material handling operation, which results in airborne dust, etc.
b. Combustion reactions: The combustion gas contains sulfur dioxide, NOx, CO, CO_2, etc. The waste gas from indurating strand may also contain sulfur dioxide, hydrochloric acid, and hydrofluoric acid.

11.4.2 Removal of Pollutants

Solid Pollutants These are controlled by passing the off-gas through bag filter and electro-static precipitator.

Oxides of Sulfur (SOx) The source of SOx is sulfur in the raw materials besides natural gas, oil, and coal. In the pelletizing process, all the sulfur is oxidized to generate SOx, most of which is emitted as exhaust. Unlike NOx, SOx cannot be reduced by the process itself. Therefore:

a. The usage of raw materials with less S in the pelletization process may be required.
b. Semidry desulfurization may be followed by dedusting.
c. Waste gas scrubbing is a possible approach. The installation of a desulfurization unit may be required. Typically, a lime-slurry method can be applied. This involves mixing lime with water into a slurry that then reacts with SO_2 to form gypsum ($CaSO_4$. $2H_2O$).

Nitrogen Oxide (NOx) The hot air used in the grate kiln system for induration is supplied from the cooler. Temperature of this combustion air is 1,000–1,100°C, while the flame temperature is 1,600°C or higher. Thus, the NOx generated is, for the most part, thermal NOx. The generation of thermal NOx is greater as the (burner) flame temperature and temperature of combustion air are increased.

The pollutant, NOx, can be reduced by:

- improving the process and adopting low NOx burners. This approach includes:
  - Decreasing the amount and/or temperature of the combustion air
  - Reducing the fuel combustion of the kiln burner by providing additional burners in the grate
- Installation of NOx removal system.
- Selective catalytic reduction of the NOx in the flue gas.

Space for Environmental Protection A comprehensive environmental protection facility, in the case of an agglomerating unit, may require space—sometimes more than the main unit itself. In this respect, a pellet plant has an advantage over a sinter plant as it can be installed at the mine site itself where spaces are available with more easiness. A sinter plant with its environmental protection facilities has to be accommodated in the integrated and steel works premises and would require much more space.

11.4.3 Energy Conservation

The energy consumed in the pelletizing plant can be conserved through:

a. Recovery of sensible heat in the induration process in a straight-grate machine (see **Sec. 14.4.4**) and grate-kiln-cooler system (see **Sec. 14.5.5**)
b. Adopting cleaner technology as follows:
   - NOx abatement by water injection into strand burners
   - Improvement to the firing equipment process gas recuperation systems

- Use of exhaust gas from other processes as combustion air and indirect water injection in the cooling system
- Cold bonded pellets and briquette technology

11.5 SPECIFICATIONS OF PELLET PLANTS

Table 11.1 lists the specifications of typical pelletizing plants constructed by Kobe Steel Ltd.[2]

TABLE 11.1
Reference List of Pelletizing Plants (after Ref. [2])

Plant	FMO Venezuela	Chador Malu Ardakan/Iran	GHC No.2 Bahrain	Vale Sohar/Oman
Capacity, MT/year	3.3	3.1	6	4.5*2 lines
Start-up, year	1994	2007	2009	2010
Feed material	hematite	magnetite-hematite	hematite	hematite
Product	DR pellets	DR pellets	DR pellets	DR pellets
Balling disc, dia., m	7.5*5 units	7.5*6 units	7.5*9 units	7.5*7 units (2 lines)
Traveling grate	4.716 m (w) 66.388 m (L)	4.716 m (w) 63.975 m (L)	5.782 m (w) 83.020 m (L)	4.716 m (w) 68.801 m (L) (*2)
Effective area, m^2	284.5	273.1	440.8	295.9
Wind box, No.	25	24	25	26
Length (m)/bay	2.413	2.413	3.050	2.413
Rotary kiln	6 m (dia.) 46 m (L)	6 m (dia.) 46 m (L)	7.2 m (dia.) 50 m (L)	6.9 m (dia.) 45 m (L) (*2)
Annular cooler	18.5 m (dia.) 2.8 m (W)	20.5 m (dia.) 3.1 m (W)	22.0 m (dia.) 3.7 m (W)	22.0 m (dia.) 3.7 m (W) (*2)
Effective area, m^2	145.5	177.5	234.0	234.0

11.6 COMPOSITE PELLETS

Several research studies carried out in the past aimed at the development of composite pellets with various organic as well as inorganic binders through a cold-bonding approach.[3,4]

Interest in iron ore-coal composite pellets has grown because of the following advantages:

- The cheaper material resources, such as iron-bearing fines, coal fines, and coke breeze, can be utilized.
- Cold-bonded composite pellets do not require firing; hence, a savings in energy can be expected by adopting this methodology.

- Cold-bonded pellets can serve as feedstock in the iron-making process at smaller scale with less capital investment, such as direct reduction. By this way, the prospect for diversification of natural resources gets enhanced.
- The pellets possess uniform size and have consistent product quality as their chemical composition does not change much.
- A fast reduction reaction can be attained in the case of composite pellets due to intimate contact between reductant and iron oxide particles.

However, the composite pellets, in their original form, have lower cold crushing strength (CCS); hence, they cannot be charged in a larger blast furnace as feedstock. Since composite pellets cannot be fired, these are cold-bonded in order to improve mechanical strength. The composite pellets should have sufficient compressive strength to withstand stresses during handling, transportation, and thermal stresses in reduction furnaces.[5]

A process, called the "Orcarb process" was developed in the mid-1950s in which the ore is coated with 10%–25% carbon to produce micropellets 0.25–1.7 mm in size that are suitable for fluidized reduction or open arc reduction.[6]

The process of pelletization as well as the quality of produced ore–coal composite pellets depends on various factors, such as feed particle size, amount of water added during pelletizing, disc rotating speed, inclination angle of the disc bottom, and residence time of materials in the disc. The effect of these parameters on the mechanical properties of green, dried, and fired iron ore pellets has been the subject of much research.[7] These have been detailed in **Chapter 13** and **Chapter 16**, respectively.

Composite pre-reduced (CPR) pellets have been developed mainly

a. To utilize plant waste
b. To prepare a cheap, metallized feed for iron making[8]

They consist of metallized sponge iron ore encased in a partially reduced and sintered iron oxide shell.

CPR pellets are produced by firing composite iron ore-coal char pellets prepared by two-stage pelletization; namely, iron ore and coal char are mixed in a desired proportion to form ore pellets that are further coated with iron ore only during second-stage pelletization. The reducing gas CO is generated on firing under atmosphere. However, the continued firing reoxidizes the reduced pellets.

REFERENCES

1. Cappel, F., A. Geock, and G. Samant, Projects, developments, environmental protection costs and prospects of pelletizing, in: *National Seminar on Role of Iron Ore Pellets in Iron & Steel Industry*, Iron & Steel Davison of Indian Institute of Metals, Ranchi, April 15–16, 1988, org. by Steel Authority of India Limited, Kudremukh Iron Ore Company Limited, 7 Mandovi Pellets.
2. Yamaguchi, S., T. Fuji, N. Yamamoto, and T. Nomura, KOBELCO pelletizing process, *Kobelco Technology Review*, 29, (Dec. 2010), 58–68.
3. Dutta, S. K. and A. Ghosh, *Trans. IIM*, 48, (1995), 1.
4. Sastry, K, V. S. Hosten, and P. Dontula, Prepr. Soc. Min. Engrs. *AIME*, 93, (1993), 275.

5. Sah, R. and S. K. Dutta, Effects of binder on the properties of iron ore-coal composite pellets, *Mineral Processing Extractive Metall. Rev.*, 31, (2010, 73–85.
6. Leshar, C. E., *Agglomeration fine sized ores—temperature coke*, Min. Eng., 7, (October 1955), 944–948.
7 Ahmed, Y. M. Z., F. M. Mogamed, N. A. El-Hussiny, and M. E. H. Shalabi, Pelletization of heterogeneous iron concentrates for blast furnace operation. *Canadian Metallurgical Quarterly*, 44(1), (2005), 95–102.
8. Mishra, S. N. and R. C. Gupta, Development of schedule to prepare composite pre-reduced pellets (CPR), *ISIJ Inter.*, 34(6), (1994), 468–476.

12 Green Pelletization: Process and Mechanism

12.1 BALLING PROCESS

The process of transformation of fines into agglomerates is referred as "balling." The balls serve as input material for induration in the pelletization process to get the final products, which are called pellets. The balling stage is the most important part of the overall pelletization process as this influences the strength, size, compactness, and other properties of the indurated pellets significantly. The balling operation is performed in rotating devices like drums (**Figure 12.1a**), discs (**Figure 12.1b**), cones, etc.; however, the former two are mostly used. In industrial practice, the balling devices are continuously operated by making a constant supply of new feed available to previously formed seed pellets. Thus, it is convenient to conceptualize that the agglomerating system consists of two classes of species: the well-formed species and the added moist feed fines.

12.1.1 PELLETIZING UNITS: DESCRIPTION AND OPERATION

Drum Pelletizer In an industrial practice, the pelletizing drum has a 9–11 m length and 3–3.6 m diameter for the production of green pellets, with a capacity of 90–120 tonne per hour (TPH) (Figure 12.1a). It is inclined to 6° to 10° from horizontal. The metallic smooth liner wall is coated with moist fines in the drum to improve adhesive strength and roughness. Rotating spiral scrapers are used to control the height of the pellet mix being pelletized. The prepared pellet mix is charged in the upper end of the drum. The drum has a water spray arrangement at several points across the drum length. Water is sprayed into the pellet mix being balled from these distributed points to achieve optimum ball formation. The pellets of a certain size are produced depending on the length, slope, rotational speed, and the extent to which the drum is filled with the solid material. The size and shape of the drum should be such as to obtain the most favorable conditions of motion. The drum has limitations in its operation as the green pellets are not sorted (according to size) inside the drum. The pellets are screened, and the oversize ones are crushed, combined with fresh feed and smaller sized pellets, and charged again (**Figure 12.2**). The new seeds are formed as the fresh feed material is taken up during the progression of charge through the drum in the first couple of meters of the drum.

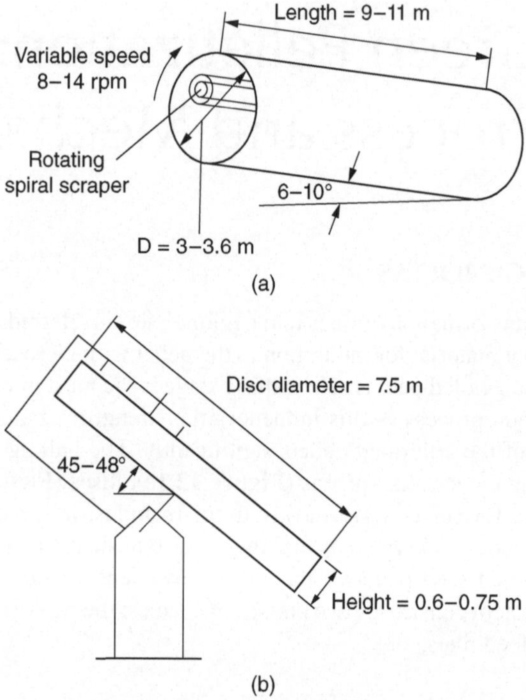

FIGURE 12.1 Pelletizing units for the production of 90–120 tonne per hour (TPH) green pellets: (a) pelletizing drum with dimensions; (b) pelletizing disc with dimensions.

Disc Pelletizer The pelletizing disc, used in the industry, is an inclined pan that varies in size from 6 to 7.5 m in diameter for capacities of green pellet production between 90 and 120 TPH (Figure 12.1b). About 70% of the disc space is actively used in balling action. The disc pelletizer has a water spraying arrangement to maintain the moisture level of the pellet mix (**Figure 12.3**). The disc rotates slowly: around 6–8 rpm. The inclination of the disc is approximately 45°. Material to be pelletized is fed to the device at a controlled rate. The pellet mix is lifted upward until the friction is overcome by gravity and material rolls down to the bottom of the disc. In case water is needed for ball formation, the same (water) can be added at points where the pellets are still nascent.

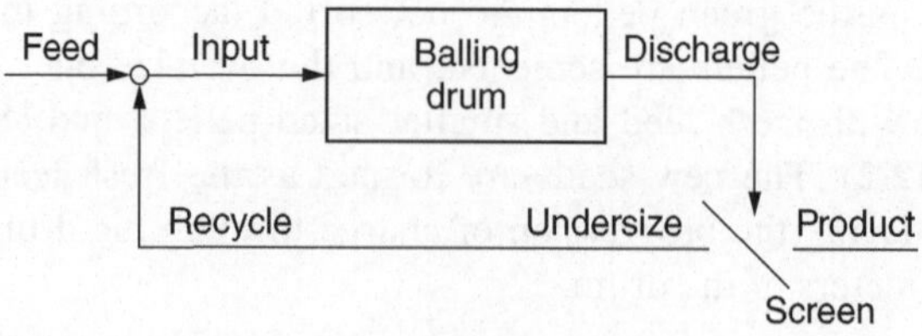

FIGURE 12.2 Pelletizing operation using balling drum in closed circuit (after Ref. [1])

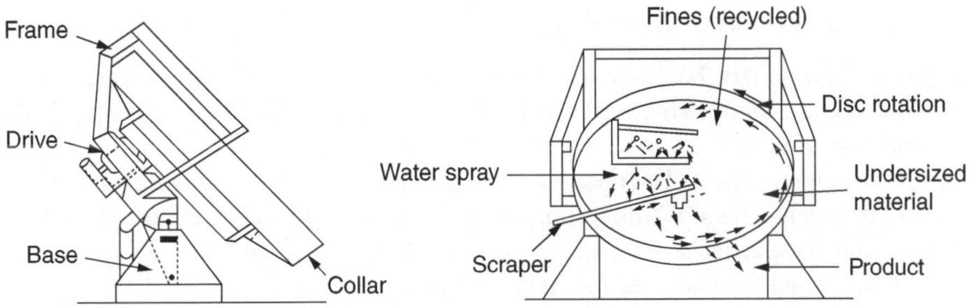

FIGURE 12.3 Framework of disc pelletizing unit and water spray arrangement

The pelletizing devices are usually connected to roller screens. Usually a double-deck roller screen is provided downstream from the pelletizing unit (ahead of the furnace) to screen out the upper size and under-sized materials (green pellets). The provision of double deck roller screens improves the screen efficiency; however, the green pellets are subjected to an additional drop from the upper to the lower deck of the roller screens. In view of this, the quality of green pellets, especially their drop number, has to support its installation. The provision of double-deck roller screens is to be seen as a compromise between plant investment costs and improved screen efficiency. Undersized pellets (less than 9 mm) are separated in the roller screen and returned to the balling unit as seeds. The pellets sized 9–16 mm in diameter go for induration through the conveyor. The oversized fraction is usually crushed and returned (to the balling unit) (**Figure 12.4**).

A narrow size distribution in the produced green pellets is an important criterion to ensure a reasonably good bed permeability during the heat hardening process for the pellet production. The pellet size distribution is regulated by:

a. Mechanical means: adjusting the screen openings for the recycling load or for the on-size fraction
b. Varying the moisture content or the binder dosage

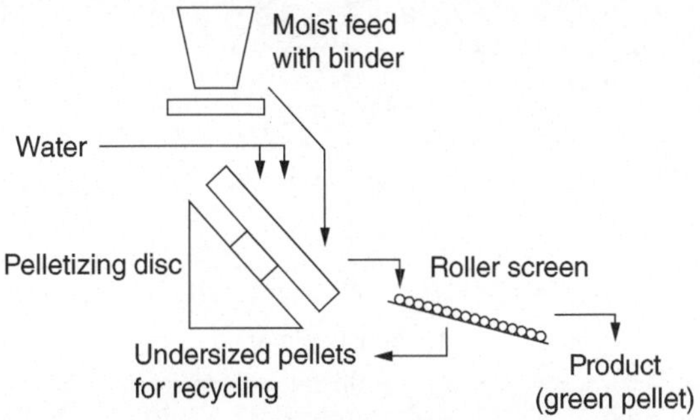

FIGURE 12.4 Pelletizing operation in closed circuit using disc (after Ref. [1])

12.1.2 Operation of Rotating Device

Figure 12.5.I shows the operation of a pelletizing disc.[2a] The raw material is fed into the disc and sprayed by water. With continuous rotation of the disc, fine particles are gathered and formed into larger pellets in the stable area, finally falling out of the disc as green pellets. The operational parameters of the balling unit do affect the quality of green pellets. While rolling down, the pellets are subjected to compacting forces, and the particles get densely packed. More rolling ensures more effective collisions among the particles and improves green strength.

Disc Speed and Angle of Inclination The movement of the pellet charge in the rotating device is considerably affected by the angle of inclination and rotational speed. Increasing the circumferential speed has limitations as it should be within three-fourths of the critical speed of rotation. At the lower speed, the charges cannot be lifted, while at the higher speed the charge is almost centrifuged.

(I)

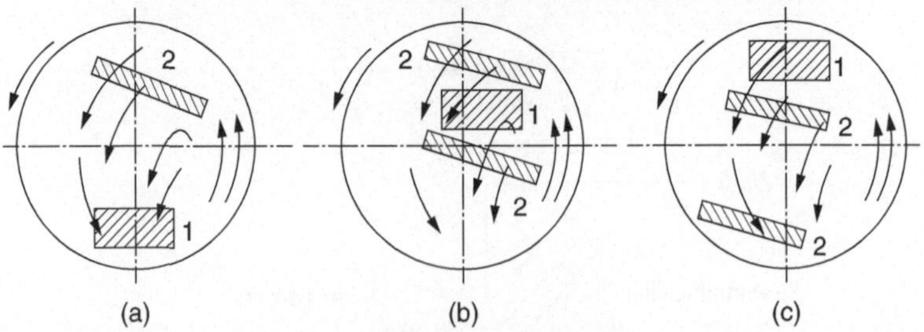

1: Feed; 2: Water

(II)

FIGURE 12.5 I: Operation of a pelletizing disc (after Ref. [2a]); II: Arrangement of material feeding and water spray on the disc pelletizer surface (1) feed and (2) water (after Ref. [2b])

The rolling of material improves when the disc speed increases. This results in higher strength of green pellets and increases the ball growth rate. An optimum of the disc speed is achieved so long as the (improvement in) rolling of material is supported by the disc speed. The excessive speed may result in breaking of pellets by impact.

The angle of the disc determines the time of residence, pellet size, and productivity of the pellets being formed. Green pellet size can be precisely adjusted by varying the disc inclination, circumferential speed, feed rate, or water addition. At a disc inclination of around 45°, the charge could be pelletized within a narrow range of speed, N = 25 to 30 rpm. When the disc inclination is increased, the balls get more compacted during rolling. This results in higher strength of green pellets. The disc loading possesses a limitation when the inclination is more than 60°. With a disc inclination of 55°, a wide range of speeds could be utilized (15 to 45 rpm) and the resulting pellets had extremely good sphericity.

For a given disc inclination and feed rate, the residence time can be adjusted to some extent by varying the height of the peripheral.

The method of feeding the pellet mix and the location of water addition across the disc surface are also important in deciding the ball growth (refer to Figure **12.3**). The basic arrangements of feeding and wetting systems are shown in **Figure 12.5.II** to produce the green pellets of (a) much larger sizes (20–30 mm), (b) normal sized pellets (10-20 mm), and (c) much smaller sized pellets (less than 10 mm).

The disturbances in balling operation increase:

a. The amount of materials being recycled to the pelletizing operation
b. The variations in the production rate of the sized green pellet fraction (surging)

Excessive surging causes problems not only in the balling circuits but also in the induration machine.

12.1.3 Material Movement in a Balling Unit

The material movement in a rotating pelletizer is highly complex and guided by the parameters related to:

a. The process (material characteristics, moisture, type and content of binder, etc.)
b. The operation (rotating device—whether drum or disc, speed)

Pelletizing Drum The movement of material in a pelletizing drum according to speed is given in **Figure 12.6** and is subjected to the following:

a. Circumferential speed of the rotating drum: the material is carried along the periphery of drum on account of friction between the material and the steel surface of the drum.
b. Gravitational force, which acts on the particles being rolled
c. Centrifugal force, which acts along the radial direction of rotating material

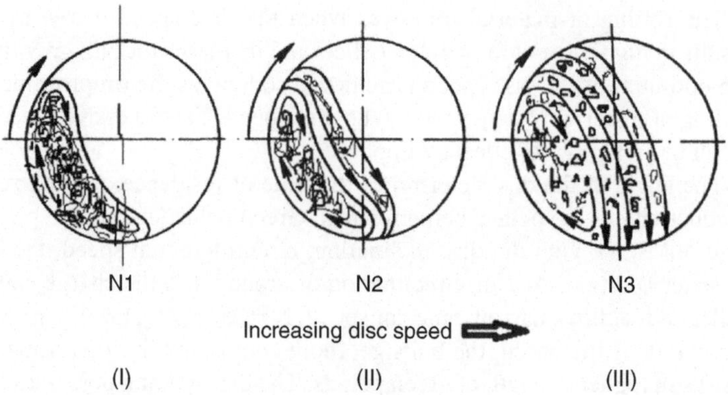

FIGURE 12.6 Characteristic of material movement in balling drum according to its rotational speed

The behavior of pellet mix material changes with the rotational speed of the drum as follows.

a. At much lower speeds (N1), the mix slides on the drum surface on a swinging motion. As the material does not roll, the formation of balls is absent (Figure 12.6.I).
b. At much higher speeds (N3), the centrifugal force acting on the material exceeds its limiting value. As a consequence, the mix material is pressed on the drum surface and lifted to a much higher circumferential distance from where these fall without rolling. As a result, the granulation does not proceed to form the agglomerate (Figure 12.6.III). This speed is around three-fourths of critical speed of rotation.
c. At an intermediate speed, the mix material is carried by the rotating balling drum to a reasonable distance due to the frictional resistance between the material and the drum wall. The agglomerate begins to form when the particles begin to roll. The optimum speed drum rotation is achieved when the material movement is of a cascade type (Figure 12.6.II)

Pelletizing Disc Unlike the movement behavior of pellets in a drum, the pellets are classified according to their sizes during their movement in the pelletizing disc. By an adequate arrangement of scrapers, the charge is so guided that, during its downward rolling, a classification is promoted. The balls of a desired size migrate to the periphery and to the top of the bed in the disc until these overflow the rim. These are discharged from the pelletizer: the quantity is balanced by an equivalent amount of feed (**Figure 12.7**). Pellet growth is controlled by the small amount of water sprayed in the disc and the adjustment of disc rotation speed. The feed point of raw material should therefore be located where seeds are formed (Figure 12.7).

The material movement of the pellet mix on the disc surface during the course of balling is complex. The influencing parameters are:

a. Proportion of the pellet mix (as a fraction of the total disc volume)
b. Angle of disc inclination

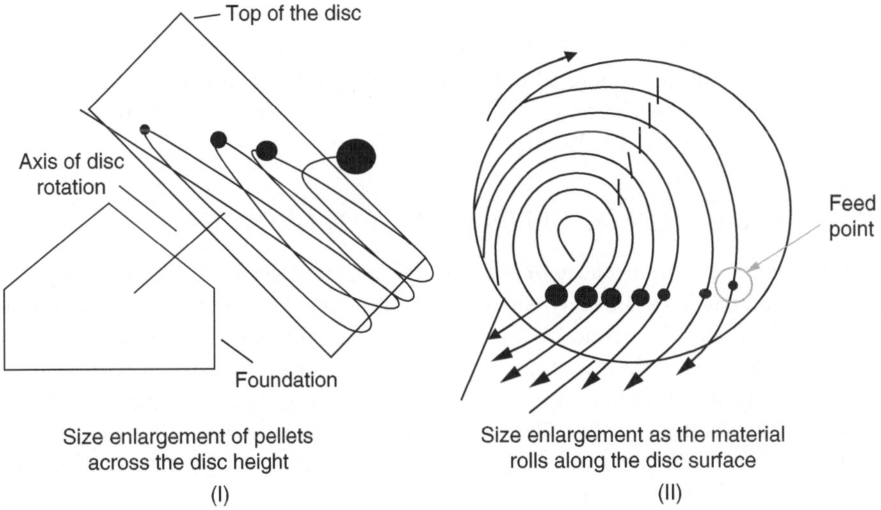

FIGURE 12.7 Schematic of disc pelletizer showing the profile of material flow. I: cross-section (development of different layer), and II: surface (movement of feed in various stages (after Ref. [3]).

c. Rim height
d. Speed of the pelletizing disc (in rpm). The critical speed of rotation is an important guiding factor in this regard: the speed at which the material is subjected to the sufficient centrifugal force that just compensates for the gravitational force. Beyond this speed, the mix material is slung to the disc wall and does not detach from the wall to roll. In view of this, the speed of disc rotation should be a fraction of the critical speed of rotation, say 65–75% (see Figure 12.6). If α is the disc inclination, and D is the disc diameter, the critical speed of rotation ($\eta_{critical}$) is approximately given by: $\eta_{critical} = [42.3/\sqrt{D}] * \sqrt{\sin \alpha}$. As a rule of thumb, $\eta_{critical} = [22.5/\sqrt{D}]$.

12.1.4 Drum versus Disc Pelletizer

When choosing a balling unit, in most cases, the disc is selected. This is due to:

a. Ease of their operation: The handling of materials, pellet mix (charge), and the produced green pellets is easier (in the case of a disc).
b. The level of recirculation (amount of undersized pellets against the total feed) is high in pelletizing drums. This high level of circulation makes pelletizing drums less sensitive to variations in feed material properties. Discs operate practically without circulating an undersized load or with only small amounts. Consequently, no recirculation of an undersized load is required, and the disc quickly comes into balance (i.e., the quantity of pelletizing material supplied corresponds to the same amount of green balls discharged).
c. Close control can be exercised over the size distribution of produced pellets through using the disc pelletizer.

d. Pelletizing discs need only a single process step to form pellets. Discharging the pellets from the disc rim within a very narrow size range is possible.
e. The discs may be operated in an open circuit, although it is more common to screen the discharge of the disc pelletizer, recycle the undersized load after crushing, and utilize the oversized load only as the product.

12.2 FORMATION OF GREEN BALLS AND GROWTH

Ball formation occurs because of the surface tension forces of water and the collisions among particles. Initially, small nuclei of pellets are formed on the addition of water. The nuclei surrounded by water film grow by colliding with either dry or wet particles, and their frictional drag against the disc surface decreases. These nuclei then turn into balls as they pick up loose grain particles during the rolling action in the rotating device. The movement of balls increases with their sizes, until these are deflected downward by a scraper attached to the disc. During the movement, the balls acquire fresh feed; thus the growth (of balls) takes place. As the balls grow in size, the centrifugal force that acts on these (agglomerates) increases due to the rotating action of the disc/drum. The balls now have come out from the nucleation zone.

12.2.1 Formation of Green Balls

The formation of balls and their growth are two-stage processes, namely,

- Nucleation or seed formation
- Growth of balls

These take place in two different portions of the pelletizer, as follows.

- Where water is sprayed into the material, the seed is formed (nucleation)
- In the remaining part of the pelletizer, the growth of balls (and breakage of pellets) occurs.

Figure 12.8 shows the schematic of the formation of balls and their growth when subjected to mechanical treatment in a rotating disc as water is added.

12.2.2 Growth of Balls

The agglomerates are formed due to the combined effect of (1) the added water, and (2) the movement of particles inside the pelletizer. The steps involved are (Figure 12.8):

a. The formation of green balls initiates as water is added into the ore concentrate. The surface of the particles gets wetted and, initially, a water film is formed that envelops the solid particles (Phase a). The thickness of the film depends on the wettability.

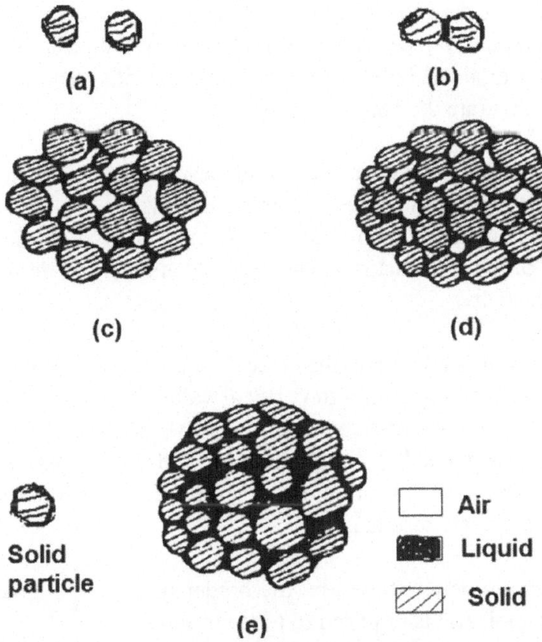

FIGURE 12.8 The formation of green balls and their growth as the material rolls and the influence of water addition

b. The wet particles come in contact with each other in many places due to the surface tension of the water films. Consequently, liquid bridges are formed (Phase b).
c. The wetted fine particles (on account of liquid bridges) coalesce and form a single unit (Phase c). At the initial stage of rotation, the balls are loosely held as their network has a large number of voidages. Water filled in the voids is less.
   - With further addition of water and on rotation of the device, the agglomerates become denser and change to a ball (Phase d). Their strength increases on impact. The capillary force of the individual liquid bridges is essentially active at this stage of green ball formation.
   - The optimum of this ball formation phase is attained when all the pores inside the balls are (completely) filled with liquid, but later this does not yet uniformly coat the whole agglomerate (Phase e). The wet strength of green pellets reaches a maximum (at this stage), and concave surfaces are formed in the pore openings due to the capillary forces; (**also see Figure 12.12**).

12.2.3 Role of Moisture in Granulation

The presence of water in the pellet mix is essential to roll the concentrate particles into balls. The formation of balls is due to the rolling of moist material, and surface tension of water in contact with particles is crucial in this respect. The extent of green ball formation is dependent on moisture content. The pellet mix should have critical moisture content for seed formation and ball growth. Maximum strength of green balls will be attained by compacting material with minimum porosity and sufficient water to saturate the voids.

- Moisture content less than a critical value leads to its nonuniform distribution, leaving behind some fines that are relatively dry.
- Moisture content more than the critical value leads to a larger growth rate (a type of lump formation). The pellets can be easily deformed due to increased plasticity.

Seed formation is favored only at the critical moisture level. Seeds contain moisture that could be a little more than the critical value. This leads to an increase in plasticity and, hence, better growth by layering or assimilation.

The amount of water needed for granulation is decided approximately by the wettability of particles. The solid-liquid contact angle is important in this respect. Lowering the contact angle results in:

- The particles are better wetted by the binder liquid. This means that less interparticle moisture is required to form granules.
- The amount of liquid that soaks into the pores of porous particles will increase. This means that less added liquid remains at the surface to aid granulation. Therefore, in fact, there may be a nonzero optimum contact angle.

Content of water in the pelletizing mix affects the capillary action:

- When it is lower, air bubbles inside the balls are partially closed and capillary action is restricted.
- When it is higher than the critical limit, water film is formed on the ball surface, which destroys the capillary forces.

12.2.4 Forces Involved in Formation of Green Balls

The driving force in the formation of balls from the ore concentrate happens to be the lowering of total surface free energy of the system through reduction of the effective air-water interfacial area. For instance in case of the taconite concentrate having specific surface area 1,800 cm^2/g, the particles on an individual basis are surrounded by water of 0.6 μ thickness. Only a few layers of the adsorbed liquid (water in this case) behave as bulk liquid. Hence, the surface free energy of the moisture layer is $72*10^{-7} J/cm^2$ (the value of surface tension) ($72*0.001$ kg/cm^2). When the nuclei are formed, the external surface of air-water film is effectively reduced (**Figure 12.9**).[4]

The free energy of nuclei formation ($\Delta G_{agglom.}$) is then given by: $\Delta G_{agglom.} = (S_{nuclei} - S_{feed}) * \gamma_{(air\text{-}water)}$

where:

S_{nuclei} = the external area of the nuclei agglomerate,

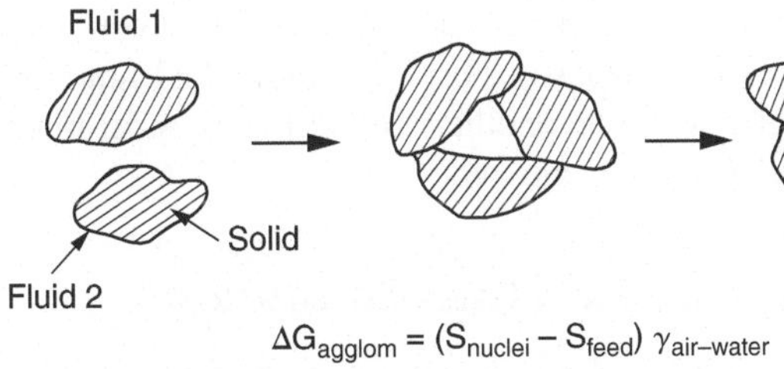

FIGURE 12.9 Schematic showing the driving force for pelletization (after Ref. [4])

S_{feed} = the surface area of feed particles, and
$\gamma_{(air\text{-}water)}$ = surface tension of water.

The forces responsible for the formation of green balls and imparting the strength can be categorized into: (a) physical forces and (b) applied forces.

a. **Physical Forces** These arise from:

- *Interfacial effects* giving rise to capillary and surface tension forces due to the presence of water. For the solid particles that are completely wetted by a bridging liquid, the capillary and surface tension forces are the most predominant. The surface tension of the binder liquid and dissolved salt, if any, do affect the bonding. The mineral salts, if present in the environmental system, crystallize out, which promotes bonding. That is why seawater is used in balling technology.
- *Van der Waals interaction* is the attraction of intermolecular forces among molecules. The van der Waals forces fall off rapidly with distance of separation and, for such forces to be effective, the interacting surfaces must be as close as possible, typically, 10 nm or less. There are two kinds:

  London dispersion forces: These are weak. Since the molecule is polar, all the electrons are concentrated at one end; hence, the molecule is partially negatively charged on that end. This negative end attracts the surrounding molecules' positive ends to also give an instantaneous dipole.

  Dipole-dipole forces: The molecule of H_2O, being polar, attracts the positive end of another polar molecule with the negative end of its dipole. The attraction between these two molecules is the dipole-dipole force. These are stronger.
- *Electrostatic interactions:* This refers to the clustering of particles due to the exchanges of charges. This interaction is of lesser importance than the van der Waals forces.
- *Immovable bridging bonds:* This is solid-solid bonding and can occur only when the particles are closely packed. The bonding occurs if a binder such as bentonite or lime is used. This type of bonding appreciably strengthens the balls when they are dried.

b. **Applied Forces** These forces arise due to the tumbling motion of drums or discs, which serves the purpose of bringing individual liquid-wetted particles into proximity with one another so that the physical forces become operative and pelletization is initiated.

12.2.5 Stages of Granulation in Green Pelletization

The balling process undergoes different stages of granulation having different degrees of liquid filling. Liquid filling degree or liquid saturation (S) describes the portion of the pore volume that is filled with the binder liquid and is calculated from Eq. 12.1.

$$S = [100\ F/(100 - F)] * [(1 - \varepsilon)/\varepsilon] * [\rho_S/\rho_L] \quad (12.1)$$

where:

S = liquid filling degree,
F = moisture content by wet weight,
ε = fractional porosity, and
ρ_S, ρ_L = density of solid particles and liquid, respectively.

The saturation degree (S) of pores strongly influences the strength of green pellets. The degree is determined by bound moisture having two components:

a. Moisture retained in the capillary
b. Moisture absorbed on the particle surface

The liquid (water in this case) acts as a binder in wet agglomerates. The formation of agglomerates as water is added and their growth due to the combined effects of water addition and their movement inside the pelletizer undergo different states (see **Figure 12.8**) depending on the extent to which the intergranular spaces are filled with the binder liquid.[5,6] **Figure 12.10** schematically shows these different states as the material is saturated with liquid, during its rolling.[5]

Pendular State: This state is at low saturation. Water presents just at the particle contact point. The particles are held together by liquid bridges (liquid trapped) and the surface tension of the water is crucial in this respect.

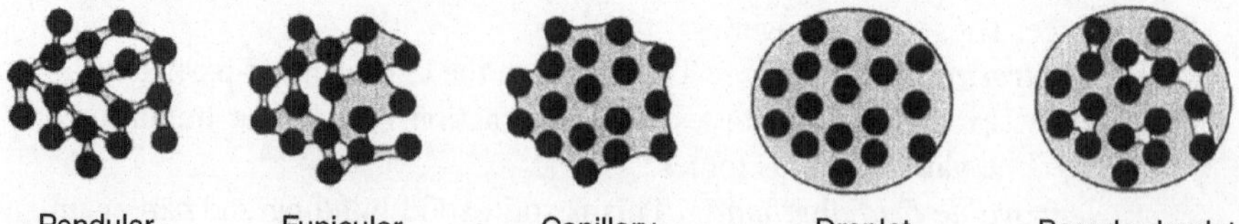

FIGURE 12.10 Schematic of different states of saturation in liquid-bonded granules (after Ref. [5])

The pendular regime is the suitable statute for water content with regard to the granulation of iron ore in the process of sintering (see **Secs. 6.5.6**).

Funicular State: Some pores of material are fully occupied by water. In this state, liquid-filled capillaries and liquid bridges coexist.

Capillary State: In the capillary state, the optimum of this phase is attained and pores are filled with liquid bridges. But water has not yet coated the whole agglomerate uniformly. Concave surfaces are formed in the pore openings due to the capillary. No water film is produced on the ball surface.

The characteristic feature of the capillary state vis-à-vis that of the funicular state can be described in terms of their relative strength to hold the agglomerates. The capillary forces are much stronger than the pendular bonds. In the funicular state, bonding takes place by both the pendular bonds and the capillary forces.

Tensile strength of the bond is contributed by these forces: the pull due to surface tension at solid-liquid-gas contract line directed along the liquid surface and the negative capillary pressure existing in the liquid bridge in an additive fashion (**Figure 12.11**).[7]

Droplet State This is the final stage of agglomeration and is attained when the solid particles are fully coated with water film. At this point, the capillary action ceases and the surface tension among water droplets becomes fully active. Finally, droplets of liquid completely surround the granule. At a fully saturation level (S = 100%), flooding takes place and the agglomerate deforms under its own weight.

In the droplet state of agglomerates, free superficial water over the whole agglomerate (outer surface) exists. Also, the agglomerate is held together by the cohesive force of the liquid. The capillary state, on the other hand, is characterized by the concave capillary openings.

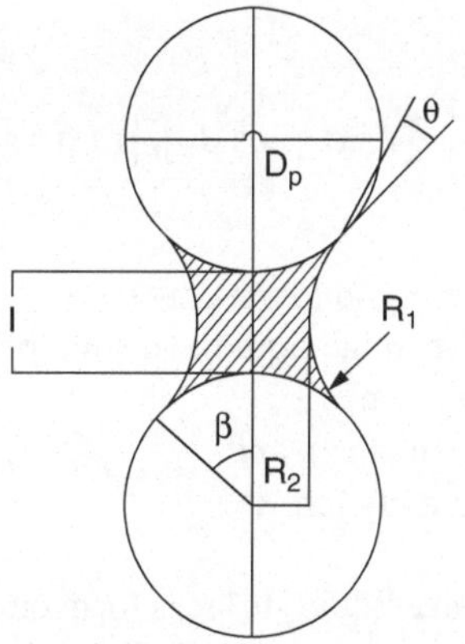

FIGURE 12.11 Pendular bonding between two spheres (after Ref. [7])

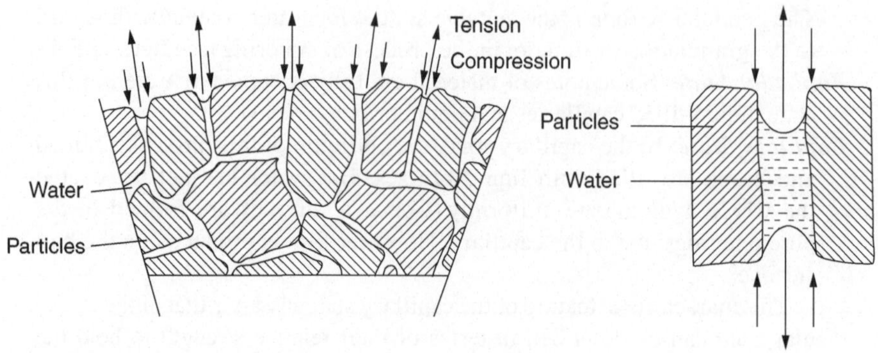

FIGURE 12.12 Influence of capillary force on binding of particles (after Ref. [8])

Tigerschiold and Ilmoni[8] have demonstrated how the particles are bound by the action of capillary forces in relation to the pelletization of the ore concentrate. Concave liquid surfaces form on the outer pores and capillary suction holds the ore particles together (**Figure 12.12**).

12.3 STRENGTH OF WET AGGLOMERATES

12.3.1 Capillary Theory for Wet Agglomerate Strength

The capillary force theory for wet agglomerate strength is well established.[9a,9b]

It applies to the particle systems with a freely movable binder, like water. The basic wet granulated unit may be visualized as two phases of solid particles and a liquid bridge that serves to hold the particles together by surface tension at air-liquid interface.[8]

The main binding force happens to be a capillary (force) as long as part of the capillaries are still to be filled with water.

The height of the liquid column, h (in cm), in a capillary (**Figure 12.13**) is given by:

$$h = [2 * \sigma * \cos \theta_{sl}] / [r * \rho * g] \quad (12.2)$$

where:

σ = surface tension of the liquid (dyne cm^{-1}),
θ_{sl} = angle of contact between the liquid and solid phase (degrees),
r = radius of the capillary (cm),
ρ = density of the liquid, (g cm^{-3}), and
g = acceleration due to gravity (cm s^{-2}).

In case of pelletization, concave liquid surfaces form on the outer pores and capillary suction holds the ore particles together. The ball strength is directly proportional to the value of h (**Figure 12.13**). In the pelletization process, balling of ore concentrate

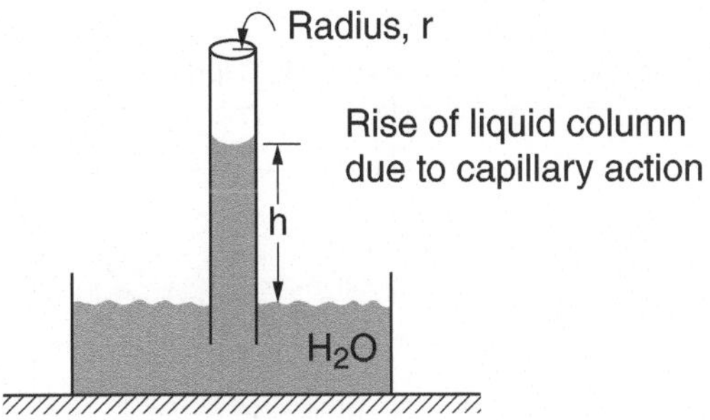

FIGURE 12.13 Surface tension versus height of liquid column

is carried out using water. We may safely assume that the solid particles are fully wetted by water (see **Figure 12.15** also):

$$\Theta_{sl} = 0;\ \text{hence}\ \cos\theta_{sl} = 1,\ \text{and}\ \rho = 1\ \text{g cm}^{-3}\text{at } 20°\text{C}; \sigma = 72\ \text{dyne cm}^{-1}$$

The previous equation is simplified to:

$$h = \left[2 * 72 * 10^{-5} * 1\right] / \left[r * 1 * 980\right] = 0.15/r\,(\text{cm})$$

During the process of green pelletization, capillaries of the balls are not entirely filled with liquid in actual practice. The extent to which the capillaries are saturated with liquid is important in this regard. The agglomerate strength can be estimated, following the theory, from the relative amount of filled capillaries.

Figure 12.14 schematically shows the tensile strength of wet agglomerate in terms of the liquid filling degree, according to the capillary theory.

According to the capillary theory, the green pellet wet strength reaches a maximum at the capillary state when the pores are completely filled with water and concave surfaces are formed in the pore openings due to the capillary forces. This takes place at around 90% liquid saturation. The strength of wet agglomerate, when it reaches maximum (in the capillary state) is given by Rumpf's equation (Eq. 12.3).[11]

12.3.2 Rumpf's Equation for Agglomerate Strength

The green pellet wet tensile strength due to the capillary forces, σ_C, is represented by the following equation:[11]

$$\sigma_C = \grave{a} * \left[(1 - \varepsilon)/\varepsilon\right] * \gamma * \left[1/d\right] * \text{Cos}\,\theta_{sl} \tag{12.3}$$

where:

- $\grave{a}$ = constant,
- ε = fractional porosity,
- γ = liquid surface tension,
- d = average particle size, and
- θ_{sl} = liquid-solid contact angle.

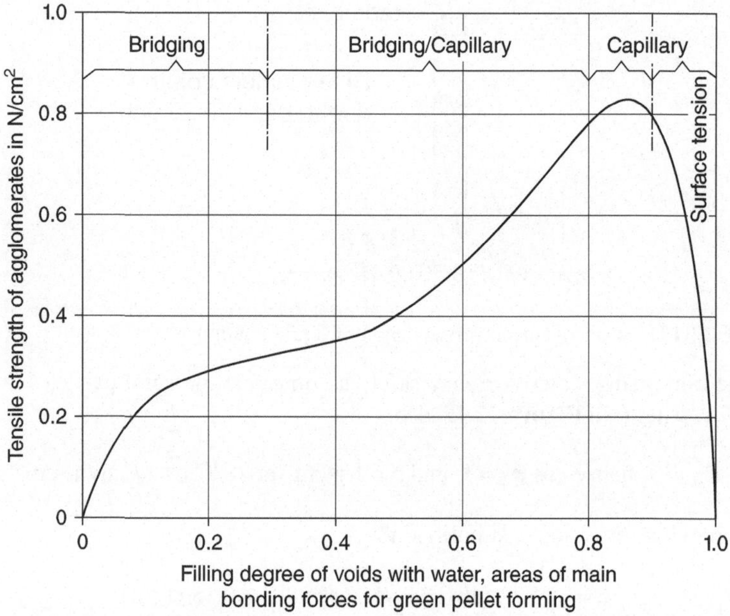

FIGURE 12.14 Schematic showing how tensile strength of green pellets changes with the degree of voids filled by binder liquid (raw material: limestone, $\varepsilon = 0.4$) (after Ref. [10])

The equation was originally verified by Rumpf,[11] who studied agglomeration made of narrowly sized limestone powders using water as a binder. The best ball strength appears when the material is perfectly wetted, $\theta_{sl} = 0$ ($\cos \theta_{sl} = 1$), and the pores are completely filled ($\varepsilon = 1$).

The constant, à, is given by à = C* S,

where:
C = geometric constant, and
S = the granule interparticle saturation.

Combining the previous two, we obtain:

$$\sigma_C = C * S * \left[(1-\varepsilon)/\varepsilon\right] * \gamma * [1/d] * \text{Cos}\ \theta_{sl} \qquad (12.4)$$

Hence, the minimum level of liquid saturation required for granulation of nonporous particles is inversely proportional to Cos $\theta_{LS.}$ That is, less liquid is needed to granulate materials with a low contact angle.

The measurement of the parameters involved in Rumpf's equation (Eq. 12.3) is, however, time-consuming to compute the tensile strength and hence cannot be applied to a large number of pellets in a similar manner as is applicable in the measurement of compression strength. The value of compression force, as it needs to overcome the friction among particles, is more than that of tensile strength. According to Rumpf,[11] a nonlinear relationship between these two forces prevails.

12.3.3 Influencing Parameters for Agglomerate Strength

The compressive strength of green pellets is influenced by the characteristics of pellet feed as well as those related to the binder liquid. The strength of wet agglomerate, according to Rumpf's equation (Eq. 12.3) is affected by the following parameters:

Moisture Content: Forsmo et al.[12] computed the liquid saturation value (S) from the batch balling (in drum) experimental data using Eq. 12.1. The liquid saturation was 100% at around 8.2% moisture content. Below 8.2% moisture, the green pellets were not fully saturated. It is important to note that the saturation points, with the addition of 0.5 wt.% binder, almost merge with those without the binder addition in the curve, which infers that the degree of saturation depends exclusively on the added moisture (not on the binder dosage). At much higher levels of the moisture content (beyond 8.2%), the balling process becomes self-regulating and strives to keep the liquid saturation constant at around 100%.

Particle Size: The compressive strength of green pellets increases with decreasing particle size of ore concentrate (Eq. 12.4). An increase in wet-CS of green pellets with increasing raw material fineness has been also reported in relation to hematite and magnetite concentrates, respectively.[13a,13b] Struve[14] studied the effect of specific surface area on compressive strength of green pellets in case of hematite concentrate and observed a linear correlation between these two. The ranges of their variation were specific surface area: 1,200 to 2,800 cm^2/g and compressive strength: 8 to 20 N/pellet.

Porosity: With the lowering of pellet feed porosity, the compressive strength of green pellet increases (Eq. 12.3). The pellet mix with a wider particle size distribution has closer packing, which yields its lower porosity. This leads to stronger agglomerates. The porosity of mix material consisting of large (L) and small (S) particles when varied in different proportions in the mix has the minimum value when the mixture contains about two-thirds of the large particles (see **Sec. 3.6.5** and **Figure 3.7**, **Sec. 6.4.2** and **Figure 6.6b**).

Interfacial Tension: The strength of wet agglomerate (Eq. 12.3) decreases with decreasing the surface tension of the binder liquid. The surface tension of water decreases when alcohol in its varying proportions is added to it. Hence, the influence of surface tension on the strength of wet agglomerate can be verified by balling iron ore with water-alcohol mixtures.

However, it is not only the surface tension of binder liquid that contributes to the agglomeration as other factors are also involved. The binder liquid has to penetrate through the pores and also spread on the surface. According to Rumpf's equation, complete wetting of material is necessary

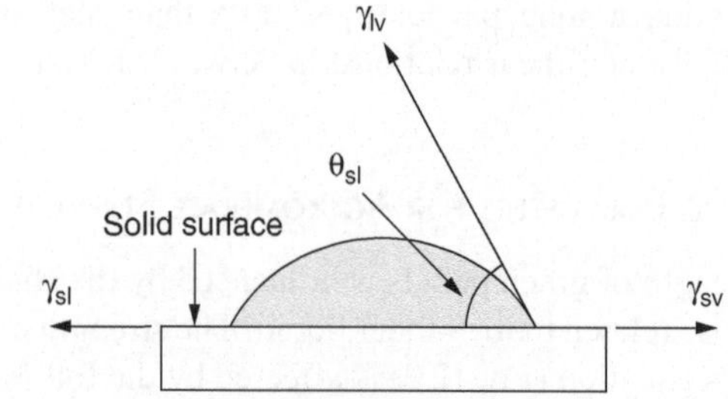

FIGURE 12.15 Contact angle of liquid/solid interface

for fully developed capillary forces. The wettability is usually identified through the contact angle, the work of adhesion and cohesion, as well as the spreading coefficient. Adhesive force refers to the binding force at the interface between the binder and ore particles. Maximum adhesion will occur when the work of adhesion, W_A, is maximum.

Contact Angle and Adhesion Forces When a liquid comes into contact with a solid surface, the liquid assumes a characteristic drop shape; this is shown in **Figure 12.15**.

The interfacial tension in the aforementioned system gets balanced according to Young's equation:

$$\gamma_{sl} + \gamma_{lv} \cos \theta_{sl} = \gamma_{sv} \tag{12.5}$$

Here, Θ_{sl} is defined as the liquid-solid contact angle; γ_{sl}, γ_{lv}, and γ_{sv} are interfacial tensions of solid/liquid, liquid/vapor, and solid/vapor interfaces, respectively. The work of adhesion, Wa, is the reversible work required to separate a unit area of liquid from a solid. It is defined by Eq. 12.6.

$$Wa = \gamma_{lv} + \gamma_{sv} - \gamma_{sl} \tag{12.6}$$

Substitution of Eq. 12.5 into Eq. 12.6 gives Eq. 12.7:

$$Wa = \gamma_{lv} (1 + \cos \theta_{sl}) \tag{12.7}$$

Hence, the work of adhesion is directly related to the surface tension of the liquid and the contact angle. This equation states that the maximum work of adhesion occurs when $\theta_{sl} = 0$.

Another important parameter is the work of cohesion, Wc, which is the reversible work required to separate a column of liquid and create two liquid/vapor surfaces. It is a good measure of molecular interactions for symmetrical molecules. It is defined by Eq. 12.8.

$$Wc = 2\,\gamma_{lv} \tag{12.8}$$

The spreading coefficient, S, is defined by:

$$S = Wa - Wc \qquad (12.9)$$

If a liquid is to spread on a solid, S must be greater than zero. That is,

$$S = (Wa - Wc) = (\gamma_{sv} - \gamma_{sl} - \gamma_{lv}) > 0 \qquad (12.10)$$

Hence, for spreading the binder liquid onto the solid surface, γ_{sl} and γ_{lv} should be made as small as possible.

Substitution of Eq. 12.5 into Eq. 12.10 gives Eq. 12.11:

$$S = (Wa - Wc) = \gamma_{lv}(\cos\theta_{sl} - 1) \qquad (12.11)$$

It can be seen from Eq. 12.11 that in the case where γ_{lv} is kept constant, the binder liquid will spread more on the surface with the smaller (but nonzero) contact angle. Therefore, the binder liquid should be made to reduce the contact angle on an iron ore surface as much as possible. In view of this, it may be necessary to introduce hydrophilic functional groups into the binder molecules in order to reduce the contact angle.

Porosity Effect: Iron ore is a porous material; in this case, water is required to penetrate through the pores. The rate at which binder liquid (water) is absorbed into the pores is influenced by the solid-liquid contact angle. When the contact angle is lower, a higher penetration of binder liquid into pores takes place (Eq. 12.2). However, the binder liquid (water) that is absorbed into the pores is obviously unavailable to aid in the granulation process. The water that has penetrated into the pores is required to be removed by evaporation during the sintering step, involving an associated energy cost. On the other hand, if the contact angle is more, the spreading decreases (Eq. 12.11), and granulation is adversely affected. In view of this, the solid-liquid contact angle has two opposing effects on granulation of porous materials and need to be optimized.

Difficulty in the Measurement of Contact Angle The agglomerate wet strength is influenced by the liquid-solid contact angle. However, the effect is difficult to measure. This is because of the fact that, in practice, it is difficult to vary the contact angle independently without influencing the surface tension. Iron ores are complex, having two or more mineral phases and associated gangue minerals, and the ores vary in both composition and morphology from pure/synthetic minerals. In view of this, the contact angle value in the case of iron ore remains uncertain. Despite uncertainties, the contact angle in the case of iron ores has a correlation with its associated minerals: the contact angle increases with increasing hematite content (or decreasing goethite content) (**Figure 12.16**). The following empirical relation holds good for the contact angle.[15]

$$\Theta = 45x + 8\,(x = \text{vol. fraction of hematite})\,(R^2 = 0.68) \qquad (12.12)$$

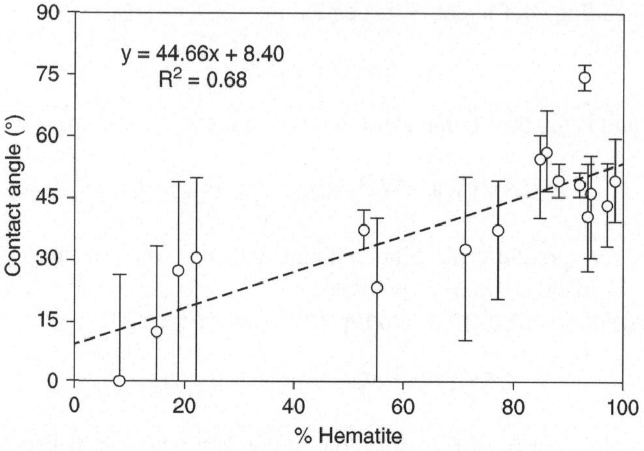

FIGURE 12.16 Contact angle versus content of hematite in ore (after Ref. [15])

12.4 VISCOSITY EFFECT AND BINDERS

12.4.1 Viscosity Effect

According to the capillary theory, the crushing strength of green pellets is chiefly determined by the capillary forces of liquid (water) and reaches its maximum value when the liquid filling degree in the capillaries is around 0.9 (Eq. 12.3 and Figure 12.9). The capillary theory, so far discussed and stated in Eq. 12.3, pertains to the particle systems with freely movable binders, like water. However, the capillary theory alone may not be sufficient to explain the balling mechanism. This is on account of the fact that the capillary forces, which are important in moist pellets, disappear once the green pellets are dried and before these are heat hardened. The process of drying involves the complete removal of moisture, resulting in destruction of capillary cohesive forces. This should decrepitate the balls had the capillary force been the only one responsible for balling. We know that the balls are not broken (on removal of water), which means that other forces (besides the capillary one) are also responsible for keeping the particles together. The forces responsible to facilitate the binding mechanism are a combination of the capillary force (easily movable liquid) and the cohesive force (viscous binder) at a low binder dosage.

During the process of green pelletization, the viscosity of the binder liquid, which is "squeezed" to the pellet surface, influences its balling rate. As the viscosity in the binder liquid increases, the superficial network of viscous fluid among the particles becomes stronger

12.4.2 Binders

The viscosity effect is apparent in pelletization if the pellet mix being balled contains even a slight amount of microfines (less than 5 micron in size). These behave as colloids when in contact with liquid (water) and act as binders in the granulation process.

In practice, binders (largely bentonite) are used in iron ore pelletization. The roles of the binder in relation to the iron ore pelletization are as follows:

- The binder makes the moist ore plastic so that it will nucleate seeds that grow at a controlled rate into required size pellets.
- During drying, the binder holds the particles in the agglomerates together while the water is removed and continues to bind them together until the pellet is heated sufficiently to sinter the grains together (see Figure 14.1).

  The basic conditions for the dry pellet strength differ from those of the green pellet strength. During drying, liquid bonds are reduced and gradually replaced by the solid bonds, which become effective among the ore particles.

The bonding efficiency of bentonite during pelletizing is influenced by the properties of bentonite, water, and iron ore concentrate. The bentonite binder, when mixed into the iron ore concentrate with water, becomes wet; the clay layers expand and disperse from the hydration of exchangeable interlayer cations throughout the concentrate, transforming into a matrix that bonds the ore particles together. The granulation of ore concentrate results in a supporting "network" of viscous liquid that is formed on the green pellet surface. The total binding force is on account of:

- The cohesive force of the network
- The compression strength as explained by the capillary force theory

The green pellets do have surface irregularities; hence, both fully developed concave pore opening and saturated areas occur simultaneously in the pellets.[13]

Thus, the bonding efficiency of the bentonite is not only related to its own quality but also to the medium water and the properties of the concentrate. So, the ternary system (bentonite, water, and concentrate) affects the bentonite bonding efficiency (**Figure 12.17**).[16]

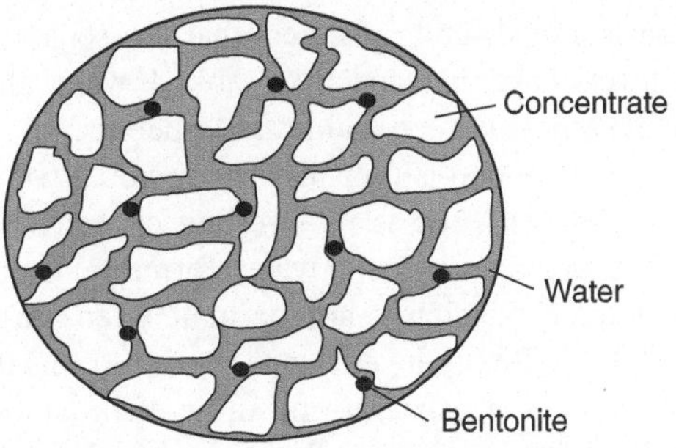

FIGURE 12.17 Ternary system in pellets (after Ref. [16])

12.4.3 Requirements of Binders

Experience of pellet production over the past decades has led to the identification of a number of criteria that a binder must satisfy. These include:

a. *Mechanical Properties.* A good binder, when incorporated in the pellet mix, should provide good mechanical strength of pellets during its various stages of production: that of green, dry, and fired pellets. That is, the pellets must have adequate resistance against: (a) their deformation under load, (b) their fracture by impact and by compression, and (c) their abrasion.
b. *Chemical Composition.* A good binder should not bring elements into the product pellets that are environmentally and metallurgically harmful, such as P, S, As, etc. It should not markedly reduce the iron grade or increase impurities such as silica.
c. *Metallurgical Performance.* The characteristic of a binder should be such that the produced pellets do possess excellent metallurgical properties, such as reducibility, swelling, and pressure drop during reduction.
d. *Processing Behavior.* The pellet mix, when binder is added to it, should neither be difficult to process in stages of pelletization: mixing, dispersion of binder, green ball preparation, pellet drying, etc., nor should it essentially change the conventional pellet production circuit.
e. *Cost Factor.* The price of the binder should be within acceptable limits for iron pellet production.

12.4.4 Bentonite (as Binder)

Bentonite is the most commonly used binder in the pelletizing industry. It meets most of the necessary criteria of a binder.[17–19] Being an effective binder, bentonite is found to improve the pellet formation and growth process. The green balls, formed using the bentonite binder, have superior properties in terms of drop strength and compressive strength compared to other binders. Additionally, the use of bentonite in the pellet mix provides better mechanical and metallurgical properties of indurated pellets.

Bentonite is basically an aluminosilicate clay mineral, montmorillonite. Its chemical formula is: (Na,Ca) (Al Mg) $Si_4O_{10}(OH)_2$. It is formed by the colloidal nature and layer structure. It consists of discrete platelets that are separated by exchangeable cation layers. These layers have the capacity to absorb water and expand, thus controlling the moisture contents of green and finished pellets.

The amount of bentonite added is typically between 0.5 wt.% and 0.7 wt.%. The most important feature of bentonite is its swelling capacity, which differs for its various types. Bentonite, when mixed with water, forms gel at about five to six times its weight, with its volume expansion to approximately ten times, and increases the viscosity of the water phase. Bentonite envelops the ore grains together with water.

Besides its water adsorption capacity, the mineral provides dry film strength. Hence, the mineral plays a dual role in pelletization: (a) It controls the moisture in the pellet mix that undergoes balling, and (b) it provides resistance to spalling during the drying of green pellets.

Additionally, bentonite is used as a moisture-controlling agent in balling circuits [4] as it possesses a moisture retention property.

Activation of Bentonite The addition of bentonite, is not without any drawbacks as it contaminates the ore with gangue (silica and alumina) constituents. Acidic constituents of the pellets do have an adverse effect on the iron-making process (see **Sec. 3.6.1**). The addition of 1 wt.% bentonite in the pellet mix lowers the Fe content of acidic pellets by approximately 0.6% and increases the silica content by 0.6–0.7%.[20,21]

One of the possible ways to reduce the consumption of bentonite is through activating the bentonite mineral. Properties of bentonite are affected by the nature of exchangeable cations in its interlayers. Due to lattice substitution, the mineral acquires a negative charge that is neutralized by the absorption of cations like sodium, potassium, calcium, and magnesium. Depending on the type of absorbed cations, the bentonites are referred to sodium—or calcium—bentonites.

The swelling properties of bentonite depend on the type of exchangeable cations, whether Na^+ or Ca^+. Calcium ions have a higher charge and smaller diameter than sodium ions, and as a result, they tend to interact more strongly with the aluminosilicate platelets, making them less prone to swelling. Sodium ions hydrate more readily causing the bentonite to swell more, so any effort to increase the concentration of Na^+ ions will improve the swelling properties of bentonite. This fact forms the basis of the activation of bentonite using an alkaline compound.[16]

A dynamic adsorption and desorption reaction is encountered, when Na^+ and Ca^{++} coexist in the bentonite system as follows: Ca bentonite + 2 Na^+ = Na_2 bentonite + Ca^{++}; Ca^{++} combined with bentonite is more stable than Na^+. In case Ca^{++} on reaction with anions forms sediments, the result would be to decrease the concentration of $Ca^{++.}$ As a consequence, the formation of sodium bentonite is promoted. In this respect, NaF, Na_2CO_3, and NaOH provide a better reaction compared to NaCl.

The activation of bentonite by substituting a Ca ion with an Na/K ion, mentioned in the preceding paragraph, reduces its specific consumption. However, such substitution (by Na/K) increases the alkali load in the blast furnace iron making, also the cost of pelletization increases.

Substitution of Bentonite The bentonite could be substituted partially by inorganic and organic additives thus, reducing the dosage of bentonite.

CaO Lime (CaO) in the pellet mix reacts with water on moistening during the green pelletization according to the exothermic reaction: $CaO + H_2O = Ca(OH)_2$ + heat.

Hydrated lime is formed, which becomes an effective binder. During its hydration, the volume expands roughly threefold; hence, the slaking process has to be completely terminated and the hydroxide thoroughly mixed with the ore before green balls are formed. Otherwise, the burnt lime is only slaked during green ball formation, which causes a local volume increase and damages the pellet structure.

The hydrated lime combines with CO_2 to form the hard phase of $CaCO_3$ in the pellet on exposure to the atmosphere.[22]

$$Ca(OH)_2(s) + CO_2(g) = CaCO_3(s) + H_2O(g/l)$$

The voids are formed within the green pellets when these are air dried as the water gets evaporated. This facilitates the penetration of CO_2 into the pellet during the treatment. The formation of calcium carbonate is primarily responsible for imparting good strength to the pellet. The transformation of hydrated lime to calcite is associated with volume expansion by 1.52%. (The molar volumes of $Ca[OH]_2$ and $CaCO_3$ are: 33.59 and 34.10 cm^3/mol., respectively.) The volume expansion is likely to be accommodated by the internal pores of the pellets, leading to a decrease in porosity and an increase in compactness of the pellets. The transformation of $Ca(OH)_2$ to $CaCO_3$ plays a predominant role in the strength development in pellets.

Organic Binders An organic binder is an alternative to the bentonite binder. Pellets made with organic binders have lower gangue content and are generally more porous in their structure, which allows the reducing gases to permeate the burden more quickly during reduction. Owing to lower gangue concentration in the pallet, substantial improvement in energy and flux consumption can be realized in iron-making processes. In contrast to inorganic binders, organic binders have the inherent advantage of lower dosage and being eliminated during firing. However, organic binders lower the mechanical strength of preheated and fired pellets since they burn out without any residue at a temperature over 250°C. Therefore, the use of organic binders in industry is limited.

Boron has a beneficial effect at all stages of metal production, and the quality of pellets has increased. This benefit was believed to be due to low melting points of boron compounds. The low-temperature binder (i.e., the starch, dextrin, or the like), in combination with the metallic iron powder, affords considerable structural integrity to the pellets during the initial and subsequent stages of the direct reduction in a shaft-type or rotary furnace.[23]

Bentonite use could also be reduced by replacing bentonite with even lower-cost binders, such as a high-carbon fly-ash based binder (FBB). While FBBs can be used alone as binders, it was considered possible that mixtures of FBB and bentonite could exhibit superior binding properties.[24]

Water Absorption Capacity The water absorption capacity of the bentonite is measured in terms of the Enslin value (EV), which is calculated as follows.

$$EV(\%) = \left[(\text{absorbed water quantity in g})/(\text{bentonite quantity in g})\right] * 100 \quad (12.13)$$

The Enslin value of Na-based bentonite is in the range of 800–900, which is much higher than that (200–300) in case of Ca-based bentonite.

The water absorption capacity is measured by the test method, ASTM E946-92, 1996.[25]

12.4.5 Ball-Ability

Ball-ability is defined as the ability of particulate matter to form pellets. The main parameters for good ball-ability in iron ore pellet feeds are the fineness of the raw material, the moisture content in balling, binder dosage, and good wetting of the particles.

Sastry and Fuerstenau[26] introduced the concept of the ball-ability index. The index considers the water balance in terms of "free water." The ball-ability index decreases with increasing additions of bentonite.

$$\beta = Q(W - \omega * B) \tag{12.14}$$

where:

β = ball-ability index,
Q = experimental constant, which is material specific,
W = total mass of moisture,
ω = the mass of water immobilized by 1 g of binder, and
B = binder dosage.

The static ball-ability index, Kp, defined here is used to evaluate whether an iron ore concentrate can be easily pelletized or not:

$$Kp = W_m / (W_c - W_m) \tag{12.15}$$

where W_m and W_c are the saturated molecular moisture (%) and capillary moisture (%) of the iron ore concentrate, respectively. The ball-ability of the concentrate is often non-acceptable/difficult if the value of Kp is less than 0.3 and acceptable when Kp is higher than 0.6.

12.5 ELASTIC AND PLASTIC DEFORMATION OF GREEN PELLETS

Elastic deformation refers to a temporary shape change due to applied force that is self-reversing, as the object returns to its original shape after the force is removed. When a material undergoes changes of shape that are nonreversible, the deformation is a plastic one.

Elasticity in green pellets is important with respect to durability during their handling. The green pellets with high elasticity exhibit a higher resistance to breakage during their loading in the balling unit. The green pellet impact strength is usually described by drop number: the number of drops from a given height before these get broken. The drop number increases with the amount of bentonite added. It is also influenced by the water content in the green pellets.

The green pellets are encountered with the phenomenon of plastic deformation during their induration in a grate machine on account of:[27]

a. Load of pellets above.
b. The suction of process gas

Plastic deformation also takes place during rolling and facilitates the consolidation and growth of green pellets. The green pellets with higher plasticity deform more in the drying zones of the traveling grate. Besides, the pellets in the lowest layer of the straight-grate furnace experience the highest compression. These phenomena distort the pellet shape and, hence, decrease the bed porosity (and permeability). This is detrimental for the fast drying sequence and also for the oxidation of magnetite to hematite, which should be accomplished before the sintering temperature is reached.

Sportel et al.[27] report that plastic deformation was strongly dependent on both the amount of moisture and bentonite content. The moisture requirement for granulation increases considerably as the bentonite dosage in the pellet mix is increased. Increasing the moisture content in the pellet mix, at a constant binder dosage, increases the plastic deformation of the green pellets. This is more so in the case of oversaturated green pellets produced from the ores with high porosity. This leads to a lower bed permeability in a straight grate, which affects the rate of pellet induration.

The dependence of plastic deformation in green pellets, Deff. (1daN), on moisture content of the pellet mix is shown in **Figure 12.18**.[12] The amount of linear deformation when the green pellets is slowly loaded to 1 daN pressure, Def 1 (1 daN), is the distance between the piston positions at the regression line that intercepts with the x-axis and at 1 daN pressure, 1 daN = 10 N = 1.02 kg, according to ISO 4700.

Plasticity is an important property of green pellets in balling. The balling circuits should be controlled based on plastic deformation and compression strength of green pellets. These two operations are not interchangeable; even if they would compensate in growth rate, the green pellet properties would differ.

Growth of pellets is controlled by its two properties:[28a]

- Plasticity of the green pellet
- Viscosity of the superficial water layer
    a. Plasticity is controlled by moisture content. The plasticity of green pellets needs to exceed a minimum level to enable growth. The usage of bentonite in excess amounts leads to the production of pellets with very plastic properties, which is undesirable for further processing. The content of minimum moisture required to be maintained in the pellet mix is material specific.

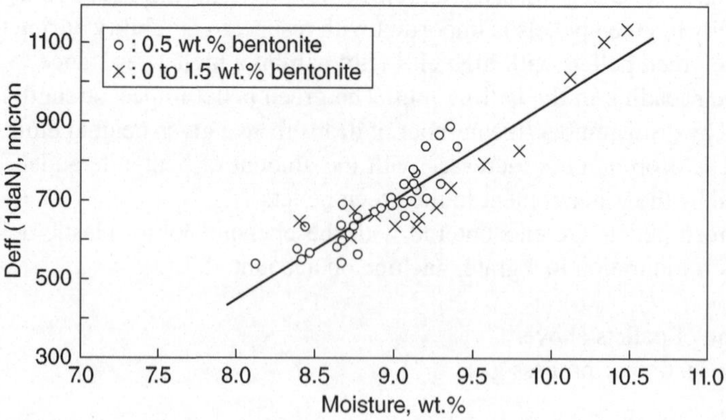

FIGURE 12.18 Deformation as a function of moisture content (after Ref. [12])

b. The growth rate is controlled by the viscosity of the superficial water layer rather than by the mobility of the pore water. The viscosity, if it is too low, leads to uncontrolled growth of balls. It has to be adequate enough for colliding pellets to combine within the time available during collision.

12.6 STAGES OF PELLET FORMATION AND GROWTH

The mechanism of green pellet growth includes:

- Crushing and layering, by Capes and Danckewerts[28b]
- Granule growth by coalescence, by Kapur and Fuerstenau[28c]

Figure 12.19(a) and **12.19(b)** shows the important stages of the agglomerate formation and size change (growth) in iron ore pelletization.[4] These steps include:

a. Nucleation: This causes the appearance of the newly born agglomerates along with discrete changes in their sizes. Nucleation refers to the formation of particles from the continuous phase due to interactions within the environment itself. A binder liquid is added to the pellet feed that wets the particles, though not uniformly. The pellet nucleation initiates due to the surface tension forces of water and the collisions among particles. When the disc rotates, it causes the wetted particles to move close together. Individual wet particles stick together to form small seed-like "nuclei" of particles that grow into pellets as they pick up loose gain particles during rotation of the disc. A weak nucleus may be disintegrated by the impact during its movement over the pelletizing surface. The disintegrated particles continue to form a new nucleus elsewhere.
b. Coalescence: As soon as the initial formation of nuclei occurs, coalescence begins to set in. Two or more smaller units (could be nuclei also) join together to form a larger unit. This stage of the balling causes discrete size changes and leads to a decrease in the number of agglomerates but not their mass. If no fresh feed material is added for balling, the rolling action may break some of the granules (mainly the smaller ones). These coalesce with the growing ones. Thus, the growth of pellets by assimilation takes place.
c. Layering: When balling proceeds with the addition of fresh feed material, there is a continuous supply of fresh material. Balls pick up material while rolling on a layer of fresh feed. The amount of material picked up is directly proportional to the exposed surface area of the rolling balls.

   The layering corresponds to continuous size changes of the particles. The particles envelop the nuclei and the agglomerates grow on their sizes. The stage is affected by interaction of particulates with the environment (continuous phase). Accordingly, layering brings changes in the mass of the agglomerates but not their number. Clearly, layering can occur only when both the agglomerates and new feed are present.

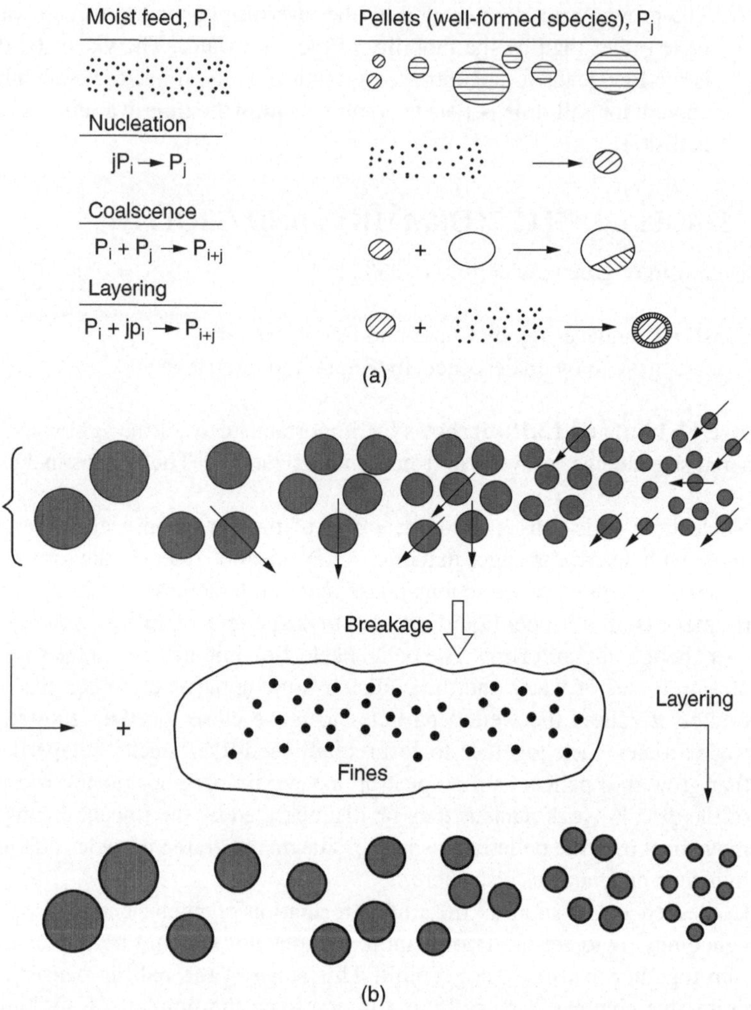

FIGURE 12.19 Schematic representation of various stages of green pelletization and pellet growth: (a) size enlargement and (b) breakage and layering (after Ref. [4])

d. Breakage and Attrition: Breakage of pellets occurs in the case when (a) the pellet size has reached its equilibrium size and/or (b) the binding force can no longer maintain the load. The phenomenon of attrition of pellets, in particular with respect to their sharp edges, smoothens their surface.
e. Consolidation and Growth: This stage of ball growth is through the process when the nuclei are joined by more fines through collision and further "sticking together" to form an agglomerate. Two micropellets, surrounded by liquid films, are brought into contact with each other.
   - A seed pellet is formed that captures dry and wet particles until the desired pellet size is obtained.
   - Pellet porosity decreases due to the continual impact, and the pellets are consolidated.

The growth of pellets in the initial stages of batch pelletization proceeds first by nucleation and then mainly by the coalescence mechanism. In the later stages there is evidence that a limited amount of breakage of pellets occurs. In industrial pelletizing operations, where a continuous supply of moist feed is charged to the pelletizing device, the growth mechanism of layering (or snowballing) of added moist feed onto the pellets becomes important.

12.7 KINETICS OF BALL GROWTH

12.7.1 Stages of Kinetics

Sastry and Fuerstenau,[29] using the batch (drum pelletizer) experimental technique in relation to taconite concentrate, have presented their results with respect to the size distribution of green pellets at various stages of pellet growth: the pellet diameter increases with an increasing number of drum revolutions (**Figure 12.20**). They established that the average diameter of pellets when plotted against the number of drum revolutions has a characteristic S-shaped curve (**Figure 12.21**).

Kanetkar et al.,[30] using a batch disc pelletizer, have also observed similar shapes of kinetics in the case of iron ore fines (blue dust) from Chowgule & Co. Ltd., Goa, India. The average diameter ($D_{av.}$) of pellets has been considered as an index of the growth rate in batch balling. It refers to the arithmetic mean:

$$D_{av.} = \Sigma ni * D_i / \Sigma n_i \qquad (12.16)$$

where n_i is number of pellets of diameter di.

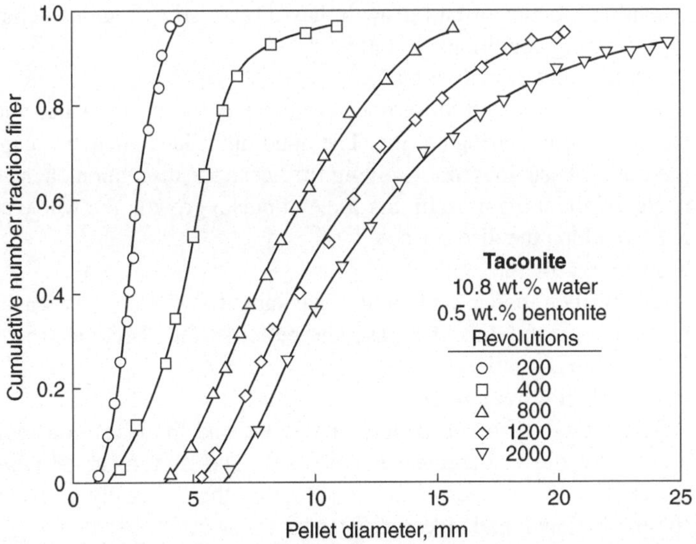

FIGURE 12.20 Size distribution of green pellets at various stages of pellet growth (after Ref. [29]

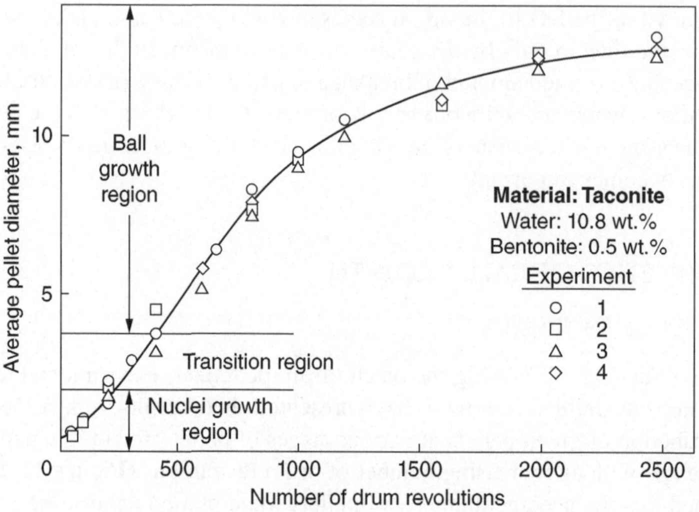

FIGURE 12.21 Typical pelletization curve: Average green pellet diameter as a function of the number of drum revolutions (after Ref. [29])

The size distribution of the resulting pellets has/had a self-preserving nature, which implies that irrespective of the operating conditions and the feed material, the size distribution, when normalized with respect to the average size of the distribution, is represented by a single curve. This characteristic size distribution is called the similarity size distribution.

Stages of Ball Growth The characteristic S-shaped curve (Figure **12.21**), representing kinetics of ball growth, has three distinct regions depending on the operating conditions, as follows:

i. Formation and growth of nuclei:

Highly porous and loose flocs are formed on mixing water with the ore concentrate during balling. The bonding is due to negative capillary pressure. These flocs, on revolving further in the disc, burst out as stable nuclei of about 0.5 mm. In this stage of nuclei growth, the whole charge is lifted along the disc surface.

ii. Transition growth:

On further increase of pellet size (about 4–5 mm), the cascading movement in the disc changes. The pellet surface becomes shiny and growth rate is rapid.

iii. Pellet or ball growth stage:

The growth of green pellets in size beyond the transition region is mainly by the coalescence mechanism. The movement of pellets is affected by three types of forces acting on them, namely (a) frictional, (b) gravitational, and (c) centrifugal.

The pellets are lifted along the disc surface up to a certain point till the centrifugal and frictional forces overcome the gravitational force.

Beyond this limit, gravitational force exceeds the other two and pellets roll down the inclined surface.

According to Kapur and Fuerstenau[31], the growth of pellets in the initial stages of batch pelletization proceeds first by nucleation and then mainly by the coalescence mechanism. In later stages of pelletization there is evidence that a limited amount of breakage of pellets occurs. In industrial pelletizing operations where a continuous supply of moist feed is charged to the pelletizing device, the growth mechanism of layering (or snowballing) of added moist feed onto the pellets becomes important.

Effect of Moisture and Rolling The growth of balls in terms of the characteristic S type is applicable within a narrow range of moisture content in the pellet mix. With lower moisture content (than the theoretical saturation volume of voids) in the pellet mix, the expulsion of moisture on the pellet surface would be slower at a particular collision frequency in the balling device. As a result, the growth of pellets remains sluggish and the transition region may extend to much higher levels of disc rotation (period). The inner structure of the pellets remains porous for a longer time. Once the limiting porosity is reached by repeated collision with other nuclei in the disc, the excess moisture is made available on the pellet surface for coalescence to occur with other wet or surfaces of dry granules, thus increasing the growth rate. The expelled water from inside is partly pulled back into the voids by negative capillary pressure to prolong the transition stage in the pellet growth curve, Therefore, the transition region gets delayed for charges containing less moisture than the theoretical saturation volume of voids. Beyond the transition region, the excess moisture is available at the pellet surface, and pellets grow further by the coalescence mechanism.

12.7.2 Kinetic Equation for Pellet Growth

Figure 12.21 shows that the average diameter of pellets increased very rapidly in the initial stages of pelletization (lower number of revolutions) and then at a slow rate. This indicates that the rate of increase of (average) diameter was a logarithmic function of the number of revolutions. The growth kinetics can be represented by a differential equation:

$$d\,D_{av.}/dN = K' * d\ln N/dN, \text{ or}$$

$$D_{av.} = K \ln N + C \tag{12.17}$$

where:

$D_{av.}$ = arithmetic average of the green pellets (Eq. 12.16),
N = total number of disc revolutions, and
K = overall rate constant.

Hence, a straight line is obtained when the average diameter (Dav.) of the pellets is plotted against the logarithmic value of the number of disc revolutions. The

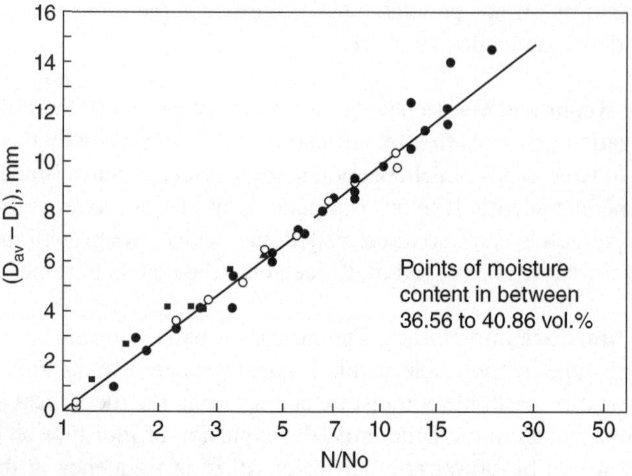

FIGURE 12.22 Plot of $(Dav - D_O)$ as a function of (N/N_o) in logarithmic scale (after Ref. [30])

straight lines at different moisture contents within the pellets' applicable range (see **Sec. 12.2.3**) are parallel—that is, the values of the slopes remain the same. The range has been observed, typically, between 38.06 and 40.19 vol.% in the case of blue dust from Goa, Western India.[30]

If D_o is the diameter of the balls attained during the green pelletization in the case of typical pellet mix at a given moisture content and N_o is the corresponding (total) number of revolutions at a point beyond the transition region (see **Figure 12.21**) where the growth occurs by coalescence. Introducing these values in Eq. 12.17 and combining these two:

$$D_{av.} - D_o = K * Ln(N/N_o) \tag{12.18}$$

Eq. 12.18 represents the kinetic equation for pellet growth beyond the transition region.

The value $(D_{av.} - D_o)$ when plotted against Ln (N/N_o) should result in a straight line for a particular value of moisture content. Kanetkar et al.[30] have observed that all the points, corresponding to different moisture content, within the applicable range (38.06 and 40.19 vol.% in this case), lie on a single straight line (**Figure 12.22**).

12.7.3 Factors Influencing the Ball Growth

The extent of these growth regions, hence, the kinetics of pellet growth, depends strongly on:

Moisture content of the feed: The influence of moisture content in the pellet mix on the kinetics of taconite pellet growth in a batch balling drum is shown in **Figure 12.23(a)** and **12.23(b)**, respectively, in the case when no binder was used and at 0.5 wt.% bentonite. The average diameter of the

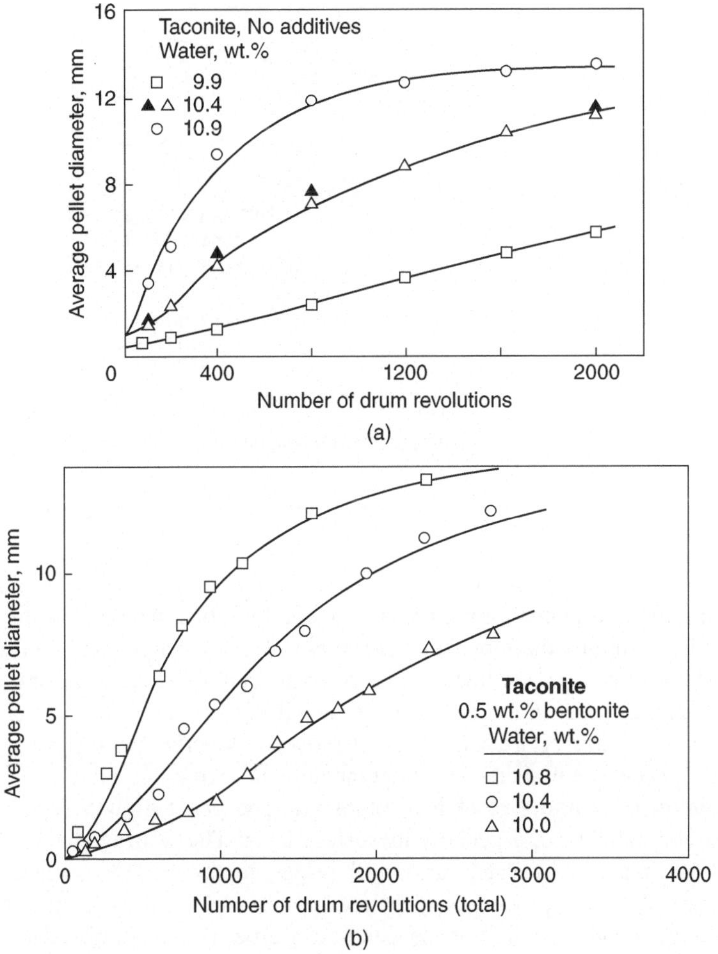

FIGURE 12.23 Effect of moisture content on the ball growth of taconite concentrate: (a) without additive (after Ref. [32]) and (b) with 0.5% bentonite (after Ref. [31])

pellets at a fixed number of drum revolutions increases with increasing the moisture content (within its applicable range). At a particular value of D_o, say 4 mm, it is apparent from Figure 12.23(a) and 12.23(b) that the value of N_o (Eq. 12.18) decreases very rapidly with increasing moisture content.

The naturally occurring capillary force predominates at a theoretical saturated volume of moisture content, which results in the highest compressive strength (see **Figure 12.14**). With the content of moisture higher than the theoretical saturated volume, the rate of moisture expulsion onto the surface is faster and this in turn increases the coalescence rate.[5]

Method of feed preparation: The kinetics of green ball growth depend on how the pellet concentrate is prepared: whether grinding is through wet circuit or following dry grinding. The concentrates generated from wet

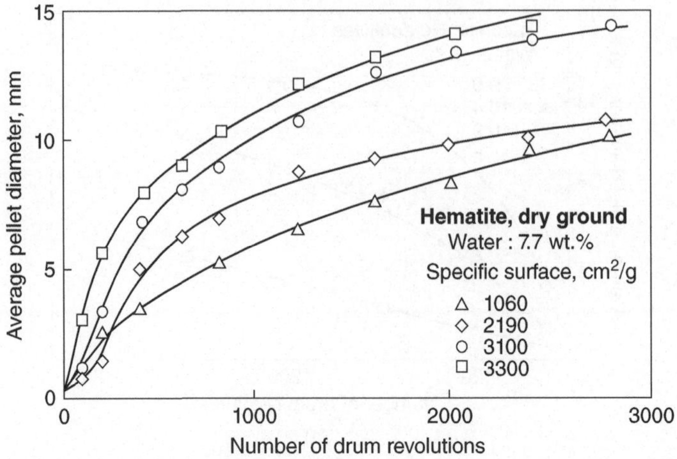

FIGURE 12.24 Effect of specific surface area of dry ground hematite concentrate on the ball growth (after Ref. [33])

comminution appears to exist as individual entities, whereas in the case of dry grinding, the forces that are responsible for size reduction of larger particles can cause the fine particles to compact into hard pre-agglomerates, which may exist in the feed during the balling operation. In view of this, though the size of ore concentrate is reduced, the effective surface area of the concentrate that take part in granulation is decreased.

Characteristics and size of feed material: The pellet growth depends on the nature of feed, especially its surface layer. The wetting characteristic of feed material by water varies with respect to its type. The contact angle data and surface properties of the feed materials indicate that water forms stronger bonds with iron ore concentrate or limestone as compared to silica.

The growth of green pellets should increase for the coarser-sized concentrate. However, in the case of dry ground concentrate, an accelerated growth of green ball formation has been observed with increasing fineness of dry ground material (**Figure 12.24**).[33] This anomalous behavior has been explained by the fact that the effective specific surface area actually decreases with increasing fineness due to the compacting effect of the fines when subjected to the grinding forces.

The pellets from coarser ores with lower bulk density lead to a lower rate of disintegration, resulting in an increased rate of growth. Also, the yield of green pellets is higher with coarser iron ore feed than with a finer one. On the other hand, the slower ball formation with finer ore results in more compact and stronger balls. These finer ores, therefore, are better suited for the production of quality pellets.

The addition of binders: The addition of bentonite in the pellet mix immobilizes the excess moisture,[26d] decreasing the growth rate of ball formation.

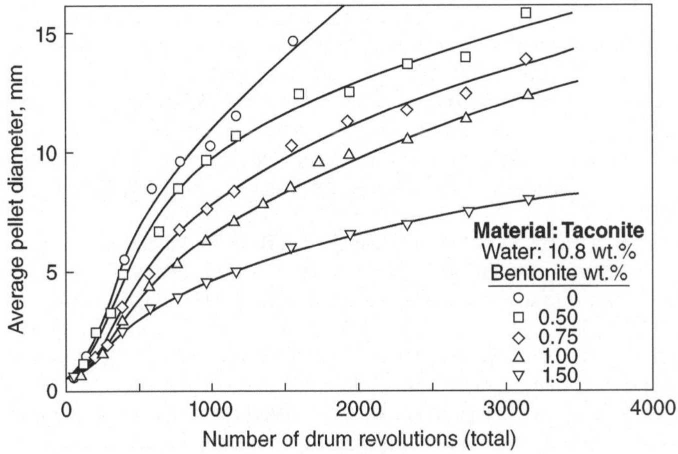

FIGURE 12.25 Influence of bentonite addition on ball growth kinetics (after Ref. [31])

The effect of binder on the growth rate of green pellets is not in isolation as the growth rate is also influenced by the pellets' moisture content. Increasing moisture content of the green pellets increases their growth rate, whereas increasing the bentonite dosage has an opposite effect (**Figure 12.25**). The presence of bentonite tends to slow the growth rate of balls compared to balling without bentonite. The addition of bentonite in the mix, though, reduces the pelletizing capacity, which leads to more uniformly sized and smooth balls than would have been produced without a binder.

REFERENCES

1. Sastry, K. V. S. and M. Cross, Basic and applied aspects of pelletizing of fine particles, in: *Proc. Inter. Seminar on Fine Particle Processing*, Las Vegas, Nevada, Feb. 24–28, 1980, Somasundaran P. (ed.), Pub. by Amer. Inst. of Mining, Metallurgical & Petroleum Engineers, New York, Vol. 2, pp. 1463–1493 (Chap., 74).
2. (a) Liu, Xi, C. Mao, W. Sun, and X. WU. Image-based method for measuring pellet size distribution in the stable area of disc pelletizer, *ISIJ Inter.*, 384, (2018), 1–7. Advance Publications by J-Stage, DOI: 10:2355/ISIJ Inter.
(b) Jaroslav, S. R. B., Z. Ruzickova, and Z. Ruzickova, Pelletization of fines, in: Furestenau D. W. (ed.), *Development in Mineral Processing*, Elsevier Science Pub. Inc., New York, Vol. 7, (1988), p. 69.
3. Meyer Kurt, *Pelletizing of Iron Ore*, Springer-Verlag, Berlin, (1980), p. 204.
4. Sastry, K. V. S. and D. W. Fuersteau, Kinetics and process analysis of agglomeration of particulate materials by green pelletization, in: Sastry K. V. S. (ed.), *Agglomeration 77*, AIME, New York, (1977), pp. 381–402.
5. Newitt, D. M. and J. M. Conway-Jones, A contribution to the theory and practice of granulation, *Trans. Inst. Chem. Eng.*, 36, (1958), 422–442.
6. Iveson, S. M., J. D. Litster, K. Hapgood, and B. J. Ennis, Nucleation, growth and breakage phenomena in agitated wet granulation processes: a review, *Powder Technol.*, 117, (2001), 3–39.

7. Xuewei, L. V., C. Bai, G. Qiu, S. Zhang, and M. Hu, Moisture capacity: definition, measurement, and application in determining the optimal water content in granulating, *ISIJ Inter.*, 50(5), (2010), 695–701.
8. Tigerschiold, M. and P. Ilmoni, Fundamental factors influencing the strength of green and burned pellets made from fine magnetite–ore concentrates, in: *Proc. Blast Furnace, Coke Oven and Raw Materials, session on ore and agglomeration*, 9, (1950), pp. 18–45.
9. (a) Ross, J. S., Bentonite—A challenge for Canadian industry, *CIM Bull.*, 57, (1964), 648–652.
(b) Sargent, D.H., N. O. Johnson, J. Boyer, and R. B. Frahme, Marketing and transportation of western bentonite, *Min. Eng.*, 34, (1979), 1314–1319.
10. Tarjan G., Der EinfluB von Bewegungskraften auf die Granulation, *Aufbereitungstech*, 1, (1966), 28–32.
11. Rumpf, H., The strength of granules and agglomerates, in: Knepper W. A. (ed.), *Agglomeration*, Interscience, New York, (1962), pp. 379–418.
12. Forsmo, S. P. E., A. J. Apelqvist, B. M. T. Bjorkman, and P. O. Samskog, Binding mechanisms in wet iron ore green pellets with bentonite binder, *Powder Technol.*, 169, (2006), 147–158.
13. (a) Urich, D. M. and T.-M. Han, A progress report on the effect of grind, temperature, and pellet size upon the quality of specular hematite pellets, in: Knepper W. A. (ed.), *Agglomeration*, Interscience, New York, (1962), pp. 669–719.
(b) Tapia, P., J. Palacios, F. Parra, A. Rodriguez, and B. Flores, Effect of grinding grade on green pellet quality, in: *Proc. of 3rd IAS Ironmaking Seminar*, Buenos Aires, Argentina, Oct. 29–Nov. 1, 2001, pp. 67–72.
14. Struve, V. G., Grundlegende Betrachtungen uber das Pelletieren von Eisenerzen, *Chem. Ing. Tech.*, (1964), 1019–1027.
15. Iveson, S. M., S. Holt, and S. Biggs, Advancing contact angle of iron ore as a function of their hematite and goethite content: Implication for pelletising and sintering, *Inter. J. Miner. Process.*, 74, (2004), 281–287.
16. Fan, X. H., M. Gan, T. Jiang, X. L. Chen, and L. S. Yuan, Decreasing bentonite dosage during iron ore pelletizing, *Ironmaking and Steelmaking*, 38, (2011), 597–601.
17. Ball, D. F., P. R. Dawson, and J. T. Fitton, Agglomeration in iron ore pelletizing, *Trans. Inst. Min. Metall. C*, 79, (1970), C189–C196.
18. Ross, J. S., Bentonite—A challenge for Canadian industry, *CIM Bull.*, 57, (1964), 648–652.
19. Sargent, D. H., N. O. Johnson, J. Boyer, and R. B. Frahme, Marketing and transportation of western Bentonite, *Min. Eng.*, 34, (1979), 1314–1319.
20. Mohamed, O. A., M. E. H. Shalabie, N. A. El-Hussiny, M. H. Khedr, and F. Mostafa, The role of normal and activated bentonite on the pelletization of barite iron ore concentrate and the quality of pellets, *Powder Technol.*, 130, (2003), 277–282.
21. De Souza, R. P., C. F. de Mendonca, and T. Kater, Production of acid iron ore pellet for direct reduction using an organic binder, *Min. Eng.*, 36, (1984), 1437.
22. Pal, J., S. Ghorai, M. C. Goswami, S. Ghosh, D. Ghosh, and D. Bandyopadhyay, Development of fluxed iron oxide pellets strengthened by CO_2 treatment for use in basic oxygen steel making, *ISIJ Inter.*, 49, (2009), 210–219.
23. Osaman, S., I. Arol, T. Eisele, and S. K. Kawatra, The effect of calcined colemanite addition on the mechanical strength of magnetite pellets produced with organic binders, *Miner Process Extr. Metall. Rev.*, 34, (2013), 210–222.
24. Ripke, J. S. and S. K. Kawatra, Can fly-ash extend bentonite binder for iron ore agglomeration? *Inter. J. Miner. Process*, 60, (2000), 181–198.
25. ASTM E946-92, Standard test method for water absorption of bentonite by the porous plate method, in: *Annual Book of ASTM Standards*, (Metals: Test Methods and Analytical Procedures), Vol. 3.06, (1996), pp. 276–277.

26. Sastry, K. V. S. and D. W. Fuerstenau, Ballability index to quantify agglomerate growth by green pelletization, *Trans. AIME*, 252, (1972), 254–258.
27. Sportel, H., P. Beenjes, J. Rengersen, and J. Droong, Quality of green iron ore pellets, *Ironmaking and Steelmaking*, 24 (1997), 129–132.
28. (a) Sastry, K. V. S., C. Hosten, and P. Dontula, Soc. Min. Eng. *AIME (preprint)*, 93, (1993), 275.
    (b) Capes, C. E. and P. V. Danckwets, *Trans. Inst. Chem. Eng.*, 43, (1965), 116.
    (c) Kapur, P. C. and D. W. Fuerstenau, *Trans. AIME*, 229, (1964), 348.
29. Sastry, K. V. S. and D. W. Fuerstenau, A laboratory method for determining the balling behavior of taconite concentrates, *Trans. AIME*, 250, (1971), 64–67.
30. Kanetkar V. V., S. K. Gupta, and R. P. Krishna, Studies on kinetics of green pellet formation from industrial iron ore in the disc pelletizer, *Trans. Indian Inst. Met.*, 30, (1977), 50–55.
31. Kapur, P. C. and D. W. Fuerstenau, Size distribution and kinetic relationships in the nuclei region of wet pelletization, *Ind. Eng. Chem. Process Des. Dev.*, 5, (1966), 5–10.
32. Sastry, K. V. S., and A. Negm, Role of binders during pelletization of iron ore concentrates—Pellet growth and strength aspects, in: *Proc. 4th Inter. Symp. on Agglomeration*, Toronto, Canada, 1985, Capes C. E. (ed), Iron & Steel Society of AMIE CIM, Iron & Steel Society, AIME, pp. 41–48.
33. Sastry, K. V. S., S. C. Panigrahy, and D. W. Fuerstenau, Effect of wet and dry grinding on the batch balling behavior of particulate materials, *Trans. AIME*, 262, (1977), 325–330.

13 Quality of Green Pellets

ABBREVIATIONS

CS *Compressive strength*
DN *Drop number*

13.1 PREAMBLE

The process of green pelletization is characterized by:

a. The growth of pellets and its rate (productivity)
b. Quality parameters of green pellets:
   - Average size and porosity
   - Drop resistance (drop number, DN)
   - Compressive strength (CS): wet as well as dry

These affect the transportation of green pellets from the pelletizing unit to the furnace and subsequent induration in the furnace. Drop number is of greater significance, more than compressive strength, in material handling of the green balls; whereas, the green balls must have sufficient compressive strength so that these do not break and retain their identity (spherical shape and size) on preheating and firing.

The quality parameters of green pellets depend on:

a. Feed particle size
b. Amount of water added during pelletizing
c. Disc rotating speed
d. Inclination angle of the disc bottom
e. Residence time of materials in the disc

The effect of these parameters on the mechanical properties of green, dried, and fired iron ore pellets, as being material specific, has been subject of research.[1]

The factors that affect the process of green pelletization of the ores vis-à-vis the quality of produced green pellets are:

a. Material characteristics: Particle size (specific surface area or Blaine number) of the feed (ore concentrate in particular) and ore porosity
b. Process parameters: Types and quantity of binder(s) and other additives, amount of water added during pelletizing, and pellet basicity
c. Operational parameters: Disc rotating speed, inclination angle of the disc bottom, and residence time of materials in the disc (total number of disc revolutions)

13.2 SIZE AND POROSITY OF PELLETS

The size of green pellets produced in a rotating device is decided by the growth kinetics during the granulating stage and is influenced by the process and operational variables (see **Sec. 12.7.1**). The growth of pellets as a function of disc revolutions follows a typical S-type curve indicating the nucleation of green pellets, transition region, and pellet growth (through coalescence mechanism) for different levels of moisture content (see **Figure 12.21**).

Disc Loading and Duration of Pelletization Figure 13.1 shows, schematically, the average diameter of green pellets ($D_{av.}$ in mm) as a function of disc loading (expressed as percent of maximum loading) for (total number of) disc revolutions, 500 and 1,000 when the disc speed was 25 rpm. The $D_{av.}$ increases with an increasing number of disc revolutions (see **Figures 12.23A and 12.23B** also). The diameter also increases with increasing disc loading (Figure 13.1), but the curve flattens at much higher loading.[2]

The collision frequency and, in turn, the growth rate are lower with smaller loading. On the other hand, weaker pellets are formed, without much compaction, if the disc is heavily loaded.

Speed and **Inclination** of **Pelletizing Disc** The average diameter of green pellets ($D_{av.}$) is also influenced by the pelletization speed (rpm) and the disc inclination. Newitt and Conway-Jones[3] have shown that the value, diameter of green pellets, at different pelletization speeds coincides when plotted as a function of the number of drum revolutions (see **Figure 12.22**).

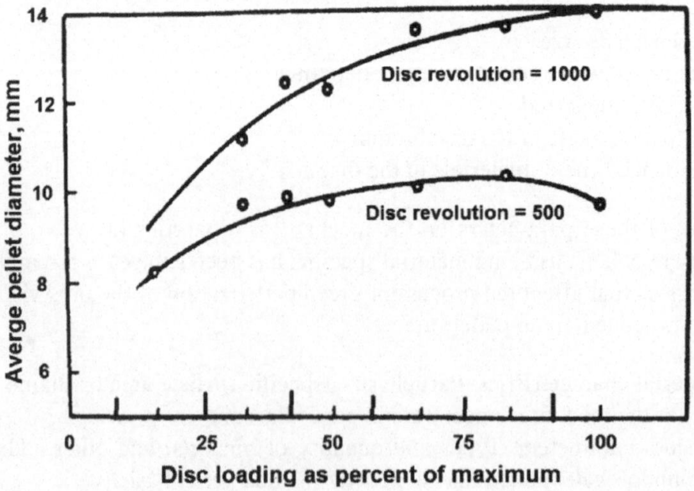

FIGURE 13.1 Dependence of pellet growth in terms of loading of pellet mix at different durations of pelletization

However, the growth mechanisms involved in the drum and disc pelletizers differ, which should be taken into account.

At lower speeds (rpm), collision frequency is less, resulting in lesser compaction of the pellets and, consequently, its structure is more porous. On the other hand, at higher speeds, the collision frequency increases and compaction is more because of the increased centrifugal force acting on the pellets. Besides, the rate of movement of moisture from the pellet interior to its surface increases with an increased speed of the rotating device (ball or disc). However, the total amount of expelled water happens to be a function of pelletization time (number of revolutions).[4]

The porosity of green pellets is contributed by:

- The internal porosity of ore concentrate: The ore concentrate has inherent porosity that remains during the pelletization process.
- The expulsion of moisture on the surface of pellet during its rolling. This is affected by the initial moisture content of the pellet mix, collision frequency in the balling device, and (total) number of rotations (duration of pelletization). Having the frequency in the balling device the same/similar, the expulsion of water onto the surface of pellets would be slower with a smaller content of moisture content in the mix. The inner structure of the pellets remains porous for a longer time. Once the limiting porosity is reached by repeated collision with other nuclei in the disc, the excess moisture is made available on the pellet surface for coalescence to occur. The expelled water from inside is partly pulled back into voids by negative capillary pressure to prolong the transition stage in the pellet growth curve.

13.3 DROP RESISTANCE (NUMBER)

Drop resistance of green pellets indicates their ability to remain intact (without a breakage) during their handling and is measured as "drop number." It is measured by dropping a freshly made (undried) single green pellet from a certain height of 0.45 m onto a steel plate before it shows perceptible cracks or it crumbles. The number of drops that the pellet had undergone before it develops a fracture is recorded and expressed in a number.

A high value of drop number of green pellets indicates that the pellets undergo more plastic deformation and that is not desirable. The pellets, when deformed, reduce the bed permeability on induration in the grate furnace. Therefore, their amenability to being dried and preheated gets diminished. On the other hand, a low value of drop number is associated with dry and brittle green pellets that crumble and generate fines that fill the voidage, and thus the flow of gases and heat through the bed decreases during the initial stages of induration.

Accuracy in the measurement of drop number, however, is not much as it is subjected to human error.

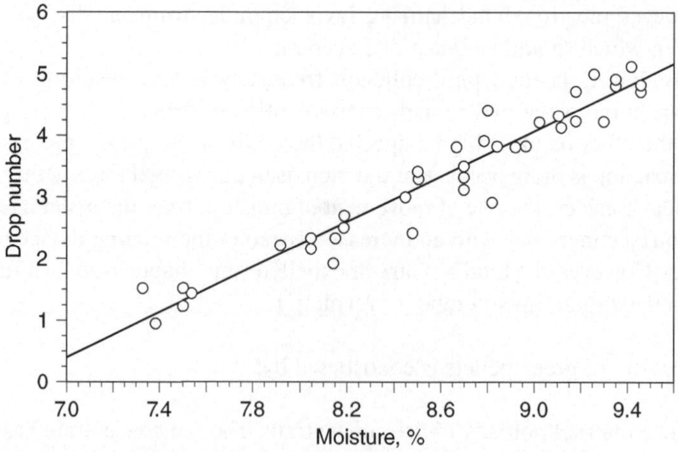

FIGURE 13.2 Effect of moisture content of the pellet mix on drop number (bentonite: 0.5%) (after Ref. [6])

Effect of Moisture Content: Water is important in green pelletization. Dry ground material should be pre-wetted to improve the balling operation. In practical operation, therefore, the green balls are seen to be slightly over-wetted. The moisture requirement for balling is material specific. The optimum moisture content for a particular feed ore depends on its particle size and its size distribution, inner porosity in particles, surface roughness, and wettability of the solids. These parameters often overlap in a complex manner. Therefore, the optimum moisture content cannot be clearly defined, though it can be worked out experimentally.

Drop number depends on the moisture content of the green pellet. It increases with increasing the moisture content within the range of 8 wt.%–10 wt.% of the pellet mix.[5]

Figure 13.2 shows the effect of moisture content on drop number of green pellets at 0.5 wt.% bentonite content. The drop number increased linearly with increasing the moisture content in the similar fashion as the plastic deformation (with increasing moisture content). However, this correlation is valid only at lower bentonite dosage (less than 0.6 wt.%) (see **Figure 13.3**).

Binder and Viscosity Effect: Resistance to fracture during the drop of green pellets is viscosity controlled. Since the fluid viscosity is influenced by the type and content of binders, these have a significant effect on drop number. The binder molasses gives a higher drop number compared to bentonite. **Figure 13.3** shows the effect of bentonite dosage on drop number of green pellets. The effect of increasing bentonite dosage, at its lower content, on drop number is marginal. The drop number increases sharply at higher, but intermediate, bentonite dosage in the 0.6–1.0 wt.% range and the relationship flattens at the bentonite dosage beyond this level.

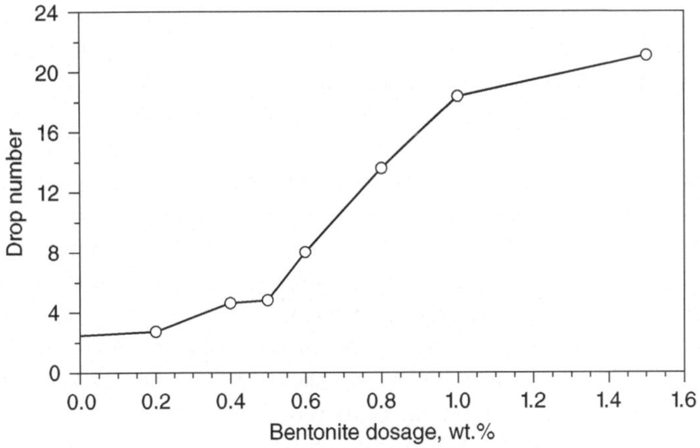

FIGURE 13.3 Relationship between drop number and bentonite dosage (moisture: 8.7%–9.3%) (after Ref. [6])

13.4 WET COMPRESSIVE STRENGTH

Moisture of Green Pellets: The maximum compressive strength of green pellets is achieved at the capillary state when the pores are filled in with binder liquid (water) adequately (see **Sec. 12.3.1**). The content of water affects the degree of its saturation in the green balls. A higher moisture content results in flooding. Accordingly, the strength of green balls increases with increasing moisture but decreases at its higher content. (See **Sec. 12.7.2.**)

Characteristics of Ore Concentrate: The green pelletization is influenced by the nature of ores and the beneficiation process adopted to produce the concentrate.

The concentrate from blue dust or flaky ore has a lower moisture soaking capacity. This results in lower green strength of blue dust pellets (due to lower saturation of water in pores).

Agglomeration of the concentrate that is enriched from iron ore by the flotation process has been recognized as a problem in green pelletization. Iwasaki et al.[7] have found that balling of iron ore concentrates in the presence of a fatty acid flotation collector reagent resulted in weaker green pellets both in wet and dry states. Gustafsson and Adolfsson[8] also report that balling of floated pellet feeds resulted in weaker green pellets (lower drop number), increased circulating loads in the balling circuits, and increased generation of fines during induration. The negative effects have been explained by a combination of decreased surface tension in the water phase and by adsorption of the collector reagent on magnetite. Complete wetting has generally been assumed.

Size of Ore Concentrate: The crushing strength of green pellets (wet-CS) becomes higher when the finer ore (higher Blain number) concentrate is balled. An increase in wet-CS of green pellets with increasing raw material fineness also has been reported in relation to hematite and magnetite concentrates.[9,10] A linear correlation has been observed between the specific surface area of ore concentrate and compressive strength of green pellets (see **Figure 16.2** [curve 3], **Sec. 16.4.1**).

The parameters, compressive strength of green pellets and drop resistance, are influenced by the specific surface area and moisture content of the pellet feed in an interacting manner. There happens to be optimum moisture content with respect to the compressive strength. The optimum value shifts toward the higher side with the increasing Blain number of the ore concentrate. In other words, the ore concentrate with a finer size (higher specific surface area) requires more water.

The control strategy for the balling circuits in green pelletization should be based on the parameters related to plastic deformation and compression strength of green pellets instead of their drop number.

Porosity of Ore Concentrate: Porosity of concentrate plays an important role in deciding the saturation degree of the pores (see Eq. 12.1), which is defined as the volume fraction of pores taken by water. The particles in green pellets are held together due to the capillary forces, which are responsible for accounting negative pressure inside the pores. The pressure in the liquid-saturated pores gives strength to the green pellets (see **Sec. 12.3.3**).[11]

Dosage of Binder: The bentonite clays form fibrous structures under compressive shear, which results in an appreciable increase in green pellet strength.

The compressive strength of green pellets, in addition to the content of moisture, is influenced by the dosage of bentonite in the pellet mix. The bentonite in the pellet mix affects the compressive strength of the green pellets as it provides a source of colloidal material that decreases interparticle distance. The wet crushing strength increases sharply with increasing the bentonite dosage; however, the effect is panned out beyond the bentonite dosage 0.6 wt.% (**Figure 13.4**).

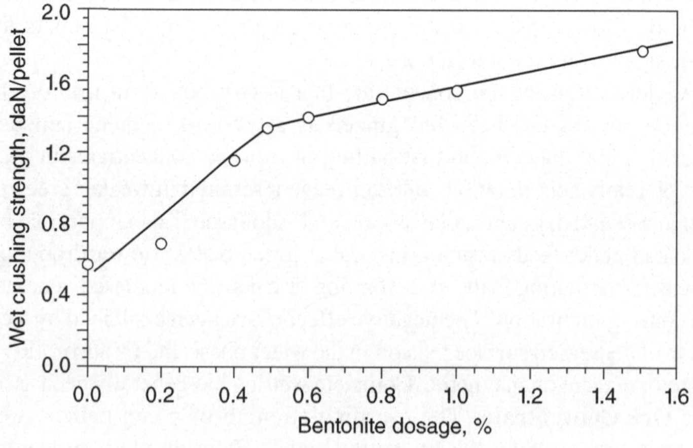

FIGURE 13.4 Crushing strength of green pellets affected by bentonite dosage (moisture content ~ 9%) (after Ref. [6])

At a low bentonite dosage, the binding force is a combination of the capillary force (easily movable liquid) and the cohesive force (viscous binder). With increasing bentonite dosage, cohesive forces from highly viscous binders dominate the binding mechanism. Increasing bentonite content in the green pellets increases their plastic deformation; when in excess, this leads to the production of pellets with very plastic properties, which is undesirable for further processing.[12]

Compressive Strength of Dried Pellets: The compressive strength of dried pellets indicates their ability to survive handling during the firing stage of induration. The strength is influenced by the binder input. With the addition of bentonite, green pellet strength changes only insignificantly; whereas the dry strength is highly influenced.

The bentonite, finely distributed in the water, concentrates during drying particularly in the residual liquid, especially at the contact points and on the contact surface of the ore grains. During drying, this viscous fluid forms a stronger gel as the moisture is removed, which gets concentrated and forms a solid bridge of hardened gel that strengthens particle contact points. Consequently, the superficial network of viscous fluid among the particles gets strengthened (see **Sec. 12.4.2, and Figure 12.17, Figure 12.12**).

The increased viscosity of fluids among the mineral grains in the pellets accounts for the production of the well-rounded, plastic pellets that can be easily handled for sizing and transport in the plant.

REFERENCES

1. Ahmed, Y. M. Z., F. M. Mohamed, N. A. El-Hussiny, and M. E. H. Shalabi, Pelletization of heterogeneous iron concentrates for blast furnace operations, *Can Metall Quart*, 144(1), (2005), 95–102.
2. Gupta, S. K., *Personal communication.*
3. Newitt, D. M. and J. M. Conway-Jones, A contribution to the theory and practice of granulation, *Trans. Inst. Chem. Eng.*, 36, (1958), 422–442.
4. Sastry, K. V. S. and D. W. Fuerstenau, A laboratory method for determining the balling behavior of taconite concentrates, *Trans. AIME*, 250, (1971), 64–67.
5. Kawatra, S. K. and S. J. Ripke, Laboratory studies for improving green ball strength in bentonite-bonded magnetite concentrate pellets, *Int. J Miner. Process*, 72, (2003), 429–441.
6. Forsmo, S. P. E., A. J. Apelqvist, B. M. T. Bjorkman, and P. O. Samskog, Binding mechanisms in wet iron ore green pellets with bentonite binder, *Powder Technology*, 169, (2006), 147–158.
7. Iwasaki, I., J. D. Zetersrrom, and E. M. Kalar, Removal of fatty acid coatings from iron oxide surfaces and its effect on the duplex flotation process and on pelletizing, *Trans. Soc. Miner.* Eng. AIME, 238, (1967), 304–312.
8. Gustafsson, J. O. and G. Adolfsson, Adsorption of carboxylate collectors on magnetite and their influence on the pelletizing process, in: *Proc. of XX International Mineral Processing Congress*, Aachen, September 21–26, 1997, 377–390.
9. Urich, D. M. and T.-M. Han, A progress report on the effect of grind, temperature, and pellet size upon the quality of specular hematite pellets, in: *Agglomeration*, Ed. W. A. Knepper, Interscience, New York, (1962), 669–719.

10. Tapia, P., J. Palacios, F. Parra, A. Rodriguez, and B. Flores, Effect of grinding grade on green pellet quality, in *Proc. of 3rd IAS Ironmaking Seminar*, Buenos Aires, Argentina, October 29–November 1, 2001, 67–72.
11. Sportel, H. and J. Droog, Influence of pore saturation on compressive strength of green iron ore pellets, *Ironmaking & Steelmaking*, 24, (1997), 213–223.
12. D. Zhu, J. Pan, G. Qiu, J. Clout, C. Wang, Y. Guo, and C. Hu, Mechano-chemical activation of magnetite concentrate for improving its pelletability by high pressure roll grinding, *ISIJ Inter.*, 44, (2004), 310–315.

14 Induration of Green Pellets

ABBREVIATIONS

CCS *Cold crushing strength*
CVRD *Companhia Vale do Rio Doce*
DR *Direct reduction*
LOI *Loss on ignition*
PH *Preheat*
RDI *Reduction degradation index*
TPH *Temperate preheat*

14.1 PREAMBLE

The induration of pellets is possibly the most important and, at the same time, critical step in the overall pelletization process affecting the profitability of the plant. The process affects:

a. The productivity of the finished pellets
b. Its quality parameters (which influence the iron-making processes)

Profitability of the induration process with regard to its operation depends on:

a. The recovery of heat in the induration process
b. The methodology adopted in the removal of gaseous pollutants and particulate materials from the induration plant and its effectiveness

A shaft furnace was used for heat hardening of the green pellets in the early days. This was discontinued as the temperature distribution within the indurated pellets was nonuniform and due to the fact that its capacity had been limited. Besides, the process was not efficient in heat recovery. Now, several technologies have been developed for heat hardening of pellets. Among these straight-grate and grate-kiln processes are the newer ones and the most widely used. This is because of their:

- Flexibility to indurate ores of different types
- High capacities and low operating costs, and ease of maintenance
- Capability to produce uniform and high quality pellets

A pellet indurating machine should be able to fulfill the following functions:

a. The green pellets must be completely dried at a rate that these should not crack in the drying or preheating zone.

b. The preheat zone should transfer sufficient heat to the pellets for strength. In the case of a grate-kiln system, the preheat zone should produce the pellets with adequate strength to survive the drop from the grate to the kiln. In case of high loss of ignition (LOI) ore concentrate or the pellets having carbonates incorporated in them, the preheating zone should be lengthened to allow for dehydration or calcination.
c. The pellets should be kept at the indurating temperature for the maximum time so that these have acceptable mechanical properties.
d. The cooling zone should be designed to recover the maximum useful heat while cooling the pellets to a temperature that prevents damage to the belt for material handling.

14.2 PELLET INDURATION: STEPS

Pellet induration consists of four main steps:

a. Drying of green pellets (temperature: 180–350°C).
b. Preheating of dried pellets (temperature: 500–1,100°C)
c. Firing of preheated pellets (temperature: 1,250–1,350°C)
d. Cooling of hot pellets (from ~900°C to ~80°C)

14.2.1 Drying

The temperature range in this step is 180 to 350°C. Free (surface and interstitial) moisture that is contained in the green pellets gets evaporated at lower temperatures. The removal of combined water from the pellets, produced from the high LOI ore concentrate, also takes place at higher temperatures. With evolving of moisture, the grains within the pellets move closely with volume shrinkage that increases the adhesion forces, resulting in the increased strength of the dried pellets.

The rate of drying influences the distribution of bentonite flakes in green pellets and the crushing strength of dried pellets. The viscous liquid forms a stronger gel as the moisture is removed during drying and, finally, solid mortar bridges are formed as the water is finally evaporated. The dehydration of bentonite gel is also claimed to be accompanied by a shrinkage that increases the adhesion forces.

The basic conditions for the dry pellet strength differ from those of green pellet strength (see **Secs. 12.3.1 and 12.4.1**). The compressive strength of green pellets is chiefly determined by the capillary forces of water. It reaches a maximum value when the liquid filling degree in the capillaries is about 0.8 to 0.9. During drying, liquid bonds are reduced and gradually replaced by the solid (bridges) bonds, which become effective among the ore particles.

Figure 14.1 schematically shows changes in the nature of the bond at different stages of pellet production. The capillary force, which is effective in green pellets, is being replaced by the solid bridges as the moisture (in the pellets) is removed in the course of drying. As mentioned later in this chapter (see **Sec. 14.2.3**), the solid bridges were fused to form slag bond at higher temperatures during induration. These bonds are responsible for the cold crushing strength (CCS) in pellets.

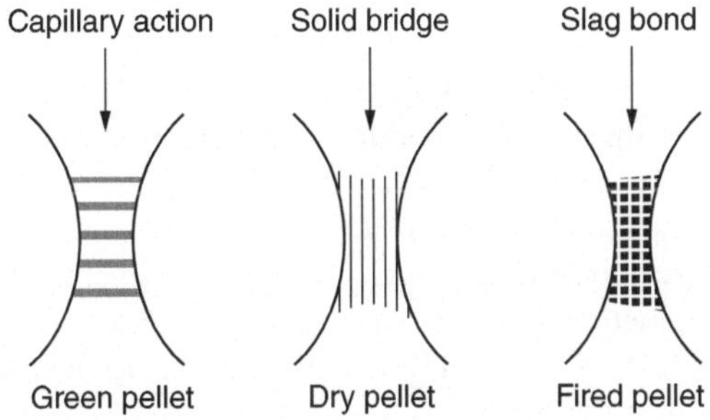

FIGURE 14.1 Bonds applicable at different stages of pellet production

Parameters, like temperature and duration of drying, and velocity of gas flow, are important to influence the total thermal treatment and the strength of dried pellet. The greater the temperature difference between drying medium and green ball, the shorter the period required for drying.

Porosity of green pellets plays an important role in their drying: the higher the porosity, the faster the drying. Besides, the porosity of fired pellets is influenced by the porosity of green pellets and the pore size distribution.

The drying rate of green pellets decreases with increased fineness of the concentrate.

Thermal spalling is the main mode of failure of green pellets during drying. It is the phenomenon of flaking off the outer layers of the pellets or even complete crumbling into small pieces. The reason for thermal spalling is built-up pressure inside the pellets, which originates mainly from the evaporation of moisture and decomposition of carbonates and hydrates.

The pressure drop due to the flow of a fluid through a porous equal-sized sphere, ΔP, is given by the Kozeny-Carman equation.[1] (See **Sec. 6.3.2** also.)

$$\Delta P = \left[K * \eta * (1 - €) * L * V\right] / \left[d^2 * €^3\right] \tag{14.1}$$

where:
K = Kozeny-Carman constant,
η = viscosity of liquid phase,
€ = porosity,
d = particle diameter,
L = depth of pellet, and
V = velocity of gas (drying rate).

Thus, for a given temperature and flow of drying gas (V): the built-up pressure inside the pellet is a function of the green pellet porosity (€), fineness of the balling feed (d), and the size of the green pellet (L). It is apparent from the previous equation that increasing the ball porosity, decreasing the fineness of balling feed (coarser sized material), and decreasing the pellet size will decrease the pressure buildup inside the pellet and hence lower the thermal spalling.

Thermal spalling depends on:

- Characteristics of the pellet feed: The heat inside the pellets builds up in the balls formed from the finely ground and lower porosity materials. This inhibits the diffusion of steam and volatilities from the interior (core) of the pellets, resulting in the pressure buildup inside the pellet, which causes spalling to occur.
- Diameter of green pellets and their porosity: A higher spalling is expected with higher diameters of pellets and/or lower porosity.
- Heating condition: A more uniform heating during drying is required to minimize the thermal spalling.

The drying rate of green pellets is reduced with the addition of bentonite. Also, the content of moisture that can be removed from the green pellets without spalling is also reduced. Both are undesirable. Nonetheless, bentonite also increases the drying temperature without spalling, which tends to cancel out the limitations of the drying process.[1]

14.2.2 Preheating

The range of temperature in this step is 500–1,100°C. Several reactions proceed simultaneously that depend on the chemistry of the green pellets, as follows:

a. **Preheating at 350°C:** Decomposition of hydrate minerals takes place. The chemically combined water from goethite ore (α-FeOOH) and/or any hydrate minerals is evolved when the pellets reach a temperature of 350°C or more:

$$2FeO(OH) = Fe_2O_3 + H_2O$$

As the moisture gets evolved from the green pellets, the osmotic suction acting among the grains is increased. This helps in pulling the grains to more close packing, resulting in the volume shrinkage and densification of the pellets.

The heating must be controlled during this period so as to remove the moisture without developing microcracks in the pellets.

b. **Oxidation of Magnetite at 980°C:** The magnetite is oxidized to hematite according to the following exothermic reaction:

$$4Fe_3O_4 + O_2 = 6Fe_2O_3 . \Delta Hr = -119 \frac{kJ}{mol} \text{magnetite}$$

Other iron-bearing minerals of a lower oxidation state, such as goethite and siderite (if present) are also oxidized to hematite.

The oxidation reaction (ferrous to ferric iron oxides) results in bonding and bridging within the minerals. This influences the mechanical strength of pellets, also their reduction behavior.

All the Fe_3O_4 present in pellets should preferably be oxidized to Fe_2O_3 in this stage before the formation of sintered bond. The oxidation reaction is dependent on the oxygen potential (availability of oxygen) at the magnetite surface. Hence, the porous form of the magnetite particles enhances the oxidation rate.

Since the reaction is exothermic, the temperature of the furnace needs to be strictly controlled. An excessive high temperature in this zone promotes sintering and reduces the pellet porosity; hence, it is detrimental to the oxidation of pellets.

c. **Other Chemical Reactions:** These constitute:
   - Decomposition of carbonates: Carbonates gets calcined and CO_2 gas is released.
   - Roasting of sulfides: Pyrites, if present in the pellet mix, get calcined and SO_2 gas is liberated.
   - Gasification of solid fuels (coal and coke).

d. **Recrystallization Bond Formation (950–1100°C):** The recrystallization of iron oxides is essentially a physical process in which smaller particles consolidate into larger ones with the loss of surface energy. Grain growth for hematite starts at around 1,100°C. Continued growth of iron oxide crystals imparts adequate strength to the pellets. The commencement of solid oxide bonding and grain growth are important physico-chemical reactions that take place in this step.

   The oxidized grains of magnetite (hematite) start forming a recrystallization bond at the grain boundaries where two grains are in close proximity to each other. This leads to the development of new hematite grains and smaller voids at the grain boundaries. The residual magnetite either forms recrystallized secondary magnetite grains, which get oxidized during the final stage of induration, or combine with the gangue forming oxides to make a slag bond.

14.2.3 Firing

The temperature range in this step is 1,250–1,350°C, which is below the melting temperature of the major oxide phases. The reactivity of gangue components and additives proceeds in this temperature range, and the final process of sintering of the grains takes place. This zone is characterized by the formation of diffusion and slag bonds. The bonding of mineral grains developed during induration of pellets is affected by the following factors:

a. **Diffusion (solid oxide) bonding:** The primary and secondary hematite grains form recrystallization bonds by the solid oxide diffusion process. Hematite crystals that are formed at lower temperatures on the edges and surface of ore grains cause the formation of initial bridges. Further oxidation of the recrystallized magnetite takes place by the diffusion of oxygen through the pores. This reaction is also associated with further shrinkage of pellet volume, which gives the bond strength.

b. **Formation of slag bond:** The gangue minerals (mainly quartz) react with FeO and other gangue minerals to form fayalite ($2FeO.SiO_2$) and complex silicates. The formation of fayalite takes place on the surface of iron oxides as well as in the form of needles or a platelet bridge. Besides SiO_2, CaO, MgO, and other nonferrous ingredients can cause the bridges: the bonds thus formed are not as strong as those caused by recrystallization of oxides.

14.2.4 Cooling

The pellets are then cooled and screened. The oversized pellets are crushed and, along with the undersized pellets, sent to the stock house bins.

14.3 PROCESS OF INDURATION USING SHAFT FURNACE

The shaft furnace is a tall, refractory lined, rectangular-shaped furnace with external combustion chambers and internal ports to give distribution to the hot gases. The external combustion chamber is provided to generate the heat required for indurating the green pellets. The green balls, charged from the furnace top, come into sufficient contact with the hot gas as they descend and exchange heat to increase their temperature. Breaker rolls are used to help support the bed and break up any clusters that develop. The heated pellets pass a cooling chamber that is placed at the lower part of the shaft before being discharged outside the furnace (**Figure 14.2**).

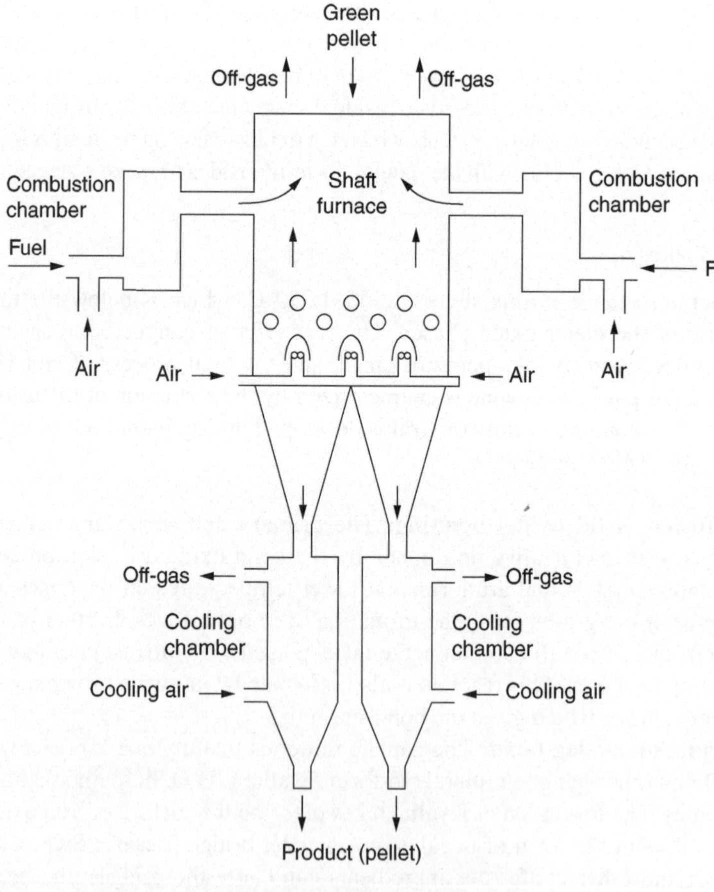

FIGURE 14.2 Flow of materials and gases in a shaft furnace

14.4 STRAIGHT-GRATE PROCESS OF PELLET INDURATION

14.4.1 Straight-Grate System

Machine The straight-grate system, used for induration of green pellets, was originally developed by Lurgi Metallurgie. This system emerged in the pellet-making industries soon after shaft furnaces. The straight-grate machine is essentially a single unit that moves as a static layer. The machine includes an endless grate car consisting of grate bars with side walls. A straight-grate traveling machine resembles the well-known sintering machine of Dwight Lloyd's design. **Figure 14.3** shows the straight-grate indurating machine and flow of materials during induration in the machine. The grate is partitioned into drying, preheating, firing, and cooling zones. The grate cars move through these zones at the same speed. The machine has several wind boxes beneath the traveling grate for updraft and downdraft flow of gases. The burners are provided in preheating and firing zones and controlled for a set of temperatures. Each burner control zone has a thermocouple in the furnace hood that is connected via a temperature transmitter to the burner zone temperature controller. The temperature controller compares the temperature signal with the set point and accordingly regulates the gas flow control valve.

Material Handling The green balls, which are sized (to 8–16 mm) in the screen, are charged onto the machine through the belt conveyor, which is upstream (ahead) of the indurating machine. The pelletizing devices

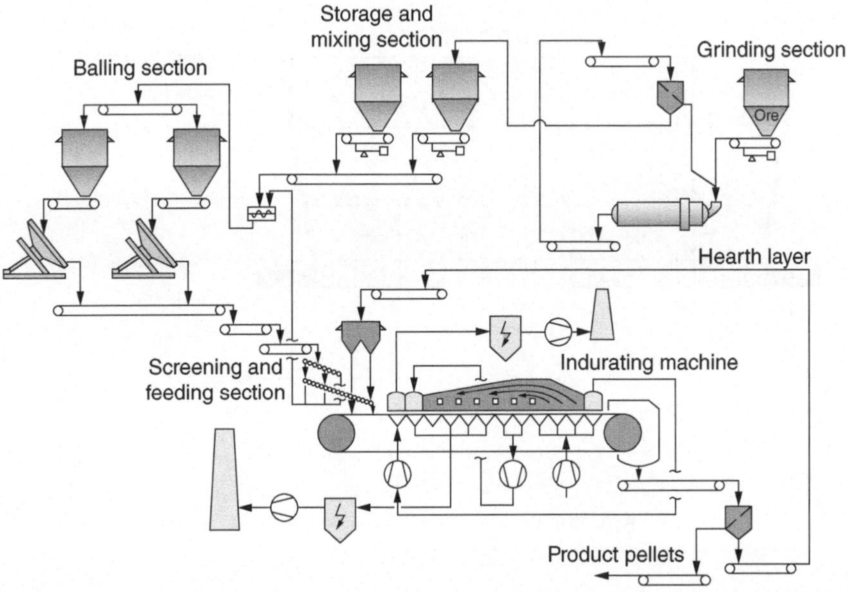

FIGURE 14.3 Straight-grate system for pellet induration and flow of materials (after Ref. [2])

are usually connected to roller screens. Undersized pellets (less than 9 mm) are separated in the roller screen and returned to the balling unit as seeds. The oversized fraction is usually crushed and returned (to the balling unit) (see **Figure 12.4**). The green pellets, which are produced in the pelletizing unit, have to be transported with utmost care by: (a) decreasing the number of transfer points wherever possible and (b) reducing drop heights as much as possible. The transfer points in green pellet handling are reduced through the incorporation of the reciprocating belt conveyor system. This controls the generation of green fines, which have to be recycled. The system also ensures that the green pellets in the size range 8 to 16 mm are evenly distributed across the width of the traveling grate.

Hearth Layer A hearth layer (about 0.05-m thick) of fired pellets is placed on the grate bars (upon which the pellets are laid) and also on the side walls. The layer of fired pellets protects the grate bar and side walls from high temperatures; it also prevents the green balls from being fired inhomogeneously. This reduces the variations in pellet quality. **Figure 14.4** shows cross section of straight-grate pallet. H1 and H2, respectively, are width of hearth layer and that of pallets which undergo induration. Typically, these values are 100 mm and 300 mm.

Bed Permeability A reasonably good permeability in a bed of pellets is beneficial in its subsequent heat hardening for the pellet production process. This minimizes the energy consumption on account of lower pressure drop over the charge. Besides, this makes an even heat treatment of the pellets, resulting in an improved quality. A higher bed

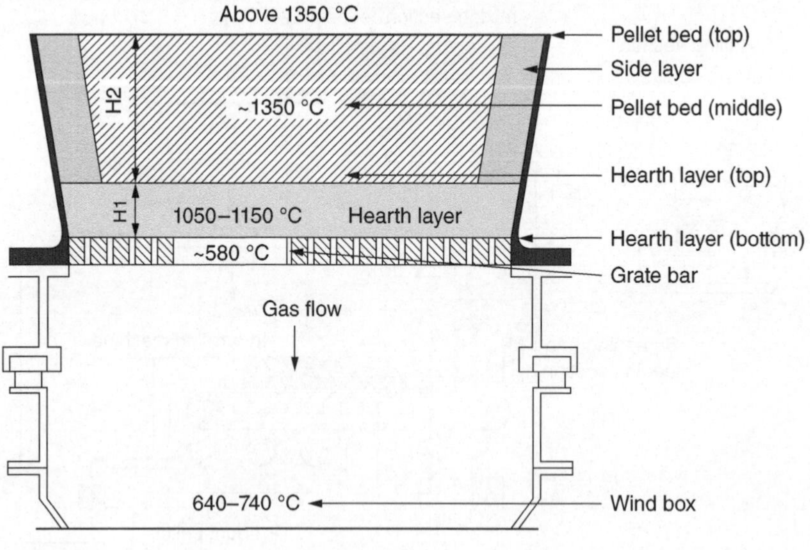

FIGURE 14.4 Cross section of the straight-grate pallet showing different layers

permeability can be achieved with the formation of a homogeneous pellet charge on the traveling grate. Besides, a narrow size distribution in the produced green pellets is an important criterion to achieve high bed permeability. Disturbances in the pellet size distribution are regulated either mechanically, by adjusting the screen openings for the recycling load or for the on-size fraction, or by varying the moisture content or the binder dosage.

14.4.2 Zones of Induration

The straight grate carries the green pellets on a bed of 500 mm to 550 mm height sequentially through a furnace through the steps of drying, preheating, firing of green pellets, and subsequent cooling. These induration steps are carried out in different sections (zones) of the straight-grate machine. Each zone is held at a predetermined temperature, and heat exchange occurs via hot air and/or combustion gas to fire the pellets (**Figure 14.5**).

The different zones associated with induration using the straight-grate furnace are mentioned in **Table 14.1** with their indicative portion in terms of percent of suction area and the maximum temperature maintained in that zone. However, these parameters may vary, and they depend on the characteristics of materials—such as whether the ore concentrate is hematite or magnetite, and their LOI content, type, amount of additives (fluxes, coke, etc.), and so on.

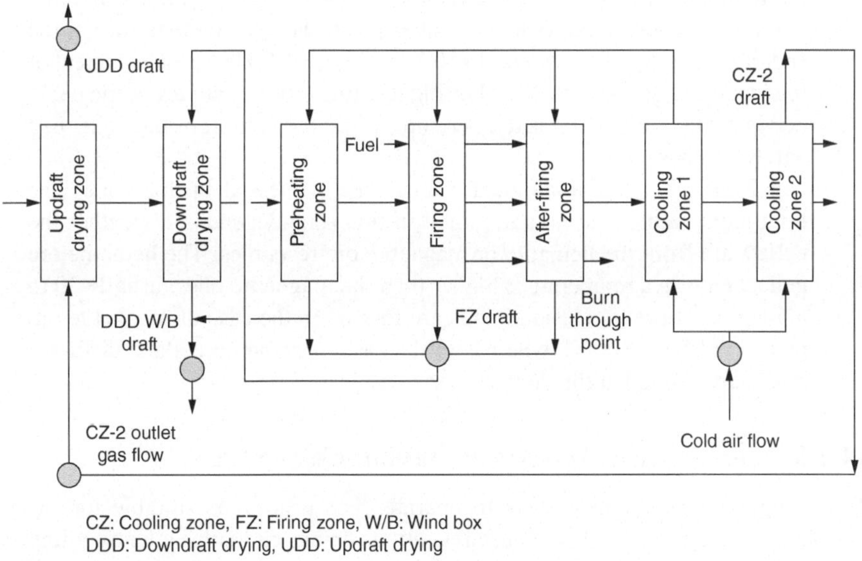

FIGURE 14.5 Schematic diagram of straight-grate induration process for production of iron ore pellets and recuperation of heat through recycling of hot air/gases (after Ref. [4])

TABLE 14.1
Different Zones of a Straight-Grate Furnace with their Indicative Areas and Temperatures

Zone	Suction area (approx.) %	Temperature, (maximum), K
Updraft drying (UDD)	25	550
Downdraft drying (DDD)	14	630
Preheating (PH)	7	1370
Firing (FZ)	32	1,520–1,580
After-Firing (AFZ)		1270
Cooling I (heat recuperation)	11	1,070–1,170
Cooling II	11	570

Interconnected process fans are provided to circulate air throughout the different zones of an indurating machine.

Drying Zones The bed of wet green pellets is transported through the drying zones: the updraft drying (UDD) and downdraft drying (DDD) zones. After the drying zones, the upper part of the pellet bed is dry and warm (temperature ~250°C), while the bottom of the bed is still partly humid.

Preheating Zones After drying, the bed of green pellets is transported through the temperate preheat zone (TPH) and the preheat zone (PH), where the main part of magnetite oxidation takes place. The gas temperature at the end of the PH zone is between 1,150°C and 1,250°C. The oxygen content of the incoming gas is 16 to 18%. During this time, the upper part of the pellet bed is heated up to the gas temperature, while the bottom of the pellet bed barely reaches 1,000°C.

Firing Zone Major mineralogical changes are witnessed in this zone. The final temperature that is maintained in this zone depends on whether the pellets are from the hematite or magnetite ore resources. The hematite ore pellets require a temperature higher than the magnetite ones: usually 50 to 80 degrees more. Diffusion bonds are formed in the case of magnetite ore pellets at 950–1,200°C, whereas in the case of hematite pellets, diffusion bonds are limited at this low temperature.[3]

14.4.3 Characteristic Features of Straight-Grate System

The straight-grate system is easy to operate. The process is suitable for carrying out induration steps of drying, preheating, firing, and subsequent cooling of heat-hardened pellets in one single unit. Its advantage over a shaft furnace exists as control of temperature with respect to individual steps could be undertaken in a wider range. The individual thermal zones are separate and can be designed to suit the ore types by controlling flow rates and temperatures.

The system, however, suffers from the disadvantage that any change in one section affects the residence time in another. That is, a change in the grate speed changes all the conditions for the subsequent process steps. In the straight-grate system, green pellets are indurated in a thick layer (500–550 mm) that may be prone to temperature variation between its top and bottom portions. This leads to variations in the product (pellet) quality. The system recirculates part of the fired pellets to form the layers on the grate bars and side walls in order to protect the grate bars and mechanical parts (**Figure 14.4**). This diminishes the effective volume of the strand for induration. Despite this protection, the pellets are subjected to wear when passing through the process steps at elevated temperatures.

14.4.4 Heat Recovery in Straight-Grate System

About 75% of the total heat supplied in the induration process (approximately 2,000 KJ/kg of pellets) is available with the fired pellets at around 1,300°C as sensible heat. A considerable advantage is achieved in terms of plant profitability and environmental protection if this heat is recovered during the cooling stage. The heat is transferred to the air during the cooling stage; as a result, the air becomes hot. **Figure 14.5** is the schematic of the pelletization process using a straight-grate machine. Possible options for flue gas recirculation to recover waste gas are also shown.

14.5 GRATE-KILN-COOLER PROCESS

14.5.1 Grate-Kiln-Cooler System

Presently, the grate-kiln-cooler process has become the most popular method to prepare iron-making burdens from ore concentrates, particularly in China. The product (indurated pellets using the process) possesses fine mechanical properties and excellent superior metallurgical performance on its reduction.

A grate-kiln-cooler system (**Figure 14.6**) consists of three major units:

a. A grate (similar to a traveling grate)
b. A kiln (similar to a rotary kiln)
c. A cooler (e.g., an annular cooler)

Drying, preheating, and oxidation are carried out in a traveling grate while pellets are fired in a rotary kiln by radiation. The heat-hardened pellets are subsequently cooled in a rotary cooler.

Traveling Grate Machine: This machine is used for induration of green pellets and resembles the well-known sintering machine of Dwight Lloyd's design. The traveling grate is partitioned into drying zones and preheating zones. In these zones, heat exchanges at relatively low temperatures occur for drying and preheating the green balls.

The grate has two or more wind boxes to provide for the gas draft. The pellet bed on the grate is only 150–200 mm deep, and no protective layer is

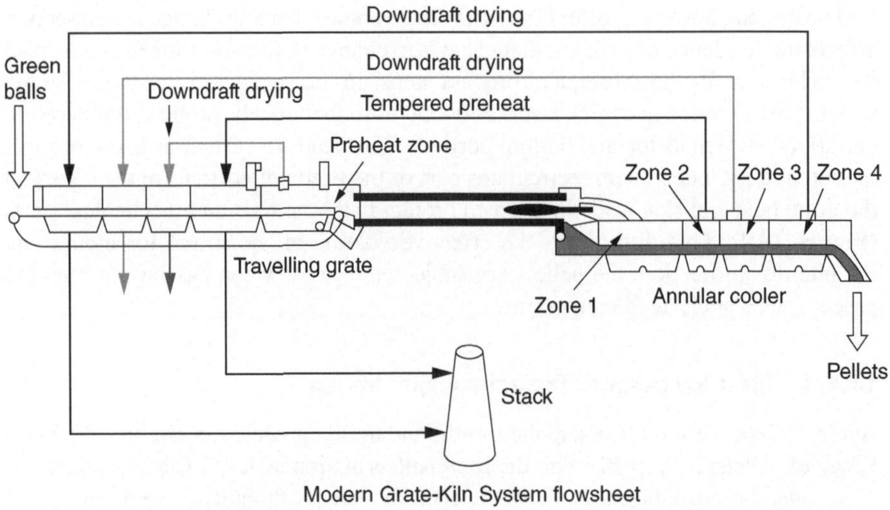

FIGURE 14.6 Grate-kiln-cooler system flowsheet for induration of pellets (after Ref. [5])

required on the grate because, in this case, the bottom of the bed does not reach a high enough temperature to damage the grate bars. The temperature of the bottom layer must be controlled around 900°C to ensure that the grate pellets are not destroyed (at a higher temperature).

The source of the heat for drying and preheating the pellets is not only the kiln off-gas but also recouped gas from the cooler: an arrangement that gives the plant, as a whole, high thermal efficiency (**Figure 14.6**). In order to ensure high thermal efficiency, the heating is executed through forced convection. The machine has a completely closed hood with many interconnected thermal zones for recirculation of gas leading to maximum heat recovery.

Kiln: The kiln is of relatively short length and is connected with the grate at its inlet and with the cooler at its outlet. The kiln is lined with refractory materials for firing at high temperature. The kiln is placed at a slight incline to the discharge and rotates at a low revolution. The pellets, after passing through the PH zone, are transferred to a rotating kiln and sintered at around 1,250°C while tumbling inside it. The tumbling action ensures the homogenous heating of all the pellets inside the kiln. This yields pellets of high quality in a consistent manner. The pellets, after firing, are subsequently transferred to the cooler.

Cooler: The hot pellets, which are discharged from the kiln, are cooled by air that is blown through the pellet bed. The bed depth is around 700 mm. The preheated air from the cooler is thereafter used as process gas in the kiln and in the different zones of the grate (see **Sec.** 14.5.5).

Figure 14.7 is a schematic representation of the material flow and heat recuperation practiced in a typical plant owned/constructed by Kobe Steel.

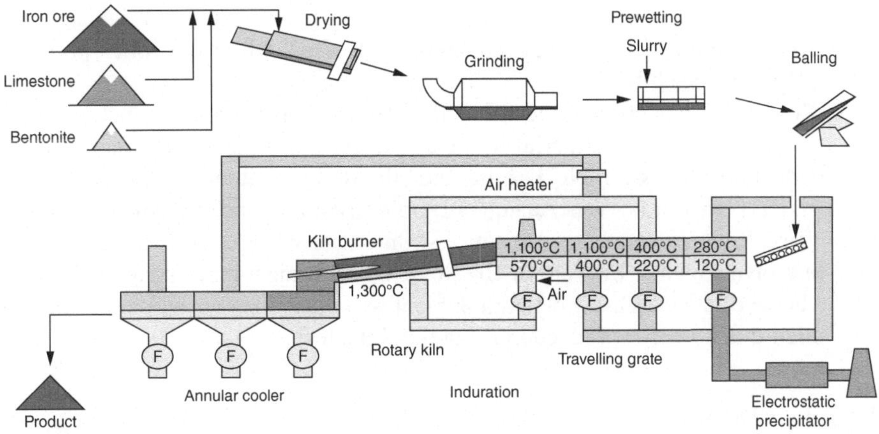

FIGURE 14.7 Material flow in grate-kiln-cooler system for the induration of pellets (after Ref. [6])

The pellet mix constituting iron ore, limestone and bentonite, is withdrawn from bunkers, dried and ground. The mix is pre-wetted and balled in disc pelletizer. The green pellets, thus produced, are sized in roller screen and charged into traveling grate. The pellets pass through drying and preheating stages. The preheated pellets are heat-hardened in rotary kiln and subsequently cooled in annular cooler.

14.5.2 Design Parameters

The design parameters of pelletizing plants incorporating the grate-kiln-cooler system vary according to their capacities, material handling, fuel used, and the manner in which the sensible heat from off-gases is recuperated.

a. The pelletizing plant of Vale Mining Company, formerly known as Companhia Vale do Rio Doce (CVRD) in Sao Luis, Brazil, happens to be one of the biggest single strand pelletizing plants in the world: 6 MT/year.[7] It consists of:
   - A 105-m-long collecting transport system that is equipped with a reciprocating head to distribute the pellets directly onto the wide belt conveyor ahead of the indurating machine
   - A single deck roller screen that serves to feed the pellets onto the pellet cars of the indurating machine
b. The Sao Luis pelletizing plant is the first one in the world where pellet feed is produced only by roller press communition.[7] This is possible mainly due to the favorable grinding behavior of the iron ores from the Carajas mines. Thus, the thickening and filtration operations are avoided. Hence, not only the capital costs but also the operation costs are significantly

lower. As a comparison, a ball mill grinding plant including thickening and filtration requires approximately 34 KWh/T while the roller press plant requires 6 KWh/T.

c. Kobe Steel has constructed the pelletization plants for the production of pellets used as feed material in the extraction of iron following the direct reduction process. Table 11.1 outlines the design features of the pelletization plants in (a) Venezuela, (b) Iran, (c) Bahrain, and (d) Oman (Vale) using Kobe's technology of the grate-kiln-cooler system.[6] All the pelletizing plants owned by Vale in Brazil have/had straight-grate systems. The choice of a KOBELCO pelletizing plant was made in consideration of the high-quality pellet production using the Kobe technology.

14.5.3 Operation

Drying and Preheating The green balls are fed uniformly onto the straight moving grate. The material moves on a straight traveling grate that passes sequentially through the steps of updraft drying, downdraft drying, and the preheat zones (below 1,050°C).

The pellets are continuously preheated in the preheat zone. The range of temperature in the preheat zone is 800°C to 1,000°C, which is adequate to oxidize magnetite ore concentrate into Fe_2O_3 in an exothermic reaction. The oxidation of magnetite leads to strong bonding in the contact points. The solidification and induration processes are partially finished, which could significantly increase the pellet strength for the following process. The pellets acquire sufficient strength to endure the tumbling and heating that occur in the subsequent kiln step. This decreases the generation of dust when pellets are loaded into the kiln.

The sources of heat in the preheating step of pellet induration include:

- Liberated heat from exothermic reaction of magnetite to hematite oxidation (see **Sec. 14.2.2**)
- Thermal heat generated by the combustion of gaseous and/or liquid fuel, which is transferred to the pellets
- Recovered heat from the waste gases generated in the kiln (see **Sec. 14.5.4**)

In the preheating zone, partial oxidation of magnetite and calcination of carbonates and hydroxide take place. The oxidation of magnetite leads to strong bonding in contact points. Solid bonding and grain growth start at the preheating zone and are complete in the firing stage. Liquid formation by chemical combination of various acidic and basic constituents and consequently development of slag bonds occur during firing. In this (high-temperature sintering) zone, the recrystallization of fine-grained secondary hematite mainly takes place.

Firing The preheated pellets, thereafter, are transferred to a refractory lined rotary kiln to go through the high temperature sintering at 1,250–1,300°C. The firing is conducted in a counter-flow manner while tumbling. Tumbling ensures uniform firing of the pellets. Radiation heating is applied to fire the pellets, which

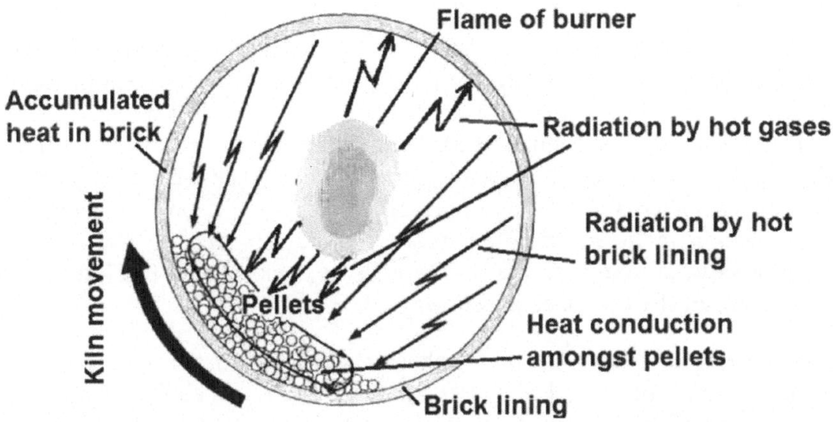

FIGURE 14.8 Various modes of heat transfer in a rotating kiln during pellet induration

makes the induration of pellets efficient and homogeneous. A major portion of the heat is supplied by hot gas and hot refractory work (**Figure 14.8**). Besides, exothermic reactions during the induration of pellets also provide heat. The highly heated pellets are pulled into the charge at the pellet bed's lower end where they give up a portion of their heat to adjacent pellets. Little or no oxidation takes place in the kiln due to the high temperature. In this (high-temperature sintering) zone, the recrystallization of fine-grained secondary hematite mainly takes place. Even some dissociation of hematite back to magnetite can occur. The firing temperature can be adjusted with ease and accuracy by controlling the fuel ratio for the kiln burner.

The burners, especially those using heavy fuel oil, are equipped with compressed air for atomization. As a consequence, air quantity is reduced by 15 to 20%. The reduction of atomizing air with ambient temperature will decrease the fuel oil consumption of the plant. Another advantage of using high-pressure burners is the expected lowering of the formation of thermal NOx in the burner flame.

Cooling (Oxidation) The pellets, after being fired at an elevated temperature inside the kiln, are cooled in an annular-shaped, horizontally rotating cooler, where the temperature of the fired pellets is decreased to a level suitable for subsequent transportation. This step employs the forced convection of air blown through a relatively deep bed (0.8–1.0 m) of (hot) pellets to produce the final product. Final oxidation of the sintered pellets takes place in the cooler. The degree of oxidation that takes place in the cooler may fluctuate in accordance with the variation in the degree of oxidation in pellets leaving the PH zone. Thereafter, the pellets are transported to steelmaking plants.

Transportation Prior to shipment, the pellets can be treated in a pellet lime-coating station. Coated pellets maintain their high iron content while displaying low sticking behavior during reduction. They offer higher bustle gas temperature in direct reduction (DR) plants and, as a consequence, the capacity of DR plants increases due to better reduction kinetics.

14.5.4 System of Heat Recuperation

The recirculation of hot gases, besides the recuperation of sensible heat for their usage in the process steps, makes the entire grate-kiln-cooler system thermally efficient. The principle of a heat recuperation system remains similar to that of straight grate system (see **Figure 14.5**). The available heat in the form of air and gases from the cooler and kiln/traveling grate are recirculated. However, depending on the characteristics of the raw materials vis-à-vis the process adopted, a minor variation in the recuperation process could be possible. A commonly adopted process, according to Metso–Minerals, is demonstrated in **Figure 14.6**. This is schematically described as follows.

Hot Gases from Cooler A part of the hot air collected from the cooler is used as secondary air for fuel combustion in the kiln. The hot air from the various zones of the annular cooler are also used in drying and preheating the green balls in the grate according to the following schemes:

- The hot air with temperatures 150–200°C, from the cooling zone III, passes through the pellet layer in the drying zone so that pellets are heated and attached water is removed.
- The hot air from cooling zone II is utilized to heat the pellets in the preheat zone (**PH I**) of the traveling grate. This is to accelerate oxidation reactions at 650°C so that the pellets can bear the temperatures above 1,000°C in PH II. The aided burners are set in this zone.

Hot Gases from Traveling Grate The hot air with temperature 400°C from PH II of the grate passes through pellet layers in the drying zone. The pellets are dehydrated and preheated to ensure that the pellets can bear 650°C in PH I.

Flue Gases from Kiln The flue gases that come from the rotary kiln have a temperature around 1,050°C and do possess enough sensible heat, which is used for the process step in drying and preheating of the green pellets in the traveling grate. Their recirculation to the preheating zone of the traveling grate needs to be controlled so that the temperature of the bottom layer does not exceed 900°C. This is to ensure that the grate pellets are not destroyed (at higher temperature).

14.5.5 Deposit Formation

The formation of a deposit or slag on the walls of the kiln is a common phenomenon in the grate-kiln process. It causes production disturbances that require planned and periodic maintenance. The deposits may also fall from the inner wall of the grate-kiln and may require unplanned stoppages.

The mechanism of deposit formation is complicated. It is contributed by:

a. The deposition of generated ash, which gets partially fused and takes place inside the kiln walls through the combustion of pulverized coal particles.[8,9]
b. Pellets disintegrate during their induration process, which generates fines. These fines, on partial or complete fusion, contribute to the formation of deposits.

14.5.6 Characteristic Features

A grate-kiln-cooler system has the following features:

- The grate, kiln, and cooler of the system are independently designed and constructed according to their respective thermal loads. This reduces the frequency of replacing parts such as the refractory material and grate plate and also consequently improves the viability of the plant.
- The system produces homogeneous products since the pellets produced are heated uniformly by the tumbling action inside the kiln during the firing step. The firing temperature can be adjusted with ease and accuracy by controlling the fuel ratio for the kiln burner.
- Each of the steps of preheating, firing, and cooling can be controlled either in conjunction with the others or independently from the others, according to the requirements. This enables heating the pellets in a pattern that is most suitable for each process step and makes the process versatile with regard to variations in raw material quality and the production rate. The independent speed control of the grate, kiln, and cooler, which is available to the operator, provides process flexibility to adjust the changes in concentrate feed. The induration of self-fluxed pellets requires strict temperature control: In this case, the ease of temperature control enables the homogenous production.
- Lower fuel and power consumption can be achieved.

14.6 COMPARISON OF STRAIGHT-GRATE PROCESS VIS-À-VIS GRATE-KILN PROCESS

The traveling grate process uses a stationery bed of pellets that are transported through the process zones for drying, preheating, firing (oxidation and sintering phenomena), and cooling stages. The grate-kiln process uses a shorter grate where the pellets are dried and preheated before these are discharged and fired in a rotary kiln. Both the straight grate and grate-kiln-cooler are thermally efficient and can make quality pellets. Hence, the choice of either of these processes is guided by the economics, and capital cost happens to be deciding factor in this respect.

The difference in the straight-grate process and grate-kiln-cooler process, in terms of their operations, are as follows.

a. **Process Flexibility:** In the **straight-grate** process, the heat-hardening cycles—namely, drying, preheating, induration, and cooling—are executed on a single grate. As these cars move continuously at the same speed through these operations, any change in one section affects the residence time in another. In the **grate-kiln process**, the drying and preheating operations are carried out on a traveling grate, while induration and cooling are performed in the rotary kiln and annular cooler, respectively, making it possible to control each of the steps independently. The independent control of the speed of the grate, kiln, and cooler that is available to the operator provides process flexibility to adjust to changes due to input material.

b. **Ability to Adapt to Various Raw Materials:** The **grate-kiln-cooler** system has process flexibility as each of the induration steps can be independently controlled. This provides the heat patterns most suitable to adapt to various raw materials: from magnetite to hematite in different proportions. The pellets made from high LOI ore concentrate are difficult to indurate as the crystal water may explosively vaporize to cause bursting of the green balls. A pelletizing system can be designed in this (**grate-kiln**) system by providing a dewatering step, performed at a relatively low temperature, to prevent rapid heating of the green balls. Besides, the **grate-kiln** system provides more flexibility in the type of fuels and pellet binders used as compared to a **straight-grate** system.
c. **Pellet Movement:** Pellets remain stationary throughout the process in a **straight-grate** machine, whereas pellets tumble continuously in a **rotary kiln.** The straight grate generates less fines due to the absence of transfer points and little relative movement/abrasion between pellets and furnace parts (**Figure 14.9**)
d. **Heat Transfer:** In the **grate machine**, some pellets may receive an inferior indurating cycle because of uneven gas flow and heat transfer. This is not the case with the **grate kiln** since within the rotary kiln all pellets receive the same treatment. The heating rate from 200°C to 1100°C is higher for the grate-kiln than for the straight-grate process.
e. **Hearth Layer:** In a **straight-grate** system, grate bars are subjected to high temperatures. Hence, a hearth layer of fired pellets is used to protect grate bars (upon which the pellets are laid). In a **grate-kiln** system, the green pellets are charged to a moving grate, but no hearth layer of fired pellets is required because, in this case, the bottom of the bed does not reach a high enough temperature to damage the grate bars before the burden is charged into the kiln, where it is fired.

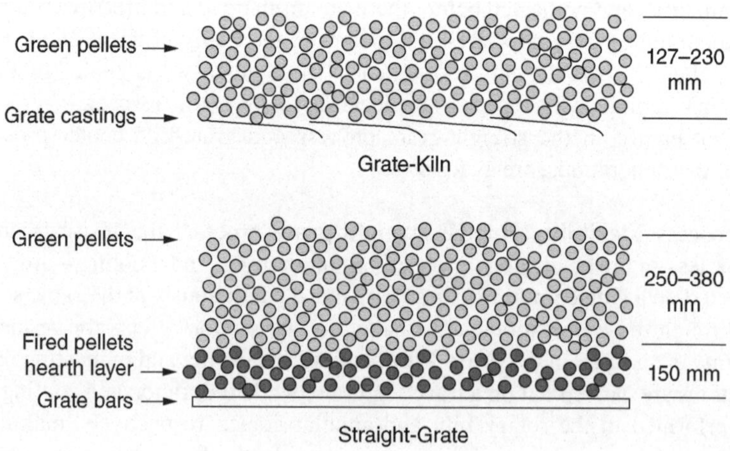

FIGURE 14.9 Schematic of pellet movement in grate-kiln and straight-grate systems (after Ref. [5])

f. **Fuel and Electricity Consumption:** Fuel consumption is lower in a **straight-grate** process compared to a **grate-kiln**. This is because heat loss due to radiation is less, and there is a better heat exchange between solids and the air due to deeper bed depths.[10] In a **grate-kiln** system, the electricity usage is lower (~20 kWh/T) compared to a **straight-grate** system (~35 kWh/T).
g. **Heat Recuperation:** It is better in the grate-kiln system as 95% of the air that is used for combustion in preheating the pellets, and also for induration in the kiln, can be recuperated from the cooler at around 1,000°C.
h. **Coal Firing System:** The **grate-kiln** system is more suitable than the straight-grate system for adopting various types of conventional pelletizing systems incorporating coal firing.
i. **Burners and Fans:** The **straight-grate** machine possesses a large number of burners along the length of the induration furnace; whereas a single burner is used in the **kiln**. The **straight-grate** machine consists of several fans operating in a series with multiple fan controls; whereas, the **grate-kiln** system requires fewer fans with a single fan control.
j. **Product Quality and Features:** Both blast furnace (BF) and DR grade pellets are/can be produced by either of the processes. A straight-grate system transfers pellets as a static layer of pellets, which is relatively thick—about 500 mm. This causes a difference in the heating profile of pellets in the upper and lower portions of the layer resulting in pellet quality, especially in compression strength and tumble strength. The **grate-kiln** system produces fired pellets with more uniform physical strength due to the nature of material flow and its tumbling in the rotary kiln, which exposes all green pellets to the peak firing temperature.
k. **Capital and Production Costs:** The grate-kiln system involves a higher capital outlay than the straight-grate process. However, for larger plants, the difference is likely to be narrowed down as the grate kiln will scale up. Overall production cost is higher with a continuous grate than with the grate kiln. The cost involved in grate bar replacement is a major contribution in this respect. The differences in production costs become smaller with a higher capacity. Hence, other factors and, in particular, required pellet qualities, are likely to influence the selection of the firing process.

The total residence time of pellets in a grate-kiln furnace is around half an hour, although this can vary depending on plant dimensions.

14.7 FACTORS INFLUENCING THE INDURATION PROCESS (AND PELLET QUALITY)

The major factors influencing the induration process are as follows.

a. Size of Green Pellets
Large-sized pellets are not desirable as thermo-chemical conditions like temperature and partial pressure of oxygen change across their cross section during the induration process and a *duplex microstructure* may

be formed (see **Sec. 15.8**). The differential contraction of the phases in bigger-sized pellets causes strains to bonds and lattices that result in the reduction of pellet strength. Besides, the propensity toward the large shrinkage gradients in the radial direction from pellet surface to the core increases with increasing the size of green pellets, resulting in the defective pellet texture.

b. Alumina Content

Alumina in the ore concentrate, due to its refractory characteristics, demands more heat during induration. Increased alumina in the pellet mix:

- Decreases the melt fluidity, which reduces the ability of bubbles to coalesce and reshape
- Results in a branch-like irregular shape that exhibits poor load-bearing capacity, resulting in lower strength

The dependence of alumina content in pellets on its strength and reduction degradation index (RDI) are mentioned in **Secs. 16.4.1** and **16.8** (see **Figure 16.7**), respectively.

c. Induration of Pellets with Volatiles

The pellet mix (or green pellet) contains certain groups of additives and/or impurities containing volatilities as follows.

- Carbonates (in the case of fluxed pellets)
- Volatilities (in the form of coal)
- Combined water (associated with the concentrate, particularly from goethite ones)

These release carbon dioxide and water when the green pellets are dried. Thermal spalling takes place during the course of drying. As a result, the outer layers of the pellet flake off or may crumble into small pieces resulting in the pressure buildup inside the pellet (see **Sec. 14.2.1** also). The gases and vapor, when released from the pellets during their induration at higher temperatures, show that pores as well as cracks are formed in the pellets, making them weak.

The LOI content of the ore concentrate is detrimental to pelletization. Each 1% LOI costs 3% in the TPH loss of itself in the indurating furnace due to the endothermic heat requirement. The loss due to thermal spalling could cost loss as high as 10%.[11]

This can be minimized by following a dewatering step at a relatively lower temperature and a slow rate so that the volatilities evolve at a slower rate. The pretreatment of goethite ore (see **Sec. 2.7.4**) also helps to improve the pellet properties following induration.

d. Porosity of Green Pellets

Porosity plays an important role in each of the stages in the pelletizing process. The feed preparation and balling processes are designed to produce green balls of adequate strength and porosity, which can then be processed to produce fired pellets of the desired quality. In the subsequent firing stage, the oxygen required to oxidize magnetite to hematite enters via the pores, and the subsequent rate of grain growth is directly related to the pore size and pore size distribution.

e. Additives in Pellet Mix

Limestone and **dolomite** are added in the pellet mix for the production of fluxed pellets. These additives have two opposing effects on the induration of pellets:

- Their calcination, being an endothermic reaction, requires a significant amount of energy (heat). The energy sink, due to calcination, will partially prevent the pellet from reaching an effective induration temperature under normal operating conditions. This is because the calcination reaction takes place at relatively the same temperature as that when the oxidation of the magnetite to hematite initiates, which is in the 600–800°C (873–1,073 K) range.

  The production capacity of a pellet plant is lower when basic pellets are produced compared to acidic ones. The calcination of flux-stone releases CO_2, which fills the pellet pores. This reduces/inhibits the diffusion of oxygen into the pellets for the oxidation of magnetite to hematite during induration and slows the oxidation kinetics.
- On the other hand, the hematite mineral dissociates at lower temperatures when basic additives are present in the pellet mix. The minerals, calcium ferrites, are formed through the exothermic reactions in the case of fluxed pellets. Consequently, the temperature increases inside the pellets. The additional melting heat would bring about an overheating in the pellet core for temperatures exceeding 1,400°C. The formation of solid solutions of additives with the hematite phase activates the dissociation of hematite back to magnetite at lower temperatures than in the case of the pure hematite phase.

Lime can be/is also incorporated to substitute for limestone for enhancing the pellet basicity. This eliminates the calcination reaction. However, it is not preferred to incorporate lime in its original form in the pellet mix. This is because it gets slaked with water that is injected into the pellet mix and generates calcium hydrate, which is not evenly distributed. Instead, the hydrated lime added during the mixing process greatly increases the specific surface area of the iron ore mixture. This in turn contributes to increased wet green balls and dried fired pellet strength, even with the more coarsely ground ores. Lime is readily available and is also less expensive than bentonite.

The addition of MgO helps in improving the high-temperature properties of the pellets. The temperature at which dissociation takes place with MgO additions happens to be the lowest among the additives.[12] Dolomite, as a source of MgO, may not be preferable as it releases CO_2 during its dissociation, resulting in the formation of pores and weakening of the strength on firing. Besides, the crystal size of dolomite is larger; hence, its calcination takes more time (see **Secs. 6.8.3** and **9.5.2**). In place of dolomite, light burned magnesite[13] and pyroxenite, a magnesium silicate mineral,[14] have been attempted as additives for pellets. Unlike dolomite, these additives do not undergo any endothermic reactions for dissociation. The pyroxenite either reacts with the surrounding iron oxide to form magnesioferrite or is assimilated in the silicate melt that acts as a bonding phase in the pellets. These pellets exhibit less swelling by forming low FeO high melting slag that gives sufficient bond strength to withstand the reduction stresses. Pyroxenite pellets result in better softening-melting characteristics compared to acid pellets due to the formation of high melting point slag and the magnesiowustite phase during reduction.

f. Thermal Treatment

The oxidation pattern and the thermal history of magnetite pellets also influences the final pellet quality. The duplex structure forms because, in the pelletizing process, only the outer shell of pellets is oxidized before the sintering temperature is reached, and the magnetite core shrinks more than the hematite shell (see **Sec. 15.8**). The degree of oxidation needs to be at least 80% in pellets leaving the PH zone to get acceptable strength in pellets after firing. Enriching the oxidizing gas in the PH zone to around 30% oxygen and controlling the speed of temperature rise in the PH zone to below 150°C/min would improve the pellet quality.[15]

High specific gas consumption is the result of low burn-through temperature, which in turn is an indication of low firing in the pellet bed. Low burn-through temperature can be attributed to poor bed permeability and resulting in low strength.

The temperature regime of pellet induration need to be optimized in order to produce quality pellets. However, this is dependent on the material characteristics and, hence, needs to be explored experimentally and also through simulation studies.

g. Bed Shrinkage

Depending on the characteristics of raw materials, the volume of pellets being indurated changes. The volatiles (carbon dioxide and water) associated with the pellets get evolved on induration, which causes the volume change, and the bed, in the case of a grate furnace, shrinks. The uniform heating on induration is essential for the production of quality pellets. Thermal conductivity of the pellets depends on the constituent minerals. Hence, the cooling of fired pellets gets affected when these contain high gangue.

14.8 FUEL SUBSTITUTION IN PELLET INDURATION

For improvements in plant profitability and material conservation, endeavors to reduce primary fuel consumption are being made by modification of the induration process and/or by using alternative fuels. Fine coke is incorporated in the pellet mix to partially substitute the gaseous/liquid fuel requirement in induration of the green pellets. Coke, when added to the pellet mix in an appropriate quantity, brings the inherent fuel value to the green pellets during induration. The solid fuel, when burned properly in the pelletizing cycle, lowers the consumption of liquid/gaseous burner fuel, while at the same time accelerating the pelletizing process. The quantity of solid fuel that can be added to the pellet mix is decided by both its proximate analysis and calorific value. However, its quantity has to be restricted in the 1.3 wt.% of pellet mix.

14.8.1 Coal Injection

The gaseous/liquid fuel required for the induration process could be partially replaced by solid fuel following any of the arrangements as follows:

a. In Japan, coal firing has been applied in the Kakogawa grate-kiln pelletizing plant of Kobe Steel.[14] Kakogawa produces dolomite pellets with feed containing 10% magnetite and 0.88% admixed coke breeze. The solid fuel is fired directly into the combustor hood kiln through suitably designed burners.[16] Solid fuel can also be fired in combination with oil in dual-type burners or as a suspension in a coal-oil or coal-water mixture.

   The solid fuel is injected into the direct recuperation header above the cooling zone of the straight-grate pelletizing system.
b. Sensible heat of (indirect) combustion of pulverized coal can also be used in firing the green pellets. The coal is ground to a fine powder and conveyed by a flow of primary air into the furnace, where it burns entirely in suspension. Secondary air supplied at the furnace also promotes more efficient and intense combustion. The burners of several designs are available with different profiles of flames.
c. Sensible heat can be generated through the fluidized bed combustion of coal, similar to that for the generation of steam in a power plant. However, the flue gas cannot attain a higher temperature through this arrangement.

14.8.2 Nature of Carbonaceous Materials

Fine coke, which has a high fix carbon (~ 81%) and low ash content, is suitable for the pelletizing process. Coke ash tends to melt at the temperature 1,200°C–1,450°C, which is near to the firing temperature in the induration process. Hence, the coke should have high ash fusion, preferably above 1,450°C, for its injection. The coal must have a good Hardgrove grindability index and good combustibility with a high content of volatile matter.

The coal is prepared in a conventional Raymond Mill grinding system. However, Allis-Chalmer's studies shows that the selection of coal for direct firing is not that simple. It is difficult to determine the best type solely on the basis of coal properties.[17]

14.8.3 Utilization of DR Coal Fines

Countries like India and China have a huge direct reduced iron (DRI) production using coal. The rotary kiln/certain shaft-based DR process for iron making predominantly uses coal as a reductant. Coal is available in 0–50 mm-sized fractions with below 3 mm fractions at load point at approximately 20%. By the time coal is brought to the kilns, it is handled multiple times and the undersize (below 3 mm) fraction goes as high as 40%. As the kilns generally cannot take more than 30% of below 3 mm of fraction, from the discharge end through injection, sponge iron manufacturers are forced to screen out 10% to 15% of below 3 mm fractions before the coal is used and must dump it in the vicinity of their setup as waste.[18] It is prudent to have a technology developed to utilize these coal chars through pelletization.

Because of the fact that the coal ash tends to melt at the firing temperature (the ash fusion point of the Indian coals varies in the range of 1,200–1,450/1,500°C), a complete replacement of fuel oil by coal could be difficult.

14.8.4 Developed Technologies

Lurgi Multiple Combustion System Lurgi[19–21] has developed a technology of coal injection while the pellets are pelletized called the Lurgi Multiple Combustion (LMC) system. The granular coal is injected in the cooling zone of the indurating machine. The temperature in this zone is below the fusion point of coal ash. Since the melting point of coal ash is not reached, this precludes the possibility of coal ash fusing with pellets. The ash is superficially deposited on the surface of pellets and gets separated out at the product screen, to be removed through the installed dedusting system. The traveling grate can be designed for the continuous use of coal as a partial substitute for fuel oil.

14.8.5 Problems and Solutions

Ring Formation The associated problem when solid fuel is injected into the kiln is its ringing due to the deposit formation on account of the ash. This could be overcome to some extent by using high-quality bituminous coal with a higher ash fusion temperature.

Pellet Structure Fine coke combusts during firing and, under oxidizing atmosphere, forms pellets with a porous structure. The hematite mineral is reduced to magnetite to a certain extent and depends on the amount of coke added during the induration process.

REFERENCES

1. Cribbs, J. D. and D. W. Kestner, Some factors influencing iron ore fired pellet quality, in: *Proc.* 2nd *Inter. Symposium on Agglomeration, Agglomeration-77,* Atlanta, GA, March 6-10, 1977, Sastry K. V. S. (ed.), Pub. by Amer. Inst. of Mining, Metallurgical & Petroleum Engineers, N.Y., 272–285.
2. ISPAT GURU, http://Ispatguru.com>understanding-pellets-and-pellet plant operations March 2015, (downloaded March 2019).
3. Ball, D. F., P. R. Dawson, and J. T. Fitton, Agglomeration in iron ore pelletizing, *IMM Transactions Section C,* 1970 (79), C189–C196.
4. Dwarapudi, S., P. K. Gupta, and S. M. Rao, Prediction of iron ore pellet strength using artificial neural network model, *ISIJ International*, 47 (2007), Fig. 1, 67–72.
5. Metso Corporation, Fablaninkatu, Fl-00101, USA https://www.metso.com, Iron ore pelletizing Brochure No. 1369-08-12-ESBL/Danville-English (downloaded Sept. 2016).
6. Yamaguchi, S., T. Fuji, N. Yamamoto, and T. Nomura, KOBELCO Pelletizing Process, *Kobelco Technology Review*, 29, (Dec. 2010), 58–68.

7. Schwalm, T. and G. Walter, New developments in agglomeration technologies, in: *Proc. Asia Steel International Conference*, Jamshedpur, India, Vol. 1, T. Mukherjee and A. S. Dhillon (eds.), Indian Institute of Metals and Zhongguo-Jinshu-Xuehui, Organisers, Indian Institute of Metals, Kolkata, Publisher, April 9–12, 2003.
8. Wall, T. F., Mineral matter transformations and ash deposition in pulverized coal combustion, *Progress in Energy and Combustion Sciences*, (1992), 1119–1126.
9. Baxter, L. L., Ash deposition during biomass and coal combustion—A mechanistic approach, *Biomass Bioenergy*, 4, (1993), 85–104.
10. Sgouris, E. E. and K. G. Oja, *Iron & Steel Technology*, (2008), 58–66.
11. Dwarapaudi, S., S. M. RaO, and N. D. Rao, Pelletization of high alumina iron ore fines: *Problems and Prospects Tata Search*, (2007), 93.
12. Ferreira, S., D. Siguin, and F. Garcia, Thermal analysis of sintering of magnetite pellets, *Ironmaking & Steelmaking*, 21, (1994), 119–123.
13. Gao, Q.-J., Shen, F., Wei, G., Jiang, X., and Zheng, H.-Y., Effect of MgO containing additive on low temperature metallurgical properties of oxidized pellet, *Journal of Iron and Steel Research International*, 20(7), (2013), 25–28.
14. Dwarapudi, S., T. K. Ghosh, A. Shankar, V. Tathavadkar, D. Bhattacharjee, and R. Venugopal, Effect of pyroxenite flux on the quality and microstructure of hematite pellets, *International Journal of Mineral Processing*, 96, (2010), 45–53.
15. Haas, L. A., J. C. Nigro, and R. Moe, Magnetite oxidation of acid and fluxed taconite pellets, in: *Proc of Ironmaking Conference*, 51, (1992), 533–549.
16. Uenaka, T., H. Isako, K. Tokutake, and K. Aketa, Coal firing in pelletizing plant developed by Kobe Steel, *Ironmaking & Steelmaking*, 10(5), (1983), 234–239.
17. AIME, A study of parameters influencing direct coal firing of grate kilns, in: *Proc. of 106th AIME Annual Meeting*, Atlanta, 1977.
18. Thawani, S., personal communication, March 2009.
19. Gielen, H. H., H. S. Heep, M. R. Holensee, and H. D. Papacek, Quality improvements and energy savings in iron ore pelletizing, in: *Proc. of SME-AMIE Annual Meeting*, Atlanta 1983.
20. Heep, H. S., Oil savings by charcoal injection in Fabrica Pellet Plant, *Lurgi Symposium*, Frankfurt am Main, June 1983.
21. Hohensee, M. R., Charcoal addition to the pellet feed, operational experiences at Ferteco's Fabrica Mine, *Lurgi Symposium*, Frankfurt am Main, June 1983.

15 Reactions and Formation of Phases During Induration of Pellet

ABBREVIATIONS

CCS *Cold crushing strength*
CF *Calcium ferrite*
NKK *Nippon-Kokam Kabushiki Kaisha (now JFE Steel)*
SFC *Silicoferrite of calcium*
SFCA *Silicoferrite of calcium and aluminum*

15.1 PREAMBLE

Induration of pellets is an extremely important stage in the process of pellet making. The induration occurs by:

a. Formation of Fe_2O_3 crystallite bonds on oxidation of the magnetite phase
b. Recrystallization of Fe_2O_3 formed by the oxidation process

Strength in the pellet is achieved through its oxidation and sintering during induration. The sintering reactions and phases formed during induration depend on whether the concentrate is magnetite or hematite.

Figure 15.1 schematically shows the minerals formed vis-à-vis the starting minerals following induration of the green pellets. The mechanism of induration in the case of hematite differs from that in the case of magnetite. The compressive strength of magnetite-bearing pellets following induration is attained by the exothermic oxidation of magnetite to hematite during cooling. Whereas in case of hematite-bearing pellets, the strength is achieved by the recrystallization of hematite minerals and grain growth. Other mineral constituents, like goethite, limonite, etc., in the drying/preheating stage form hematite; thereafter, the phases and bonds formed on firing are similar to that of hematite ores (**Figure 15.1**).

The gangue minerals form slag in the course of pellet induration. The crushing strength of fired pellets is chiefly attributed to the crystallization process and reactions of gangue constituents. For higher grade concentrates, the crystallization process, and for ores with gangue content, glassy slag formation, additionally play an important role in contributing toward the crushing strength of pellets.

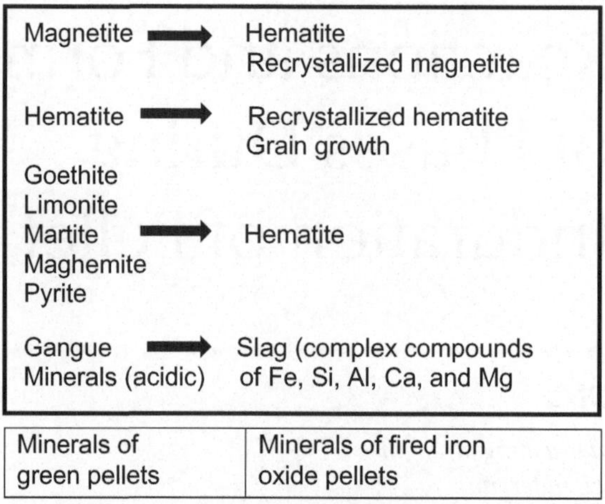

FIGURE 15.1 Mineralogical changes in magnetite/hematite pellets during induration under atmosphere

15.2 INDURATION OF MAGNETITE-BEARING PELLETS AND PHASES FORMED

15.2.1 Reactions with Magnetite Concentrate

The magnetite is oxidized to hematite according to the chemical reaction:

$$4Fe_3O_4 + O_2 = 6Fe_2O_3.\Delta Hr = -119\frac{KJ}{mol}\text{magnetite}$$

The reaction is highly exothermic. The oxidation of magnetite to hematite occurs in stages with increasing temperature and the reactions proceed as follows:

a. First, oxidation of the outer shell of pellets takes place, which is fast and controlled by the chemical reaction kinetics.[1,2]
b. Thereafter (after the fast superficial oxidation), the oxidation rate is controlled by the diffusion rate of oxygen through the product layer, which grows as the oxidation proceeds with time.[1–3] At this stage, the magnetite particles are covered by a thin layer of hematite, a consequence of which is that the oxidation rate decreases. This is due to the fact that the diffusion in the hematite phase is progressively restricted by the high-concentration hematite phase in the adjacent layer.

As the oxidation proceeds with time, temperature of the pellet bed increases, which facilitates further oxidation at a faster rate and allows for further growth of the lamellae.[4] The oxidation of magnetite to hematite is accompanied by a volumetric change, resulting in structural stress. The developed stress opens up the structure for the diffusion of oxygen.

The kinetics prevailing in the oxidation of magnetite to hematite at high temperatures (approximately 1,230°C) under isothermal condition depend on:

a. Diameter of the pellets: The time needed to obtain a given degree of oxidation in pellets is directly proportional to the square of the diameter of the pellets. If the pellet size is doubled, the oxidation time would be four times longer.
b. Porosity of the pellets: The time needed for oxidation increases with decreasing the porosity of the pellets.[1]
c. Oxidation temperature: The driving force for diffusion of oxygen is its partial pressure,[1] which decides the time needed for oxidation. This decreases (and the kinetics of oxidation reaction increase) with increasing the oxidation temperature in the range of 800–1,000°C.

   At higher temperatures, however, sintering takes place, and the oxidation reaction could be retarded. Gibbs free energy of the oxidation reaction decreases as the temperature is increased. At much higher temperatures, the hematite mineral dissociates back to magnetite.[5,6] This starts at 1,457°C in oxygen and at 1,392°C in air.[7]
d. Other Factors: The oxidation of magnetite ore pellets is influenced by (a) the fineness of the magnetite concentrate[1] and (b) the calcination of fluxes.[8]

15.2.2 Phases Formed

The induration of magnetite concentrate occurs in stages and, accordingly, the phases are formed as follows:[9,10]

a. **First Oxidation Stage:** Initially, the mineral γ-hematite (Fe_2O_3) is formed on oxidation of magnetite-bearing green pellets at low temperatures (below 400°C). This mineral phase is a kind of hematite with the same cubic lattice structure as Fe_3O_4; γ-Fe_2O_3 is unstable, and the crystal lattice is rearranged and converted into a hexagonal lattice structure at higher temperatures. Heating above 500°C converts the ϒ-hematite to α-hematite, which undergoes spontaneous nucleation following increasing structural stress. After nucleation, the low-temperature oxidation proceeds with α-hematite as the only oxidation product.
b. **Second Oxidation Stage:** Major chemical and mineralogical changes accompanying the induration do occur in this stage. The formation of Fe_2O_3 crystallite bonds and recrystallization of Fe_2O_3 occur by the oxidization process. At higher temperatures, the oxidation of magnetite particles to α-hematite takes place, and the hematite needles (lamellae) form on the particle's surface.

With reference to magnetite concentrate, the pellets contract following the oxidation and incipient sintering at 900°C. The shrinkage in the magnetite phase could be about 1.4 times greater than in the hematite phase. The shrinkage in pellets increases with increasing specific surface area in the magnetite raw material. The degree of shrinkage is dependent on the time the pellet is exposed to sintering temperatures.

15.3 INDURATION OF HEMATITE-BEARING PELLETS AND PHASES FORMED

The induration of green pellets constituting hematite concentrate is characterized by the growth and recrystallization of Fe_2O_3 grains at high temperature. The crystal growth has to be achieved in one step only. This requires a higher induration temperature and a longer firing time. The hematite particles may keep their original shape and remain separate below 1,200°C (1473 K). Only when the sintering temperature reaches 1,300°C (1573 K), can adequate strength (of the pellets) be achieved.

The degree of recrystallization (**α**) of hematite during induration is expressed, according to Tamman,[11] as follows:

$$\alpha = \left[\text{firing temperature in °C}\right] / \left[\text{fusion temperature in °C}\right]$$

For the recrystallization to take place, a minimum $\alpha = 0.33$ is required. When the firing temperature approaches the fusion temperature, the particles get fused. This occurs at around α = 80% to 100%. The intermediate range, α = 0.65% to 0.75%, is suitable for induration as adequate diffusion during recrystallization takes place in this range.

The effect of firing temperature on strength of fired pellets is schematically shown in **Figure 15.2**. The strength of pellets increases with increasing firing temperature. The compressive strength of hematite pellets rises very quickly to a maximum temperature, 1,100°C.[12] The quantum of increase, in the case of hematite pellets, is much more when the firing temperature exceeds 1,050°C and reaches the peak in the temperature range 1,250–1,300°C. Thereafter (firing temperature more than 1,300°C), the strength of the pellets decreases. In the case of magnetite pellets, such an exponential increase starts 50–70°C earlier.

The grain size of hematite minerals increases at a much higher rate during the firing of hematite pellets beyond 1,200°C (**Figure 15.3**)[13] and the pellet strength decreases.

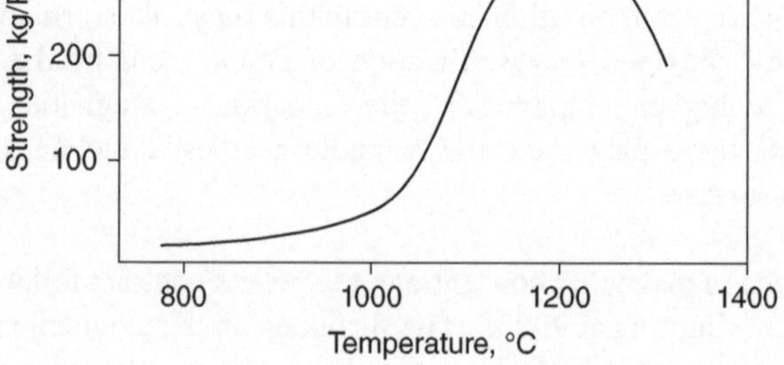

FIGURE 15.2 Effect of induration temperature on crushing strength of fired hematite pellets (schematic)

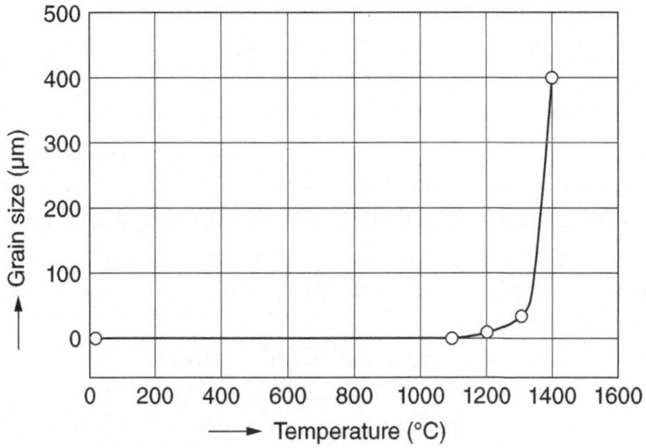

FIGURE 15.3 Effect of induration temperature on grain size of fired pellets (after Ref. [13])

The reason why the pellets fired at 1,100°C were noticeably stronger than those below that temperature has been explained by Urich and Han[12] in their studies. The mineralogical and textural transformation in the preheated pellets appears to be incipient or just beginning: at temperatures between 925°C and 980°C, the specular hematite particles retain their original size and shape, and they show no identifications of surface textural changes. At temperature around 1,040°C, while all specular hematite particles larger than 2 micron retain their original texture, some below 2 micron tend to grow into tiny granules. At temperature greater than 1,100 some larger specular hematite particles start having traces of tiny humps around the particle margins. Grain growth of fine particles appears to relatively conspicuous as compared with that in the pellets preheated between 980°C and 1,040°C.

15.4 INDURATION: MAGNETITE VIS-À-VIS HEMATITE CONCENTRATE

The induration of magnetite-bearing green pellets differs from that of hematite concentrate with respect to (a) the involved mechanism and (b) the fuel requirement:

a. **Mechanism**

Magnetite Concentrate: The mechanism of induration of magnetite concentrate is the exothermic oxidation of magnetite to hematite. Crushing strength is developed as the particles are bonded and accompanies the grain growth.

Grain bridges are formed, particles are enlarged, and recrystallization takes place. High local temperature may also result in the formation of slag, which reinforces the oxide network.

Hematite Concentrate: The mechanism of induration of hematite concentrate is the thermal recrystallization accompanied by surface-to-surface and solid-state diffusion. In the absence of any exothermic reaction, comparatively higher temperatures (than those in the case of magnetite) are required for recrystallization. The crushing strength is also not developed as early as in case of magnetite.

b. **Fuel Requirement**

The requirement of fuel in the case of hematite concentrate is higher than that in the case of magnetite. The firing temperature in the case of hematite is 60–70°C higher than that in the case of magnetite at the same compression strength.[13] However, the fuel requirement depends on (a) the ore characteristics: whether it is earthy or crystalline, and (b) the quantity and nature of gangue present.

The fuel requirement for the pelletization process varies with the type of material being processed. The most economical is magnetite, which yields exothermic heat on oxidation to hematite. The least economic are those materials with either combined water or carbonate content:[14]

The fuel requirement in case of different materials, in MJ/T, is given below.
Magnetite: 633–844
Hematite: 1,055–1,266
Limonite: 1,266–1,476

As a comparison, in the sintering process, due to insufficient fuel utilization and nonrecovery of sensible heat of the product, the fuel requirement is in the range 1,528–2,640 MJ/T of sinter. That is, pelletizing requires less than 50% of the fuel for sintering.

15.5 PHASES AND PORES IN INDURATED PELLETS

It is largely accepted that the properties of pellets are closely related to their mineralogical constituents as well as the pores and pore distribution within them (pellets).[15–16]

15.5.1 Bonding and Phases

Bonding The formation of bonds in the pellets during their induration takes place at 1,250–1,350°C (1,523–1,623 K) as follows:[17]

- Slag/glassy bonding: This type of bonding is more prominent in the pellet core.
- Crystal bonding (oxide bonding): Outer mantle and shell have this type of bonding.

The compression strength is primarily influenced by the oxide bonding of hematite grains and slag bonding among the minerals.

Phases The formation of melt occurs with a solid-state reaction. The minerals, lime, and silica present in the pellet react with each other at the contact points to form the orthosilicate phase. Also, hematite and lime react (at contact points) to form the ferrites, $CaO.Fe_2O_3$ and $2CaO.Fe_2O_3$. These phases then react with each other at around 1,000°C to form silicoferrites of calcium (SFC) in minor amounts. The remaining silicates and ferrites begin to melt around 1,200°C when SFC formation is accelerated. The mineral silicoferrites of calcium and aluminum (SFCA) at a higher temperature (1,300°C) in alumina-rich areas precipitate as a solid phase. The remaining melt solidifies as glass of ferrous-calcium silicate with the excess iron oxides migrating to nearby hematite minerals. The high strength of pellets can be attributed to uniformly distributed silicates and other gangue minerals.

15.5.2 Formation of Pores

The possible reaction steps resulting in the formation of pores are:[18]

- The carbonates present in the green pellets dissociate on heating and CO_2 is evolved; consequently, cracks are formed.
- CaO and hematite react at the contact surface to form calcium ferrite (CF).
- The CF melts at a higher temperature. The liquid (melt) due to surface tension and/or capillary force flows (disappears) away to the:
  - Surface of the CaO
  - Surface of the surrounding iron ore (hematite) particles and voids within
  - Inside of the CaO, which has developed cracks
  - Inside of the iron ore particles through open pores
- Hematite and CaO react at solid state and/or through the melt phase and, consequently, CaO is dissolved.
- The amount of free CaO decreases with the formation of more and more melt.
- After complete consumption of CaO, the formation of pores at its sites develops. When liquid flows away from its source, it completely or partially fills some of the voids among the iron ore particles as well as some of the particles' inner pores.

15.6 PARAMETERS INFLUENCING THE MINERAL PHASES IN INDURATED PELLETS

15.6.1 Pellet Basicity

The pellet mix consists of SiO_2 and Al_2O_3 constituting gangues, and their sources are ore concentrate, coke, binder (bentonite), and fluxes. Besides, the additives CaO and MgO are incorporated in the mix when the aim is to produce the nonacidic (fluxed) pellets. These fluxes—namely, limestone, hydrated lime, dolomite, and magnesium silicates—in the mix also contain silica and alumina.

Phases The types of minerals formed in the pellets depend on their basicity:

- Acidic pellets: The acid pellet is mainly composed of the hematite phase, and the spaces within the hematite phase are filled with the slag phase.

- Self-fluxed pellets: These constitute calcium ferrite and pores in addition to the hematite and slag phase. The silicate melt in fluxed pellets happens to be the predominant bonding phase up to the basicity of 1.3.
- Basic pellets: The basic pellets, like the acidic ones, are mainly composed of the hematite phase. With increasing the pellet basicity beyond 1.3, more and more amount of the liquid phase is formed during induration, which contributes toward its higher strength and lesser porosity.[19] Besides, the grown phases of calcium ferrites ($CaO.Fe_2O_3$, $2CaO.Fe_2O_3$) are formed in (fluxed) pellets as the pellet basicity is increased through exothermic reactions between hematite and lime at contact surfaces, resulting in the temperature rise inside the pellets.

The minerals, lime and silica, present in the pellet react with each other at the contact points to form the orthosilicate phase. These phases react with each other at higher temperatures (approximately 1,000°C) to form silicoferrites of calcium (SFC). However, the formation of SFC is in a minor amount at this stage, and the remaining silicates and ferrites begin to melt around 1,200°C when the formation of SFC is accelerated. In the case of the hematite concentrate with relatively higher alumina, the mineral silicoferrites of calcium and aluminum (SFCAs) are formed at higher temperatures (1,300°C) in alumina-rich areas that precipitate as the solid phase. The remaining melt solidifies as glass of ferrous–calcium silicate with the excess iron oxide migrating to nearby hematite minerals.

When carbonates are incorporated in the mix making the pellets basic ones, pores are formed within the pellets.

Figure 15.4 represents the typical microstructure of hematite-bearing basic and acidic pellets. Also, the microstructure of self-fluxed sinter is given in the figure.[20]

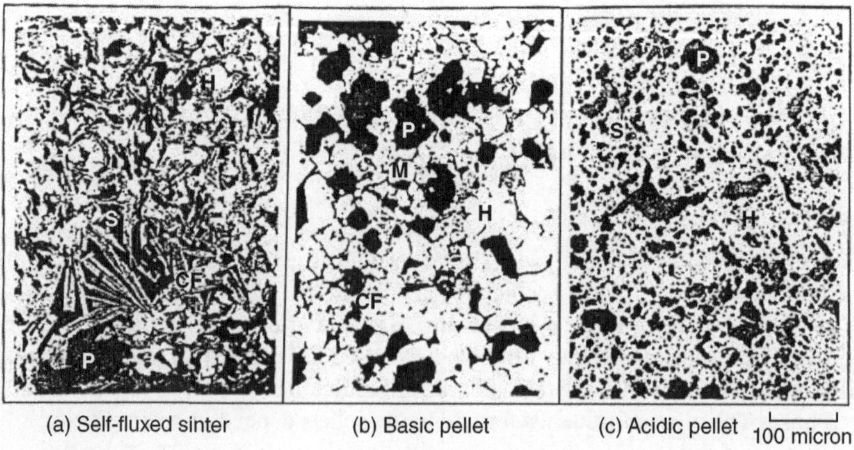

FIGURE 15.4 Typical microstructure of (a) self-fluxed sinter, (b) basic pellets, and (c) acidic pellets used in Japan (NKK data) (CF: calcium ferrite, H: hematite, M: magnetite, p: pore, s: slag) (after Ref. [20])

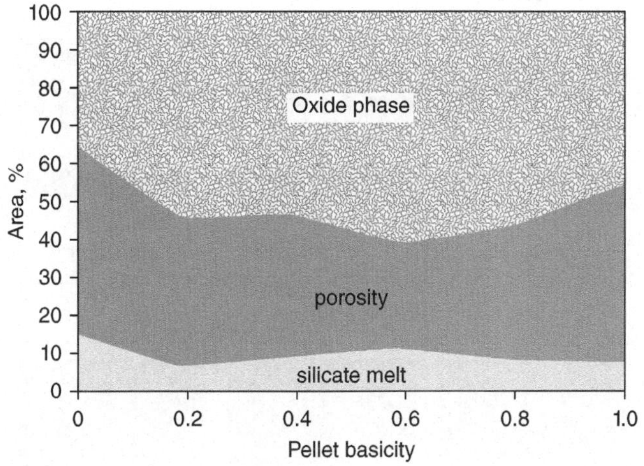

FIGURE 15.5 Variation of oxide phase, porosity, and silicate melt with increasing basicity of hematite pellets (after Ref. [21]).

The acidic pellet is mainly composed of the hematite phase, and the space between the hematite phases is filled with slag phase. The basic pellets, similar to the acidic pellet, are mainly composed of the hematite phase, while a calcium ferrite phase can also be observed. However, the structure is not as grown as the calcium ferrite sinter.[20]

The oxide phase, porosity, and silicate melt in hematite pellets change with their basicity. The maximum amount of the oxide phase, at the cost of decreased porosity, has been typically observed at around 0.6 basicity (**Figure 15.5**).[21]

15.6.2 Addition of Coke

The addition of coke to the pellet mixture influences the microstructure of the heat-hardened pellets. During the induration process, the hematite mineral is partially reduced to magnetite. The extent of reaction depends on the amount of coke added. The addition of carbon in the form of coke lowers the melting point of the slag phase, resulting in increased formation of the binding phase, which causes weakening of the pellet structure.

15.6.3 Bentonite Dosage

The hematite pellets vary in their mineralogical characteristics and are dependent on the fineness of the concentrate as well as the bentonite dosage.[22] The pellet prepared from the concentrate of very fine mesh of grind (93% below 45 micron) at lower bentonite dosage (0.4 wt.%) and with additives (coke: 1.75 wt.% and limestone: 2 wt.%) showed a variegated structural character (**Figure 15.6a**). Its core was poorly bonded; the intermediate layer showed some bonding among the iron-bearing minerals; but the outer layer was compact with better bonding. The crushing strength was lower (150 kg/P) due to the weaker inner layer of the pellet (see **Sec. 16.4.1** also).

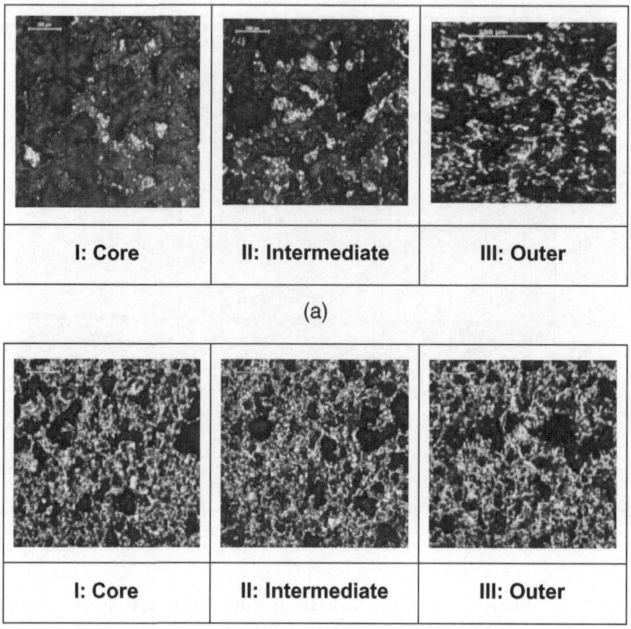

FIGURE 15.6 (a) Structure of pellets showing variegated character at lower (0.4 wt.%) bentonite dosage. I: Core, II: Intermediate layer, and III: Outer layer (underlined scale shows 100 micron) (after Ref. [22]) (b) Structure of pellets showing uniform distribution of minerals at higher (0.8 wt.%) bentonite dosage. I: Core, II: Intermediate layer, and III: Outer layer (underlined scale shows 100 micron) (after Ref. [22])

In contrast, the pellets prepared at a higher bentonite (0.8 wt.%) dosage, showed good polish surface and uniform distribution of minerals and other constituent matter. The core, intermediate, and outer layers were almost similar and quite compact (**Fig. 15.6b**). The CCS was higher (280 kg/P). The silicate and other gangue minerals, were uniformly distributed within the pellets, which contributed toward their high strength. The higher reflectivity of the iron-bearing phases indicates a much better metallization in the sample.

15.7 PARAMETERS INFLUENCING THE POROSITY IN INDURATED PELLETS

15.7.1 Effect of Pellet Basicity and MgO Content

The mean size of pores in pellets (at 1.5% MgO level) was found to increase with increasing basicity. This is indicated by lower pore density with increasing pellet basicity (**Figure 15.7**). Pore density, as defined by the number of pores per square millimeter, is an indication of pore size. Higher pore density means a greater number of pores in a given area with smaller pore size and vice versa. Pore density is maximum in the case of acidic pellets. The density decreases sharply when the

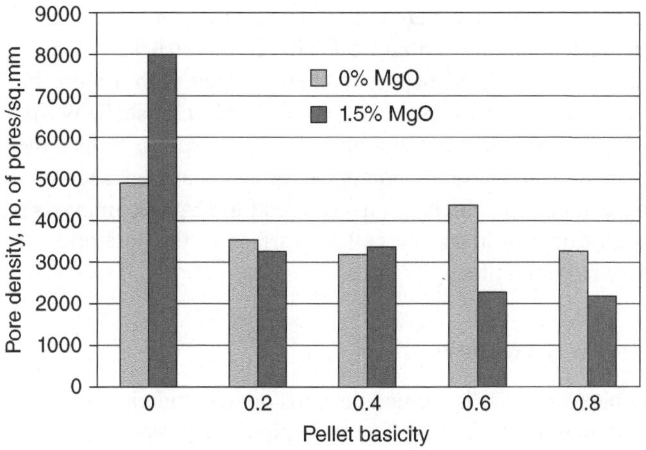

FIGURE 15.7 Variation of pore density with increasing basicity of hematite pellets at 0% and 1.5% MgO (after Ref. [21])

pellet basicity (CaO/SiO_2) is increased from 0 to around 0.2, and thereafter its value remains the same till 0.4 basicity. On further increase of basicity, the pore density decreases (**Figure 15.7**).[21]

15.8 DUPLEX STRUCTURE

The formation of duplex structure during induration of green pellets was described as early as the 1950s.[23–24] The duplex structure in pellets forms due to the difference in oxidation and sintering across their radius during induration. A distinct concentric oxidation front between the oxidized outer shell and the nonoxidized core has been reported in pellets oxidized at temperatures above 1,000°C.[23–25]

Effect on Pellet Quality The differential contraction of the phases in bigger sized pellets causes bond and lattice strains, which results in the reduction of pellet strength. Besides, the texture of a pellet deteriorates as its size is increased. This is on account of larger shrinkage gradients in the radial direction from the pellet surface to the core.

15.8.1 Formation of Duplex Structure

Oxidation An outer shell of hematite is formed on oxidation of Fe_3O_4 to α-Fe_2O_3 around the pellet before the sintering temperature is reached while the core is still magnetite. This results in the formation of a duplex structure in pellets. The lattice constant of Fe_3O_4 is: A = 0.838 nm, while that of α-Fe_2O_3 is: A = 0.542 nm. The crystal size is reduced gradually on oxidative induration, which may result in the contraction of pellet volume: The magnetite core shrinks more than the hematite shell.[24,26]

Sintering In addition, the pellets are encountered with the change in their volume on account of thermal load. This causes structural stress in pellets. The duplex structure of pellets is characterized with a more heavily sintered core pulling away from the less-sintered outer shell. Wynnyckyj and McCurdy[26] report large temperature gradients within pellets: the shell and core structure and layered homogeneity are characteristics of the severe heating conditions, which in turn can lead to large shrinkage gradients in the radial direction from the pellets' surface to the core and can result in defective pellet texture.

15.8.2 Mechanism of Formation

The way a duplex structure is formed depends on the oxidation pattern and the thermal history of magnetite pellets. The oxidation and sintering phenomena, which occur during the induration of pellets, show their interdependency in a complex manner during the formation of duplex structures.

a. **Oxidation of Magnetite to Hematite** The reaction involving oxidation of magnetite to hematite is highly exothermic. Therefore, the temperature of the outer surface of the pellet that has been oxidized increases; whereas the inner core of the pellet is still magnetite and gets heat thermally. This results in temperature differences between these layers, which could be around 100°C.[27] As the oxidation proceeds with time, more and more magnetite transforms into hematite, and the unreacted magnetite core becomes smaller. Consequently, the contact interface of magnetite and oxygen gas is gradually reduced, which results in slowness of the oxidation rate.
b. **MgO Constituents** MgO content of the silicate melt is unevenly distributed during induration. It increases from the core of the pellet constituting more magnetite to the shell constituting a higher amount of hematite. The crystal lattice of hematite does not allow any substantial amount of magnesium to enter since Mg^{2+} ions cannot replace Fe^{3+} ions in the hematite lattice. Hence, all the released MgO is distributed between silicate slag and magnesium-ferrite in the oxidized hematite shell. In the magnetite core, the bulk of released magnesium is concentrated in the magnetite grains and a minor part is in the slag phase. Magnetite contains both Fe^{2+} and Fe^{3+} ions. Mg^{2+} ions can easily substitute a part of the Fe^{2+} and form Mg-magnetite.
c. **Large-Sized Pellets** The phenomenon of the occurrence of the duplex structure is more common in large-sized pellets wherein thermochemical conditions like temperature and partial pressure of oxygen are not even and change across their cross sections during the induration process. Poor oxidation during the induration of pellets also leads to the duplex structure in the process, with hematite predominantly in the shell and unoxidized magnetite in the core.

15.8.3 Ways to Minimize the Duplex Structure

One solution to avoid duplex structures would be complete oxidation of pellets before sintering is started; this process is called double firing.[24] One possible way is to enrich the oxidizing gas in the PH zone to around 30% oxygen, and controlling the speed of temperature rise in the PH zone to be below 150°C/min. This arrangement would improve the pellet quality.

REFERENCES

1. Papanastassiou, D. and G. Bitsianes, Mechanism and kinetics underlying the oxidation of magnetite in the induration of iron ore pellets, *Metallurgical and Materials Transactions B*, 4 (1973), 487–496.
2. Pape, P.O., R. D. Frans, and G. H. Geiger, Magnetite oxidation kinetics and thermal profiles in a magnetite pellet plant cooler, *Ironmaking & Steelmaking*, 4(3), (1976), 138–145.
3. Gassel, J. B. and D. W. Rierson, Reaction rate study in the oxidation of a magnetite ore, Second International Congress on the Science and Technology of Ironmaking and 57th Ironmaking Conference, Toronto, Canada, March 22–25, 1998.
4. Bentell, L. and G. Mathisson, Oxidation and slag-forming process in dolomite-fluxed pellets based on magnetite concentrates, *Scandinavian Journal of Metallurgy*, 7, (1978), 230–236.
5. Ferreiera, S., D. Siguin, and F. Garcia, Physicochemical study of presence of Fe2+ ion in pyroconsolidated magnetite pellets, *Ironmaking & Steelmaking*, 20, (1993), 482–484.
6. Ferreira, S., D. Siguin, and F. Garcia, Thermal analysis of sintering of magnetite pellets, *Ironmaking & Steelmaking*, 21, (1994), 119–123.
7. Darken, L. S. and R. W. Gurry, The system iron-oxygen. II. Equilibrium and thermodynamics of liquid oxide and other phases, *Journal of the American Chemical Society*, 68, (1946), 798–816.
8. Firth, A. R., Interactions between magnetite oxidation, calcination of carbonate minerals and melt formation in iron ore pellets, Iron Ore Conference, Fremantle WA, September 19–21, 2005, 335–341.
9. Gillot, B., A. Rousset, and G. Dupre, Influence of crystallite size on the oxidation kinetics of magnetite, *Journal of Solid State Chemistry*, 25, (1978), 263–271.
10. Lepp, H., Stages in the oxidation of magnetite, *American Mineralogist*, 42, (1957), 679–681.
11. Tamman, G., Lehrbuch der Metallographie, Voss (Leipzig) 1923. (Loc. Cit. Ref. 13).
12. Urich, D. M. and T. M. Han, A progress report on the effect of grind, temperature, and pellet size upon the quality of specular hematite pellets. In: W. A. Knepper (ed.) *Agglomeration*-1962, Interscience, N.Y. 669–719.
13. Jaroslav, S. R. B., Z. Ruzickova, and Z. Ruzickova, Pelletization of Fines, in: *Development in Mineral Processing*, Vol. 7, Ed. D. W. Furestenau, Elsevier Science Pub. Inc., New York, (1988), p. 69.
14. Ball, D. F., J. Dartnell, J. Davidson, A. Grieve, and R. Wild, *Agglomeration of Iron Ore*, Sec 2, Chap. 11, Heinermann Education Books Ltd., London, (1973).
15. Soin, I. and R. J. Fruehan, The reduction of iron oxides by volatiles in a rotary hearth furnace process: Part I: The role and kinetics of volatile reduction, *Metallurgical & Materials Transactions*, 36, (2005), 605–612.
16. Trunshenski, S. P., K. Li, and W. O. Phiibrook, Non-Topochemical reduction of iron oxides, *Metallurgical & Materials Transactions B*, 5, (1974), 1149–1158.

17. Thomas, C. G., J. D. G. Hamilton, and K. M. Bowling, in: *Proc. of Pellet & Granules Symp. AMIM*, Parkville, Victoria, Australia, (1974), 93.
18. Firth, A. R. and J. R. Manuel, Thermal implications of phase transformations during induration of iron ore pellets produced from hematite, *ISIJ International*, 45, (2005), 1561–1566.
19. Panigrahy, S. C., P. Verstraeten, and J. Dilewijins, Influence of MgO addition on mineralogy of iron ore sinter, *Metallurgical Transactions B*, 15B, (March 1984), 23–32.
20. Suzuki, G., T. Satoch, and N. Sakamoto, Use of iron ore pellets in Japan, National Seminar on Role of Iron Ore Pellets in Iron & Steel Industry under the aegis of Iron & Steel Division of Indian Institute of Metals. Panji, Goa, April 15–16 1988, Org. by Steel Authority of India Limited, Kudremukh Iron Ore Company Limited and Mandovi Pellets Limited.
21. Dwarpudi, S., T. K. Ghosh, A. Shankar, V. Tathavadkar, D. Bhattacharjee, and R. Venugopal, Effect of pellet basicity and MgO content on the quality and microstructure of hematite pellets, *International Journal of Mineral Processing*, 99, (2011), 43–53.
22. Bhagat, R. P., B. Nayak, K. K. Bhattacharyya, and M. S. Goswami, Mineralogy of iron ore pellets and its correlation with crushing strength. *Journal of Metallurgy and Materials Science*, 56(2), (April–June 2014), 117–123.
23. Cooke, S. R. B. and T. E. Ban, Microstructures in iron ore pellets, *Transactions AIME*, 193, (1952), 1053–1058.
24. Cooke, S. R. B. and W. F. Stowasser, The effect of heat treatment and certain additives on the strength of fired magnetite pellets, *Transactions AIME*, 193, (1952), 1223–1230.
25. Papanastassiou, D. and G. Bitsianes, Mechanism and kinetics underlying the oxidation of magnetite in the induration of iron ore pellets, *Metallurgical Transactions*, 4, (1973), 487–496.
26. Wynnyckyj, J. R. and W. A. McCurdy, Causes of the shell and core structure and layered in homogeneity in iron ore pellets, *Metallurgical and Materials Transactions B* 5, (1974), 2207–2215.
27. Ruiz Sierra, J. C., J. M. Badie, and F. Chejne Janna, Non-isothermal conditions inside magnetite pellet due to its oxidation by air, *Ironmaking & Steelmaking*, 21(2), (1994), 114–117.

16 Quality of Indurated Pellets

ABBREVIATIONS

AI	*Abrasion index*
CCS	*Cold crushing strength*
LOI	*Loss on ignition*
RDI	*Reduction degradation index*
RI	*Reducibility index*
RSI	*Reduction swelling index*
SFCA	*Silico-ferrite of calcium and aluminum*
SEM	*Scanning electron microscopy*
TI	*Tumbler index*

16.1 PREAMBLE: PELLET PROPERTIES AND FACTORS INFLUENCING

16.1.1 Pellet Properties

Important properties of the indurated pellets are:

a. Bulk density and angle of repose.
b. Cold compression strength (CCS): Ability of the pellets to withstand compressive load.
c. Porosity: Pellet voidage. It decides the pellet strength and pellet reducibility.
d. Tumbler index/abrasion index (TI/AI): Resistance against tumbling of the pellets.
e. Reduction parameters: reducibility index (RI), reduction–degradation index (RDI).
f. Swelling index: Characteristic of pellets to swell during reduction.
g. Softening and meltdown characteristics: Decides the size and shape of the cohesive zone.

Some of the aforementioned properties (especially d, e, and g) are common to sinter and defined in **Sec. 9.1.1**.

16.1.2 Textural Characteristics and Pellet Properties

The properties of indurated pellets, mentioned previously, are largely governed by the form and degree of bonding achieved between ore particles and the stability of these bonding phases during reduction of iron oxides.[1]

The textural characteristics of the indurated pellets, which are influenced by the operation and process-related variables, are decisive in this respect. The new

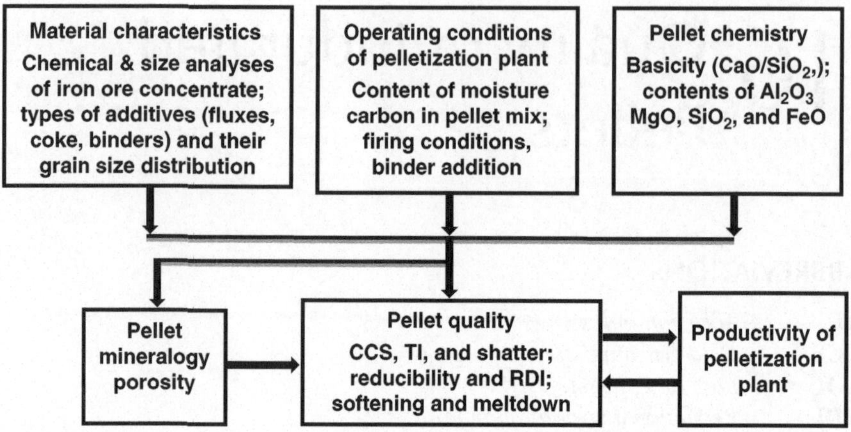

FIGURE 16.1 Schematic representation of the factors that influence the pellet qualities (cause and effect with reference to the pellet qualities and productivity vis-à-vis their mineralogy)

minerals are formed in the pellets owing to the physical–chemical reactions during the induration that initiate in the solid state, go through semiliquid, and conclude in the liquid state at higher temperature.

The microstructure of pellets is characterized by the following entities.

a. Primary and secondary iron oxide (hematite and magnetite) phases. These constitute the main minerals.
b. Slag-constituting phases of glass-like and crystalline types. These phases also act as bonds among the grains.
c. Porosity and compactness of the pellets.

16.1.3 Factors Affecting Pellet Properties

Figure 16.1 represents the factors that influence the quality parameters of pellets vis-à-vis their productivity. These are related to and categorized under:

a. Material characteristics
b. Operating conditions of pelletization plant
c. Pellet chemistry

The cause and effect are also shown in Figure 16.1. The interacting parameter, in this regard, is the pellet mineralogy.

16.2 PELLET CHEMISTRY

In modern blast furnace practice there is an increasing demand for high-grade burden materials rich in iron. The pellet should be low in alumina, silica, and sulfur as these constituents adversely affect the iron-making processes through the blast furnace as well as direct reduction routes (ref. **Sec. 4.4**).

TABLE 16.1
Bulk Density and Angle of Repose of Sinter and Pellets (for Comparison, those of Coke Are Also Shown)

Material	Sinter	Pellet	Coke
Bulk density, kg/m^3	~1,475	~3,250	~610
Angle of repose, deg.	31–34	26–28	35–38

Pellets are rich in iron and usually contain about 65% iron, with silica as the main gangue constituent, and usually with low phosphorous and sulfur contents. The ore concentrate, produced after intensive beneficiation of iron ore, consists of higher Fe (Fe around 65 wt.%), and is used as pellet feed. The additives (nonferruginous) in the pellet mix, unlike sinter, are in a much lesser amount. Hence, the pellets consist of high Fe.

16.3 BULK DENSITY AND ANGLE OF REPOSE

Pellets do have higher bulk density (3–3.5 t/m^3) than sinter (**Table 16.1**), which helps in improved heat exchange due to the longer duration of gas–solid contact. The pellets, because of their round shape, do have an angle of repose that is less than the sinter; hence, this requires wider conveyor belts, also of a lesser slope (than in the case of sinter) for material handling.

16.4 CRUSHING STRENGTH

The crushing strength of fired pellets is chiefly attributed to their crystallization during the induration processes and reactions of gangue constituents. For higher grade concentrates, the crystallization process, and for ores with gangue content, glassy slag formation additionally plays an important role. The strength is primarily influenced by the oxide bonding of hematite grains and slag bonding among the minerals. The silicate and other gangue minerals, if uniformly distributed within the pellets, contribute toward its high strength.

16.4.1 Raw Materials (Including Binders) and Their Characteristics

Ore Characteristics The crystalline ores demand a higher fineness than those of an earthy character. Consequently, one can improve the pelletizing ability of the crystalline ore by mixing it with another ore having a greater bonding power. The shrinkage in pellets is observed in the case of magnetite concentrate, which increases with increasing its specific surface area.

Alumina Content High alumina in a pellet decreases its strength. Alumina in the ore concentrate, due to its refractory characteristics, demands more heat during the induration. Increased alumina decreases the melt fluidity, which reduces the ability of bubbles to coalesce and reshape.[2]

Increased alumina results in branch-like irregular shapes that exhibit poor load-bearing capacity resulting in lower strength.

Fineness of the Concentrate Control of size distribution of the ground ore, which makes up the pellet feed, is essential in the optimization of the pellet quality. These affect the strength parameter of pellets.

The CCS of indurated pellets improves considerably with the concentrate that has undergone a very fine mesh of grind. Also, the abrasion index of the fired pellets improves with increasing the below 45 micron-sized fractions of pellet ore feed. The effect of increased fineness of ore concentrates on the pellet CCS in some cases is more than the addition of bentonite in the mix. On the other hand, increasing the fineness of concentrate has adverse effects on profitability of the pelletization operation, as follows:

a. When the ores undergo finer mesh of grinding, the cost (of operation) increases.
b. The drying rate of green pellets decreases; consequently, more time is required in drying.

In view of this, the grain size of the ore feed needs to be optimized while striking a balance between increased fineness of the concentrate to improve pellet quality and increased cost of operation.[3]

A linear correlation has been observed between the specific surface area of ore concentrate and the compressive strength of green pellets (**Figure 16.2**, Curve III)

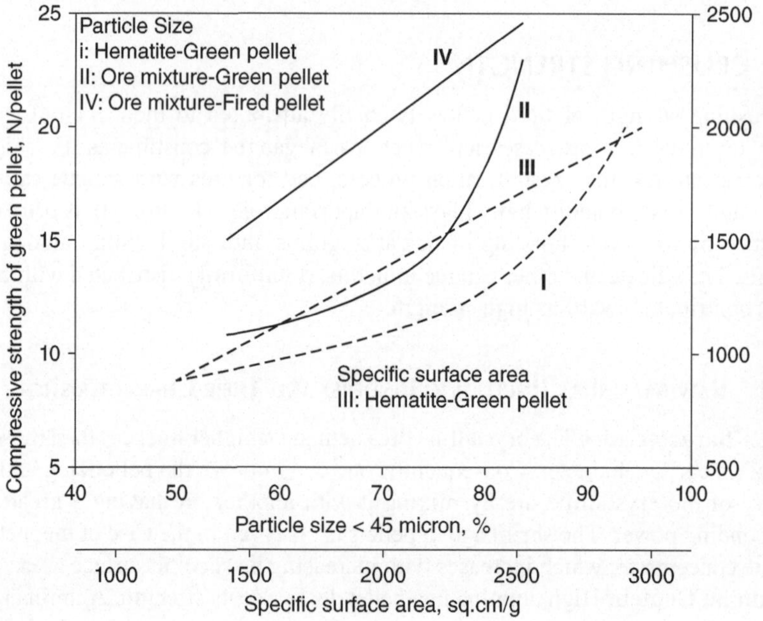

FIGURE 16.2 Effect of particle size and specific surface area on compressive strength of green and indurated hematite pellets (after Ref. [4])

(**Sec. 13.4**). Also, the strength of indurated pellets increases linearly with the increase in fineness (particle size below 45 micron) of the ore concentrate (IV).

(Combined) Water Higher moisture in green pellets creates cracks and discontinuities during their drying due to evolution of water. The green pellets after drying become less compact and more porous, which results in lowering the pellet strength on induration.[5]

The crushing strength of the indurated pellets is also less in cases where the ore (concentrate) has a higher LOI content. This is due to the emission of bonded water molecules during heat hardening. Hence, the combined water from the ore is essentially to be removed from high LOI (goethite) ore as it contains a high amount of combined water.

Binder The binder consists of low melting point alkali silicate that forms melt at high temperatures during induration. The melt phase among the particles exerts pressure to pull them together due to interfacial forces, thereby increasing bonding strength. Adequate dosage of bentonite is essential for an acceptable CCS in the indurated pellets.

The CCS increases with increasing the bentonite dosage. Typical pelletization results[6] with respect to the hematite concentrate of a finer mesh of grind (93% below 45 micron) show that with a lower bentonite dosage, the pellet CCS was much lower compared to that with a higher dosage of bentonite (see **Sec. 15.6.3**).

Limestone The addition of limestone in the mix decreases the crushing strength of pellets. CO_2 gas gets evolved on induration of the green pellets consisting of limestone. Consequently, micro- and macrocracks are developed, resulting in weaker indurated pellets.

$Ca(OH)_2$ The usage of calcined/hydrated lime in pelletization may result in crack formation in pellet, loss in bonding efficiency of bentonite. Besides, handling of hydrated lime in plant for its mixing in green pelletization is difficult. However, the usage of calcined lime in pelletization can be made by suitable adjustment of flux composition and process parameters. The slaking of lime has to be terminated and the hydroxide thoroughly mixed with the ore concentrate before green balls are formed (**see Sec. 12.4.4**).

The addition of lime in the pellet mix increases crushing strength of the indurated pellets. The abrasion value also improves (decreases) with rising calcium hydrate addition. The silicates and ferrites phases are formed with lime addition and remain predominant in the slag phase of the indurated pellets, which results in increased strength of the pellets. The glassy structure of acidic pellets (in the absence of lime) is indicated by the relatively low porosity which deteriorates the pellet reducibility.[7]

Typical study with hematite ore from Noamundi, Eastern India, has shown that the calcined lime-fluxed hematite pellet without bentonite and limestone shows a reasonably good, cold crushing strength in the range 310 kg/pellet. The bentonite constitutes of acidic gangue constituents. Hence, its elimination could help in reducing alumina and silica contents of pellet.[8]

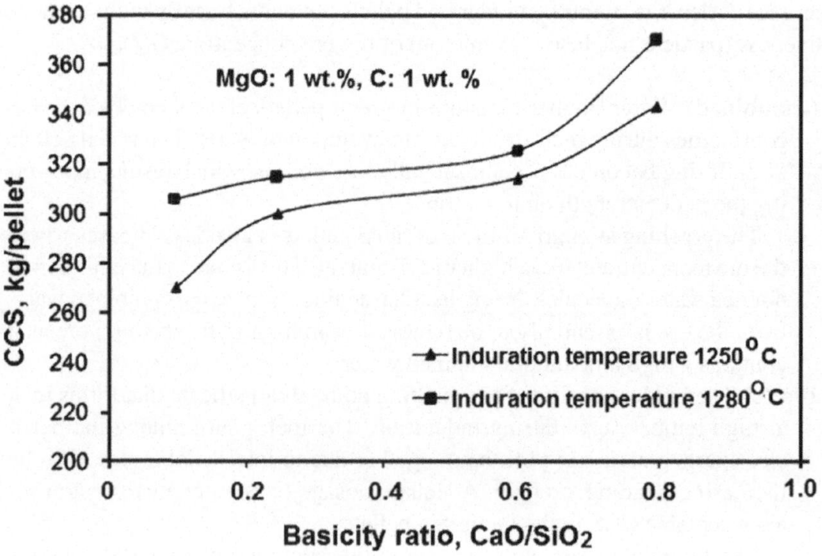

FIGURE 16.3 Effect of pellet basicity on CCS of indurated pellets (after Ref. [8])

MgO versus CaO The temperature(s) of formation of calcium compounds (ferrites, silicates, and aluminate) are lower than those of corresponding magnesium compounds. This leads to the higher stable mineral formation in the pellets with the addition of CaO (lime) compared to MgO.

Pellet Basicity Acidic pellets, in general, have good compression strength. Basic pellets also demonstrate improved quality on firing in terms of their crushing strength (CCS), reducibility, and swelling.[9] Typical study with hematite ore from Noamundi, Eastern India, has shown that the CCS of indurated (at 1,280°C for 10 min) limestone-fluxed pellets increases with increase in basicity (CaO/SiO_2) of pellets (**Figure 16.3**).[8]

The strength of fluxed pellets is increased due to the formation of calcium silicate and calcium ferrite phases in indurated pellets.[10] However, the formation of ferrites has major contribution in improving the pellet strength compared to the formation of silicates. This is because the dicalcium silicate transforms from β-phase to γ-phase with a volume change at increased basicity (more than 0.5), which causes weakening of the pellet structure.

The CaO-rich low melting slag phases $CaO{\cdot}Fe_2O_3$ and $CaO{\cdot}FeO{\cdot}SiO_2$ are formed at increased basicity: in case of former through the exothermic reaction between lime and hematite. The ferrite thus formed has the highest crushing strength, on an individual basis, next to hematite. A higher amount of Ca-rich phases contributes to the good slag bonding in indurated pellets at high basicity. The improved strength is also due to the formation of more slag in the porous structures of the pellets and its consolidation.[11] The pellet shrinkage is increased at higher basicity.

For effective results, the lime, generated during calcination of limestone, should completely assimilate with hematite to form ferrites through the solid-state reaction. The unassimilated lime, if any is present, hydrates over a period and disintegrates, causing weakening of the pellet structure.

Pellet FeO The CCS of pellets is adversely affected by their higher FeO. A higher FeO is an indication of poor oxidation during the induration of pellets, resulting in a duplex structure. The duplex structure of pellets is characterized with a more heavily sintered core pulling away from the less-sintered outer shell: The core and inner mantle show higher amount of magnetite, whereas the shell and outer mantle show high hematite content and traces of magnetite (see **Sec. 15.8**). The crystal size is reduced gradually on oxidative induration, which may result in the contraction of pellet volume: The magnetite core shrinks more than the hematite shell.[12,13] The load-bearing capacity of this duplex structure is low, resulting in low pellet strength.

Pellet MgO During induration, MgO may join either the slag phase or iron oxide grains. The components, FeO and MgO, have complete miscibility and form a solid solution (indicated by the FeO-MgO phase diagram).[14] In the oxidized hematite shell, all the released MgO is distributed between silicate slag and magnesioferrite.

FeO content of the slag phase decreases considerably with increased MgO content up to 1.5%; thereafter, no change was observed. This can be attributed to the formation of magnesiowustite, which is formed on cooling of the pellets on induration. Beyond that level (on further addition), the assimilation of MgO constituents into the pellet matrix is poor, which is indicated by the increasing amount of relict magnesium silicate phase, resulting in no further drop in FeO. Hematite contains only traces of magnesium because the crystal lattice of hematite does not allow any substantial amount of magnesium to enter. Mg^{2+} ions cannot replace Fe^{3+} ions in the hematite lattice. In the magnetite core, the bulk of released magnesium is concentrated in the magnetite grains and a minor part in the slag phase. Magnetite contains both Fe^{2+} and Fe^{3+} ions. Mg^{2+} ions can easily substitute a part of the Fe^{2+} and form Mg-magnetite.[15]

Carbon Content Higher carbon content reduces the pellet strength, which is attributed to:

a. Increased FeO content of the pellet, especially in its inner layer. This results in a duplex structure, leading to lower strength.
b. Carbon in green pellets combusts under the oxidizing atmosphere during firing and releases CO_2. The pellets formed have a porous structure.

The hematite mineral is reduced to magnetite to a certain extent, depending on the amount of coke added, during the induration process. In the case of hematite, the strength rises up to 0.5 wt.% of coal addition, then decreases, but is still sufficient up to 1.5 wt.% coal. The influence is explained by the partial reduction of hematite to magnetite and subsequent incomplete reoxidation of magnetite to hematite.

16.4.2 Induration Conditions

Feed Rate Higher feed rate decreases the pellet strength. At a higher feed rate, the pelletizing grate moves at a faster speed, which reduces the residence time of pellets in the high-temperature sintering region, resulting in low strength.

Bed Height A higher bed height results in lowering the temperature in the bottom layers of the bed, leading to poor strength of the pellets.

Firing Temperature The surface fusion occurs at higher firing temperatures that creates solid bonds among the particles. The cross section of these bonds (the neck between two adjacent particles) increases with the increase in induration temperature, which in turn improves the strength.

In addition to the (absolute value of) firing temperature, the heating rate affects the pellet quality on account of changed microstructures. Typical studies on induration of pellets prepared from mixed magnetite –35 wt.% hematite concentrate revealed different forms of pellet microstructures with varied heating rate.[16] The pellet processed at 50°C/min had compact microstructure with small grains of sintered secondary hematite (SH) that were partially surrounded by a slag phase. The driving force at high temperatures for sintering of the SH particle increases with increasing heating rate. In contrast, the microstructure from the induration process at 5°C/min shows an SH phase with faceted morphology surrounded by a relatively large amount of the slag phase. At lower heating rates, the time for the diffusion process is favored, and larger grains were formed in comparison to those observed with a higher heating rate.[16]

Figure 16.4 (I) shows microstructure of pellets processed up to 1,200°C. These constitute secondary hematite (SH), hematite (H), magnesium ferrite (MF), silicate (S), and silico-ferrite of calcium and aluminium (SFCA). The presence of hematite particles indicates its stability at 1,200°C. The SFCA phase was formed between 1,120°C and 1,160°C for 5°C/min (**Figure 16.4 II(a)**) and 50°C/min (**Figure 16.4 II(b)**), respectively. **Figure 16.4 (II)** shows the micrographs of pellets processed up to 1,400°C. The microstructure is characterized by a faceted SH phase surrounded by an abundant SCFA phase at heating rate 5°C/min (photo a). When the pellets were processed at 50°C/min, an abundant amount SH phase is formed, which is characterized by small connected grains that are partially surrounded by the SFCA (photo b).

Sometimes the pellet bed come across its low firing during induration. This is due to its poor permeability. The phenomenon of low firing is indicated by low burn-through temperature and increases the requirement of specific gas consumption; this results in poorer crushing strength of pellets.

Figure 16.5 shows the crushing strength (a) and apparent pellet densities (b) plotted against induration temperature. The plots of strength versus induration

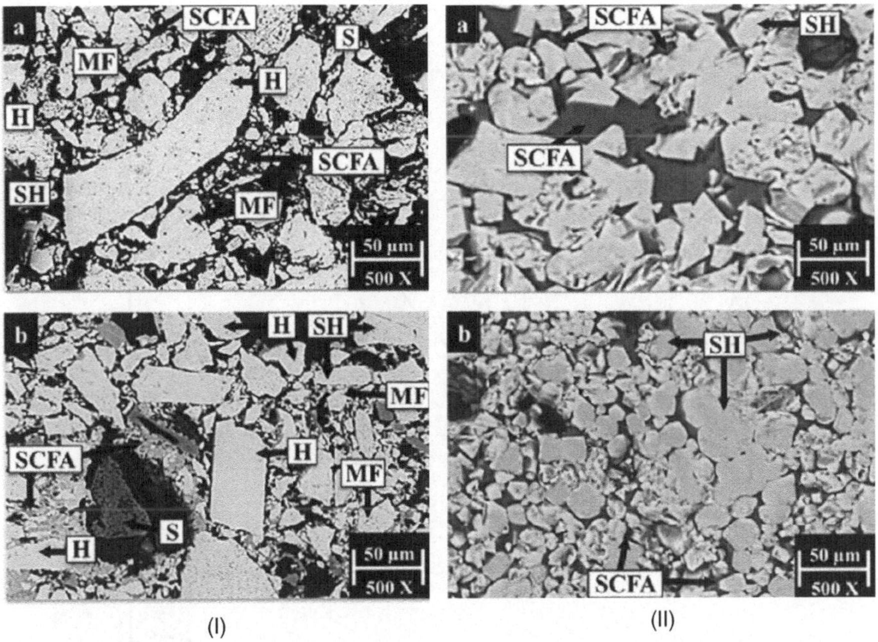

FIGURE 16.4 Microstructure of pellets: **(I)** at 1,200°C and **(II)** 1,400°C. Heating rate: (a) 5°C/min and (b) 50°C/min. Phases present: secondary hematite (SH), hematite (H), silicon oxide (S), magnesium ferrites (MF), and silico-ferrites of calcium and aluminium (SFCA) (after Ref. [16])

temperature show the characteristic (typical) "knee" curve. Initially, the strength of pellets increases slowly with the increase in temperature. This characteristic gives way at a higher temperature to an abrupt rise then levels off at about 1,270°C.[10] Strengths are higher for the finer material.

Pellet Density Above the knee, the pellets undergo significant densification (Figure 16.5 (b)), which causes the development of pellet strength. No shrinkage could be detected below the knee.

Contractions in pellets during oxidation and incipient sintering at 900°C have been reported. At 1,230°C, shrinkage in the magnetite phase was about 1.4 times greater than that in the hematite phase. The shrinkage in pellets increased with increasing specific surface area in the magnetite raw material. The degree of shrinkage was dependent on the time the pellet was exposed to sintering temperatures.

A large temperature gradient within pellets is characteristic of the severe heating conditions, which in turn can lead to large shrinkage gradients in the radial direction from pellet surface to the core and result in defective pellet texture. This causes the shell and core structure and layered inhomogeneity.[13]

Cooling Rate A slower rate of cooling of fluxed pellets during induration increases their strength.

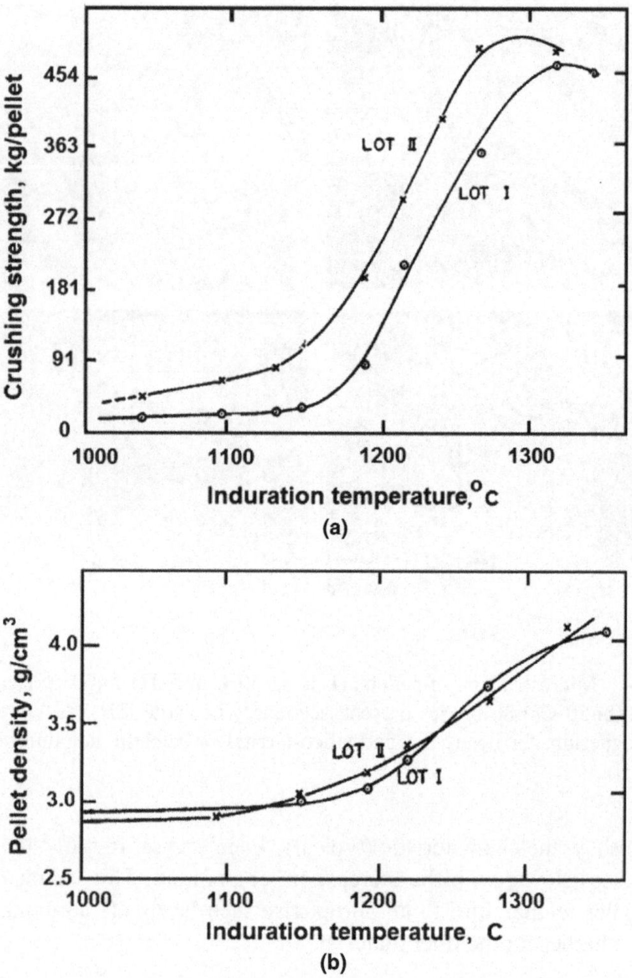

FIGURE 16.5 (a) Strength development and (b) densification at increasing induration temperature in pellets made from the magnetic concentrate (pellet diameter = 18 mm) (after Ref. [10]) (Crushing strength in lb/pellet from Ref. [10] converted into kg/pellet)

16.5 POROSITY

The porosity of pellets and its characteristics are important to the diffusion of reducing gases through fired pellets in the blast furnace. The reducing gases enter the pellets via the pores and affect the removal of oxygen. Typically, a porosity of 22–30% for fired pellets is associated with good reducibility.

The porosity of pellets is composed of microporosity and macroporosity. These are typically categorized as:

a. Microporosity (pore size less than 10 micron and up to 0.001 micron)
b. Macroporosity (pore size between 10 micron and 100 micron)

The fraction of microporosity of a pellet out of its total porosity is important in affecting the reducibility. Matsuo et al.[17] performed reduction experiments with the synthesized pellets from hematite mineral: (A) with micropores (around 15 micron) alone and (B) with both macropores and micropores. They observed that while the overall porosity was equal, pellets in (B) were reduced at lower rates than pellets in (A) Similar results also have been obtained in the case of sinter (see **Secs. 9.2.3** and **10.7.2**).[18,19]

a. Fluxed pellets exhibit more porosity than nonfluxed pellets. The carbonates in limestone/dolomite dissociate on calcination during induration and release CO_2 gas, resulting in the formation of pores and hence improving its reducibility. However, a higher melt phase, which could be generated with increased basicity, will impregnate the pores.
b. Dolomite fluxed pellets exhibit higher porosity compared to limestone pellets. MgO increases the melting point of the binding phase. The MgO, being a major constituent of dolomite, raises the liquidus temperature of slag, and consequently, a lower amount of liquid slag is formed. This helps in increasing the pellet porosity since the impregnation of pores with the melt phase in this case (dolomite fluxed pellets) is restricted.[20]

16.6 SWELLING INDEX

Swelling index is a measure of the volume change of pellets during their reduction. The pellets swell during their reduction on account of the stress developed in transformation of the hematite phase to magnetite and wustite with volume expansion. The volumetric expansion of iron ore pellets during their reduction can be mainly attributed to the increased volume requirements for the anisotropic growth of magnetite (111) planes parallel to the hematite (001) planes.[21]

The swelling decreases the voids in the charge and thus impedes the gas flow inside a blast furnace (see **Secs. 3.6.6** and **4.2.2**). When swelling occurs, pellets may crumble and gas permeability could be impaired. The maximum allowable swelling index of pellets for blast furnaces ranges from 16 to 18%.

Content of Gangue The ability of gangue and/or slag phase(s), present in the pellets, is important in affecting swelling, as it is related to the resistance to withstand the reduction stresses of oxide particles. High melting point slag would produce sufficient bonding strength to limit swelling and low melting point slag enhances swelling.

Qing et al.[22] have observed that when the SiO_2 content of pellet was 4.8 wt.%, the reduction swelling index (RSI) was 16.5% and compressive strength of reduced pellet was 423 N/pellet. When the pellets fired at 1,280°C constitutes SiO_2 at 2.8 wt.% and 1.8 wt.%, the RSI of reduced pellets increased considerably 34.5% and 55.8%, respectively. The compressive strength of reduced pellets was also lower: 200 N/pellet and 80 N/pellet, respectively. However, the usage of pellets with SiO_2 content in the blast furnace would increase the slag volume. The adverse effect of decreasing pellet SiO_2 (with regard to RSI) could be neutralized by adding some MgO in pellet.

Pellet Basicity The tendency and intensity of pellets to swell during their reduction is influenced by their basicity (CaO/SiO_2). At CaO/SiO_2 greater than 0.25, the pellets have a strong tendency to swell during their reduction, which might disturb proper blast furnace operation (see **Sec. 3.6.6**). Swelling in limestone fluxed pellets happens to be the highest at around 0.8 basicity and decreases thereafter. At increased pellet basicity, stronger bonding phases are formed in the pellets that can absorb the stress arising from hematite to magnetite transformation during their reduction.[1,23]

The maximum swelling of pellets happens to be within the basicity range between 0.2 and 0.7. A higher swelling index (over 20%) during reduction in the case of acidic pellets could be attributed to the increased mobility of the melt phase due to the formation of low melting point calcium olivines between Fe_2SiO_4 and Ca_2SiO_4, with the lowest melting point of 1,115°C in this basicity range. In these (acidic) pellets, the reduction is accompanied by the reaction between Fe^{2+} and SiO_2 to form the low melting point phase, fayalite (Fe_2SiO_4), which melts at 1,175°C.[24]

The low melting point silicate melt (slag) acts as a bonding phase in the pellets. Its plastic or mobile nature provides a medium for absorption of the reduction stresses by increased distances among the particles.[21] The amount of silicate melt, which acts as a bonding phase, increases with the increasing basicity. The number (quantity) of phases/grains per unit volume increases in case the silicate melt is more distributed within the pellets. In this case, the pellets have a high silicate melt density.

In addition, acid pellets are very sensitive to the variation of their CaO content, though in minor amounts, with respect to swelling.

Assimilation Due care must be undertaken to avoid the appearance of free lime in pellets as it promotes the rate of whisker growth and more swelling. On the other hand, the pellets, with fully assimilated lime in them (to form ferrites) exhibit less swelling by forming low FeO and high melting slag, which gives sufficient bond strength to withstand the reduction stresses.

Pellet MgO As mentioned in preceding paragraph, the pellets with higher silica content have acceptable and lower RSI. However, a higher input of SiO_2 in blast furnace through pellet increases the slag volume—the consequence of which is reduction of furnace productivity and increase in coke rate. The adverse effect of decreasing the content of silica in indurated pellets (on account of higher swelling index) could be neutralized by adding more MgO in pellets.

The suitable modification in pellet chemistry results in an acceptable level of RSI at the same time, lower content of silica in pellet. Qing et al.[22] have studied the swelling behavior of the pellets on reduction. The RSI of the pellets fired at 1,280°C with contents of SiO_2 and MgO at 2.8 wt.% and 1.7 wt.%, respectively, was less than 20%. The same level of RSI could be maintained at reduced SiO_2 content (1.8 wt.%) by increasing the pellet MgO to 2.5 wt.%. The compressive strength of reduced pellets was around

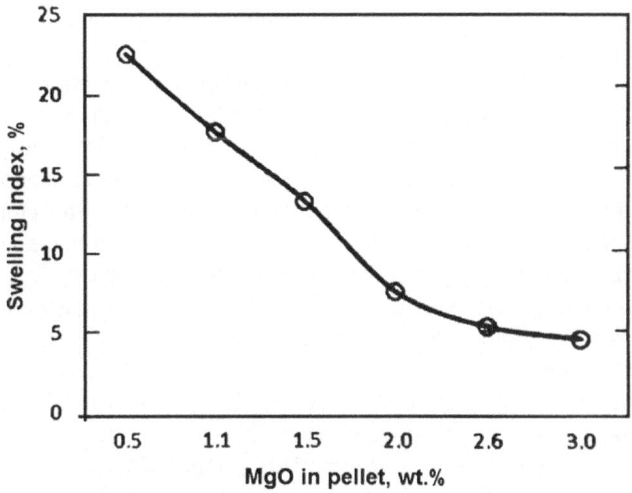

FIGURE 16.6 Relationship between the MgO content of fluxed pellets and swelling index (after Ref. [25])

300 N/pellet. The added MgO pellet with low SiO_2 content has been used in 5,500 m^3 BF in Shougang Jingtang Corporation and delivered cost reduction on sustainable basis.[22]

Figure 16.6 typically shows the effect of MgO content of dolomite-fluxed pellets produced at Kakogawa Works of Kobe Steel on its swelling index.

The pellet constituting moderate MgO exhibits less swelling by forming low FeO. This is on account of high melting slag, which gives sufficient bond strength to withstand the reduction stresses. The MgO either reacts with the surrounding iron oxide to form magnesioferrite or is assimilated in the silicate melt that acts as a bonding phase in the pellets.[20]

Pyroxenite, a magnesium silicate mineral, unlike dolomite, does not undergo any endothermic reaction for dissociation. The MgO from the pyroxenite gets distributed only in magnesioferrite and does not join the slag phase. FeO content of the slag phase decreases with increasing the pyroxenite addition up to around 5% (~1.5% MgO); beyond that, the MgO constituent (of pyroxenite) gets poorly assimilated into the pellet matrix. This is indicated by the increasing amount of relict magnesium silicate phase, resulting in no further drop in FeO.[26]

Temperature of Pellet Firing With respect to the fluxed pellet, the swelling can be reduced by adjusting the firing temperature sufficient to form calcium ferrite. The appearance of free lime in pellets promotes the rate of whisker growth.[9]

In this regard, the firing temperature of the pellets should be made sufficient to form calcium ferrite. Under this condition, the swelling value of pellets decreases as the whiskers formed are very small and random.[27]

Metallic Iron Whiskers The swelling of pellets has a direct correlation with the formation of metallic iron whiskers. The growth of metallic iron whiskers is also/mainly responsible for higher swelling value and degradation of pellets during reduction.

The formation of whiskers in pellets and theirs swelling on reduction depend on the amount gangue in the ore concentrate undergoing pelletization. Using scanning electron micrograph (SEM), microstructural examination of the pellets, produced from typical Indian iron ore concentrates, shows that the growth of metallic whiskers is affected by the gangue content of the pellet: the higher the gangue content, the lower the growth of whiskers and hence the lower the swelling value were observed.

The acidic pellets, prepared from Bailadila ore concentrate, containing higher Fe (around 68 wt.%) and lower gangue (SiO_2+Al_2O_3 around 2.0 wt.%) show a large amount of whiskers. These are large and pervade over the entire cross section of the pellets. The pellets from Noamundi ore had smaller whiskers and also a lower swelling value. The maximum swelling (about 40 %) was observed after forty minute of reduction at 1,273 K. The corresponding swelling index in case of Bailadila pellet was more than two times.[28,29]

Source	Fe	SiO_2	Al_2O_3	LOI
Bailadila	68.0	0.72	1.23	1.0
Noamundi	64.0	2.0	3.36	2.9

16.7 PELLET REDUCIBILITY

Basicity In the case of basic pellets, the limestone that is added into the mix gets calcined through the endothermic reaction during induration. The CO_2 gas, thus evolved, increases the porosity of pellets, and hence, their reducibility.

The pellet reducibility reaches a maximum at basicity 1.3 and then decreases with increased basicity. Pellet basicity of approximately 1.3 value seems to be optimum since at this basicity the hematite mineral formed on induration has a smaller grain size (before the pellet is reduced). This provides a large surface area for reduction. The optimum value of pellet basicity around 1.3 with respect to reducibility is also on account of the following:

a. In pellets having low basicity, a dense iron layer forms around the reducing pellets during the reduction of wustite, which slows down the reduction.
b. Reducibility decreases beyond 1.3 basicity due to the formation of large secondary hematite grains.

MgO/SiO_2 Ratio The ratio of **MgO/SiO_2** in magnetite pellets is an important parameter affecting its reducibility. Fluxed pellets containing approximately 1.5% MgO showed the highest reducibility and improved softening characteristics. Panigrahy et al.[1] report that the pellets fluxed with dolomite had the highest reducibility at around 1.3 basicity; though this reduces the pellet strength on account of increased porosity.

C in Pellet Mix Incorporation of coke in the pellet mix improves the porosity of pellets and, subsequently, their reducibility. However, the pellet reducibility may be adversely affected at higher content of C in pellet mix due to slagging and greater formation of adverse mineral phases, such as fayalite (FeO).

16.8 REDUCTION–DEGRADATION INDEX (RDI)

Pellet Basicity At a higher CaO/SiO_2 ratio (in the range of 1.6–2.0), low temperature breakdown, reduction degree, and strength of pellets on reduction improved. A CaO/SiO_2 ratio beyond 2.0 was not favored due to the formation of $2CaO{\cdot}SiO_2$. The magnetite pellets with basicity up to 1.3 exhibit less low temperature disintegration. The ferrite minerals in the fluxed pellets act as a bonding phase and absorb stress during their reduction in the iron-making process. This decreases the RDI and swelling index of the pellets.

Pellet Alumina The effect of alumina content on the pellets' RDI is shown in **Figure 16.7**. Alumina in the ore concentrate, due to its refractory characteristics, demands more heat during the induration. Increased alumina decreases the melt fluidity, which reduces the ability of bubbles to coalesce and reshape. High alumina in a pellet also decreases its strength.

16.9 SOFTENING AND MELTDOWN CHARACTERISTICS

The softening and melting characteristics of pellets affect the furnace cohesive zone during reduction (see **Sec. 3.6.6**).

The pellets have a softening temperature of 1,200–1,350°C, which is higher than that of lump ore or sinter. Hence, the cohesive zone becomes narrower and is pushed lower in the blast furnace. This increases the ease of reducing the gas driving rate in this zone with an ultimate effect on productivity.

The addition of MgO helps in improving the high temperature properties of the pellets.[21] The study by Nishida et al. [25] has shown that the properties of dolomite-fluxed pellets produced at Kakogawa Works of Kobe Steel improved in the high-temperature zone, contributing to better permeability in the blast furnaces.[25] Kakogawa Works has initiated the production of self-fluxed pellets (CaO/SiO_2 approximately 1.3) and

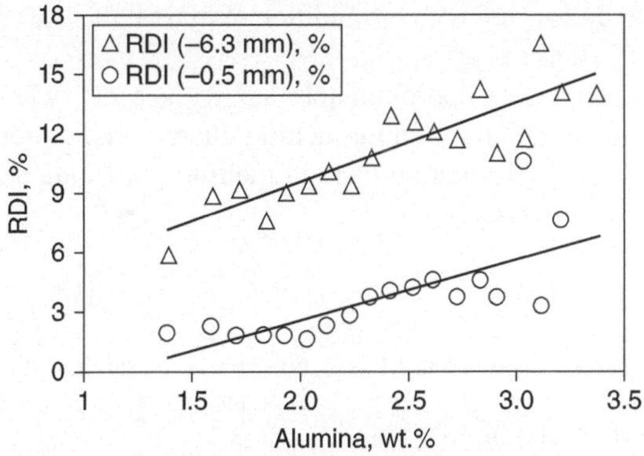

FIGURE 16.7 Dependence of sinter RDI on its alumina content (after Ref. [30])

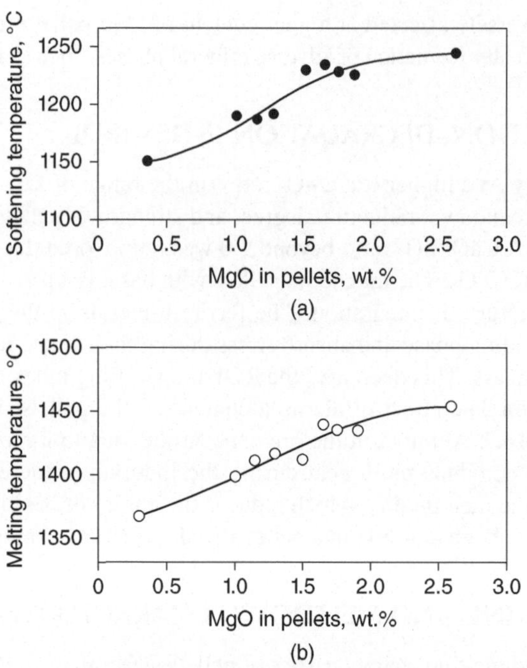

FIGURE 16.8 Dependence of (a) softening and (b) meltdown temperature on pellet MgO (temp. of 10% contraction) (after Ref. [25])

dolomite fluxed pellets since the 1970s. The self-fluxed pellets witnessed contraction (expressed in percentage) following reduction under load which depends on the MgO content. The contraction decreases almost linearly with increasing the pellet MgO and flattens at around 10%. The corresponding MgO in pellets was 1.5 wt.%. **Figure 16.8** represents the effect of MgO content in pellets on the softening and melting temperature of pellets (temperature of 10% contraction).

In the case of pyroxenite fluxed pellets, the amount of melt phase undergoes considerable variation when the firing temperature changes. The chemistry of the slag phase was found to be varying with the addition of pyroxenite. Accordingly, the CCS and RDI values of pellets vary.

Fluxed pellets containing approximately 1.5 wt.% MgO, when contributed by pyroxenite, resulted in better softening-melting characteristics compared to acidic pellets. This is due to the formation of high melting point slag and the magnesiowustite phase during reduction.

REFERENCES

1. Panigrahy S. C., B. C. Jena, and M. Rigaud, Characterization of bonding and crystalline phases in fluxed pellets using peat moss and bentonite as binders, *Metallurgical Transactions B*, 21, (1990), 463–474.
2. Loo, C. E. and W. Leung, Factors influencing the bonding phase structure of iron ore sinters, *ISIJ International*, 43, (2003), 1393–1402.

3. Cribbes J. D. and D. W. Kestner, Some factors influencing iron ore fired pellet quality, in: *Proc. of 2nd Inter. Symposium on Agglomeration, Agglomeration–77*, Atlanta, GA, March 6–10, 1977, Sastry K. V. S. (ed.), Pub. by Amer. Inst. of Mining, Metallurgical & Petroleum Engineers, N.Y., 526–537 (Chap., 29).
4. Struve, V. G., Grundlegende Betrachtungen uber das Pelletieren von Eisenerzen, *Chemie Ingenieur Technik*, (1964), 1019–1027.
5. Mayer, K., *Pelletizing of Iron Ore*, Springer-Verlag, Berlin, (1980), 265.
6. Bhagat, R. P., B. Nayak, K. K. Bhattacharyya, and M. C. Goswami, Mineralogy of iron ore pellets and its correlation with crushing strength. *Journal of Metallurgical & Materials Science*, 56, (2014), 117–123.
7. Jaroslav S. R. B. and Z. Ruzickova, Pelletization of fines, Ed. Furestenau, D. W., Volume 7, *Development in Mineral Processing*, Elsevier Science Pub. Inc., NY, 1988, 161.
8. Pal, J., C. A. Kumar, Y. Rajshekhar, G. Das, M. C. Goswami, and T. Venugopalan, Development of iron ore pelletization using calcined lime and MgO combined flux replacing limestone and bentonite, *ISIJ International*, 54, (2014), 2169–2178.
9. Fua, T. and S. Banya, *Trans. ISI Jpn.*, 9, (1969), 137.
10. Wynnyckyj, J. R. and T. Z. Fahidy, Solid state sintering in the induration of iron ore pellets, *Metallurgical Transactions*, 5, (1974), 991–1000.
11. Hamilton, J. D. G., B. F. Hoskins, W. G. Mumme, W. E. Borbidge, and M. A. Montague, The crystal structure and crystal chemistry of $Ca_{2.3}$ $Mg_{0.8}$ $Al_{1.5}$ $Si_{1.1}$ O_{20}(SFCA): solid solution limits and selected phase relationship of SFCA in the SiO_2-Fe_2O_3-CaO(-Al_2O_3) system, *Neues Jahrbuch für Mineralogie—Abhandlungen*, 161(1), (1976), 1–26.
12. Cooke, S. R. B. and W. F. Stowasser, The effect of heat treatment and certain additives on the strength of fired magnetite pellets, *Transactions AIME*, 193, (1952), 1223–1230.
13. Wynnyckyj, J. R. and W. A. McCurdy, Causes of the shell and core structure and layered in homogeneity in iron ore pellets, *Metallurgical and Materials Transactions B*, 5, (1974), 2207–2215.
14. Prestes, E. A., S. A. Chinelatto, and W. S. Resende, *Ceramica*, 55(33), (2009), 61.
15. Dwarapudi, S., T. Uma Devi, S. Mohan Rao, and M. Ranjan, Influence of pellet size on quality and microstructure of iron ore pellets, *ISIJ International*, 48(6), (2008), 768–776.
16. Martinez, P. N., M. H. Trejo, R. M. Estrella, M. de Jesús Castro, R. Rolando, M. Esparza, and M. C. Villareal, Induration process of pellets prepared from mixed magnetite – 35% hematite concentrate, *ISIJ International*, 54, (2014), 605–612.
17. Matsuo, N., T. Maeda, and Y. Ono, *CAMP-ISIJ*, 1991(4), 1081.
18. Bhagat, R. P., U. S. Chattoraj, M. C. Goswami, D. P. Singh, and S. K. Sil, Effect of size parameters of mix ingredients on the porosity and reduction characteristics of sinter, *Steel Research International*, 78, (2007), 451–454.
19. Bhagat, R. P., U. S. Chattoraj, and S. K. Sil, Porosity of sinter and its relation with sintering indices, *ISIJ International*, 46, (2006), 1728–1730.
20. Gao, Q.-J., F.-M. Shen, G. Wei, X. Jiang, and H.-Y. Zheng, Effect of MgO containing additive on low temperature metallurgical properties of oxidized pellet, *Journal of Iron and Steel Research International*, 20(7), (2013), 25–28.
21. Frazer, F. W., H. Westenberger, K. H. Boss, and W. Thumm, The relationship between basicity and swelling on reduction of iron-ore pellets, *International Journal of Mineral Processing*, 2, (1975), 353–365.
22. Qing, G., K. Wu, Y. Tian, G. An, X. Yuan, D. Xu, and W. Huang. Effect of the firing temperature and the added MgO on the reduction swelling index of the pellet with low SiO_2 content. *Ironmaking & Steelmaking*, 45, (2018), 83–89.
23. Dwarapudi, S., T. K. Ghosh, A. Shankar, V. Tathavadkar, D. Bhattacharjee, and R. Venugopal, Effect of pellet basicity and MgO content on the quality and microstructure of hematite pellets, *International Journal of Mineral Processing*, 99, (2011), 43–53.

24. Dwarapudi, S. and M. Ranjan, Influence of oxide and silicate melt phases on the RDI of iron ore pellets suitable for shaft furnace of direct reduction process, *ISIJ International*, 50, (2010), 1581–1589.
25. Suzuki, G., T. Satoch, and N. Sakamoto, Use of iron ore pellets in Japan, National Seminar on Role of Iron Ore Pellets in Iron & Steel Industry. Under the aegis of Iron & Steel Division of Indian Institute of Metals. Panaji, Goa, April 15–16, 1988, Org. by Steel Authority of India Limited, Kudremukh Iron Ore Company Limited, and Mandovi Pellets Limited.
26. Dwarapudi, S., T. K. Ghosh, A. Shankar, V. Tathavadkar, D. Bhattacharjee, and R. Venugopal, Effect of pyroxenite flux on the quality and microstructure of hematite pellets, *International Journal of Mineral Processing*, 96, (2010), 45–53.
27. Moon, J. T. and R. D. Walker, *Ironmaking & Steelmaking*, 1975 (2), 30.
28. Sharma, T., R. C. Gupta, and B. Prakash, Degradation of pellets during reduction in blast furnace, in: *Proc. Conf. All India Seminar on Extraction of Iron*, Bhilai, February 28–29, 1984, Org. by Indian Institute of Metals and Institution of Engineers (India), Paper No. III-1.
29. Sharma, T., R. C. Gupta, and B. Prakash, Swelling of iron ore pellets by statistical design of experiments, *ISIJ International*, 32, (1992), 1268–1275.
30. Dwarapaudi, S., S. M. Rao, and N. D. Rao, Pelletization of high alumina iron ore fines: problems and prospects, *Tata Search*, 2007, 82–95.

Annexures

Topics common to sintering and pelletization are described here. Annexure I lists salient transport parameters which could serve as guideline for describing the processes and their modelling. Annexure II mentions various test methods for describing the quality parameters of sinter and pellets (including lump ore). These tests have been devised following various international standards to investigate, identify, and quantify, on a relative basis, an increasing number of burden properties. The purpose is to determine the extent of their suitability in iron-making processes using blast furnace and/or following direct reduction. The rationale for these tests in order to understand the fundamentals are mentioned in brief.

ANNEXURE I

Annexure I: Transport Parameters for Sintering and Pellet Induration[1]

Property	Sintering	Pelletizing
Particle size d_p (m *10^3)	0.8–3.0	11–12.5
Sphericity- φ	0.5–0.8	0.90
Porosity- ε	0.42–0.49	0.38–0.41
Specific bed surface- S_B (m^{-1})	1,500–4,000	300–380
Mass velocity of gases- G ($kgm^{-2} s^{-1}$)	0.9–1.2	1.0–2.5
Reynolds Number- Re	30–120	240–1250
Nusselt Number- Nu	6.5–11	15–34
Heat transfer coefficient		
h ($Wm^{-2} s^{-1}$)	240–350	130–180
h_v ($kWm^{-2} s^{-1}$)	200–1,000	20–60
Mass transfer coefficient- k_c (ms^{-1})	0.8–1.4	0.25–0.82
Bed depth (height)- L (m)		
new charge	0.35–0.50	0.40–0.45
grate charge	NA	0.13–0.15
Pressure drop- ΔP		
(kPa)	9–19	5.0–6.5
(m H_2O)	0.90–1.9	0.50–0.65
Temperature front velocity v_B ($ms^{-1}\times10^{-4}$)	3.0–5.0	3.0–5.0

References 1: Wynnyakyj, J. R., and R. J Batterhamt, Proc. 4th Inter. Symp. On Agglomeration Toronto, Canada 1985, C. E. Capes (editor) Iron & Steel Society of AMIE and CIM. 957-994.

ANNEXURE II

TEST METHODS FOR QUALITY PARAMETERS OF SINTER AND PELLETS

PREAMBLE

The characterization of iron-bearing materials (ore, sinter, and pellets) is required to be carried out in the laboratory with respect to their various quality parameters to assess their suitability in the extraction processes involving direct reduction and/or blast furnace route(s). Laboratory tests have been devised to investigate, identify, and quantify, on a relative basis, an increasing number of burden properties to determine the extent of suitability of a given material as a burden. These tests identify the behavior of a material in the furnace/direct reduction operations. Based on the results of the tests, the behavior of iron ores and agglomerates as iron burden can be judged.

The procedures of determination of these qualities are described. These tests follow ISO and other international standards. However, the procedure and interpretation of test results may vary depending on the local plant conditions and requirements. An important description of the test procedures have been reported by Kortmann and Burghardt [1]

Green Pellet: The quality parameters of green pellets include their

a. Drop resistance (number)
b. Wet compression strength
c. Porosity and saturation degree

Indurated Pellets: The quality parameters that are specific to the indurated pellets are:

a. Cold crushing strength (CCS)
b. Swelling index
c. Clustering index

Sinter/Pellets: Many of the quality parameters are common in sinter and pellets; hence, the test methods are more or less common. These include:

a. Bulk density and porosity
b. Shatter/tumbler index
c. Reducibility index (reduction kinetics),
d. Reduction–degradation index
e. Softening–meltdown characteristics

SAMPLING AND SAMPLE PREPARATION

This is carried out in accordance with:

ASTM E877-13: Standard practice for sampling and sample preparation of iron ores and related materials for determination of chemical composition and physical properties.

GREEN PELLETS

Drop Test The drop test of the green pellets measures the pellets' ability to remain intact (without a breakage) during their handling.

Typically, it is measured by dropping a freshly made (undried) single green pellet from a height of 0.45 m on a steel plate until it shows perceptible cracks or it crumbles. The number of drops that the pellet undergoes before it develops a fracture is recorded and expressed in a number. This test is performed on 20 randomly selected green pellets. The result is averaged out. The mean value for the tested pellets gives their drop damage resistance, or drop number.

Wet Compression Strength This is the measure of the ability of the wet pellets to retain their shape during handling. Individual undried pellets are compressed using a compression test machine with a cross head speed of 20 mm per min. The peak load required to facture the wet pellets is recorded.

Dry Compression Strength This indicates the ability of the dried pellets to survive handling during the feeding process. The index reflects the binder strength. Dry compression strength is measured according to:

ASTM E382-97, *1998.* Standard test method for determination of crushing strength of iron ore pellets.

It is measured on pellets dried at 105°C for 1 h. Dried pellets are compressed using a compression test machine. The peak load required to facture the dried pellets is recorded and reported in N/P. Dry compression strength should be at least 22 N per pellet.

Porosity and Saturation Degree The procedures for measuring these indices are mentioned elsewhere.[2] Porosity of typical industrial wet and dry pellets should be in the range of 31–36%.

INDURATED PELLETS

Cold Crushing Strength (CCS) This is a measure of the ability of fired pellets to survive during their charging in the blast furnace and their handling. It is measured in accordance with:

JIS-M ***8718****-1973:* Method for measuring the crushing strength of iron ore pellets,.

ASTM E382-97*, 1998:* Standard test method for determination of crushing strength of iron ore pellet, Annual Book of ASTM Standards, Vol. 3.06 (Metals: Test Methods and Analytical Procedures, 56–58.

ISO 4700-1983: Iron ore pellets–determination of crushing strength.

IS 8625-1977 Amendment 1993 Method for determination of crushing strength of iron ore pellets

Typically, the crushing strength of pellets is measured by compressing at least 60-pellet samples one-after-another with size below 12.5 and above 9.5 mm diameter between parallel steel plates until the pellets break. The capacity of the load cell shall be at least 1,000 kg. The test values are reported in terms of:

a. the mean crushing strength in daN (deca-Newton)/pellet, and
b. the percentage under 200 and 150 daN/pellet.

The acceptable values are:

Average: 250 (minimum)
Below 200: maximum 10%
Below 150: maximum 5%

The compression strength can be calculated by dividing the compression force by the cross-section area of the pellet.[3] However, during compression, a small variable portion of the pellet is often squashed flat at the point of contact, forming a flat platform that distributes the load. Hence, a true point contact no longer exists. The problematic issue is generally solved by using a screened size fraction of green pellets in the compression strength measurement and by expressing the compression strength in N/pellet.

The compression strength of individual pellets varies widely from product to product, and from pellet to pellet, within a product batch. It is normally the mean value of maximum and minimum values.

Compression Strength after Reduction This can be measured according to:

IS 8604-1977 Method for determination of compression strength of iron ore pellets after reduction. Typical test is as follows:

After a specified period of reduction, the individual (reduced) pellets are lifted and cooled at the upper part of the reaction tube with a small amount of the reducing gas flow to prevent any possible re-oxidation until they are cooled completely (to room temperature). These are placed in a compressive testing machine and a load is applied at a constant speed of compressive plates until the pellets break, which is recorded in Newton. Typically an Amsler type tester with a cross head speed of 0.086 mm/s has been used[4] . The maximum value of load observed just before an abrupt fall with extensive fracture of the pellet was taken as the strength. Ten almost completely reduced pellets are tested for each set of conditions and the results are expressed by the mean values together with the ranges of data.

SWELLING INDEX (SI)

The swelling index indicates the volume change of pellets during reduction. A higher swelling index decreases the strength of the pellets after their reduction. This phenomenon causes irregularities in the blast furnace like high resistance to gas flow, burden hanging, and slipping. Swelling is mitigated by proper induration in the blast furnace. Local process conditions like temperature and gas composition greatly influence the swelling behavior. At higher reduction degrees, swollen pellets shrink.

The swelling index of the pellets is determined following:

ISO 4698: 2007 Iron ore pellets for blast furnace feed stocks- Determination of the free-swelling index,

IS 8624*:1995* (reaffirmed 2001): Determination of swelling index of iron ore pellets.

The test employs isothermal reduction of the sample at 900°C using 30% CO and 70% N_2. The volume of pellets before and after reduction are measured. The test procedure typically followed is as follows: 18 numbers of randomly selected pellets in the size range of 10.0 to 12.5 mm are used for the test. These 18 pellets are dried at 105 +/– 5°C for 60 min in an oven. These pellets are then taken in a reduction tube made up of a nonscaling heat-resistant material, INCONEL 600. An arrangement is made for holding 18 pellets, in three levels, with six pellets in each level. The internal diameter of the reduction tube is 75 +/– 1 mm. The temperature of the furnace is increased to 900°C, and inert gas passed through the reduction tube at a flow rate of 10 dm^3./min, increasing to 15 dm^3./min; when the temperature reaches 900°C, reducing gas of 30% CO and 70% N_2 is introduced in the reduction tube for 60 min at a flow rate of 15 dm^3./min. After reduction, reducing gas is replaced by inert gas at 5 dm^3./min, and the furnace is allowed to cool to room temperature. The initial and final volumes of pellets are measured by the mercury displacement method.

$$\text{Swelling index} = [V1 - Vo] * 100 / Vo$$

where:
V1 = volume after reduction, and
Vo = original volume before reduction.

CLUSTERING INDEX

The sticking among individual iron ore pellets happens to be one of the most serious problems encountered with a gaseous shaft furnace. The phenomenon of sticking occur mostly during metallization of ore, which could be mainly influenced by temperature, content of gangue, reduction step etc. The sticking (tendency) of pellets are observed above a certain temperature and this temperature is mainly affected by the characteristics ore and reductant properties. The sticking tendency happens to be lower with higher gangue content in ore. On the other side, ores with higher gangue portion might cause lower melting temperature, which could also induce sticking greatly.

The clustering index is determined by following

ISO 11256; *2015.:* Iron ore pellets for shaft direct-reduction feedstocks- Determination of the clustering index

The pellets are isothermally reduced in a fixed bed at 850°C under static load, using a reducing gas consisting of H_2, CO, CO_2, and N_2, until a degree of reduction of 95% is attended. The reduced test sample is disaggregated by tumbling, using a

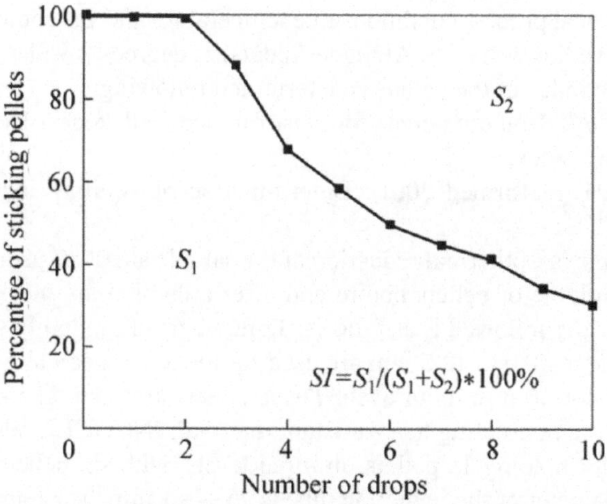

FIGURE AII.1 Calculation of clustering index of pellets following reduction (after ref. [5])

specific tumble drum. The clustering index is calculated as the mass of clustered material accumulated after specified disaggregation operations.

$$\text{Clustering index (CI)} = (100/8\ m_r)\Sigma m_{ci}$$

where:

m_r = total mass, in g, of the test portion after the reduction, and

m_{ci} = mass, in g, of the clustered material after the ith disaggregation operations.

A typical measurement adopted by Ling- yun[5] is as follows.

The pellets, in the size range of 12.5–16 mm with average crushing strength of 2973 N/pellet. were placed in a balling disc and coating was applied by spraying a suspension of coating material. These pellets were oven-dried at 105°C and isothermally reduced at 800–1000°C in synthetic gas mixtures under a load of 1.0 kg/cm^2. For each reduction experiment, 500 g of pellets were heated to the required temperature and then completely reduced in the gas mixtures at a flow rate of 0.6 m^3/h. Finally, the reduced pellets were cooled to room temperature in atmosphere of nitrogen. In the end, number of drops versus percentage of clusters was plotted. The sticking index (SI) was defined as the area proportion under curve. SI was zero when no clusters formed and 100 when all pellets remained clustered without disintegration in 10 drop tests. The calculation was illustrated by a specific example, as shown in Figure AII.1.

SINTER/PELLETS

Bulk Density and Porosity These are carried out in accordance with:

ASTM C830-00 (reapproved 2011): Standard test method for apparent porosity, liquid absorption, apparent specific gravity, and bulk density of refractory shapes by vacuum pressure.

IS 11348: *1989:* Iron ore pellets—apparent specific gravity, true specific gravity and apparent porosity—method of determination (first revision).

The pellet porosity is determined following the measurement of the true specific gravity and apparent specific gravity of the pellets. First, we calculate the apparent specific gravity and then the true specific gravity by using standard procedures, and then we calculate the porosity percent by this given formula.

Porosity (P) is expressed by the following equation:

$$P = 1 - (\rho_a / \rho_r) * 100$$

P = porosity (—).

ρ_a, ρ_r = apparent and real density of the agglomerate sample.

The real density is measured using a pycnometer (JIS-Z-8807). The apparent density is measured using the mercury method (JIS -M-8715). In case of pellets, the sample having size in between 9 and 16 mm are used for the test. The number of pellets per test shall not be less than 12. Apparent density of the sample is estimated by measuring the dry weight (D), and suspended weight (S) of each test specimen.

Apparent density of the sample (ρ_a) is given by:

$$\rho_a = [D / (D - S)] * \rho_{Hg}$$

where:

D = sample dry weight (g),

S = suspended weight of the sample (g),

ρ_{Hg} = density of mercury (Hg)

Typically, a porosity of 22–30% for fired pellets is associated with good reducibility. In order to make pellets of good physical strength and reducibility, the porosity also need to be controlled so as to have smooth furnace operation.

Shatter Index The shatter index of sinter/pellets simulates the various falls on transportation from the discharge point to the blast furnace. This is measured according to:

IS 9963-*1981*: Method for determination of shatter index of iron ore lumps, sinters and pellets: A typical test includes dropping of 20 Kg sinter sample of size range 10 mm to 40 mm/pellets of size range 9 mm to 16 mm for a predetermined number of times from a height of 2 m on steel plate. The resultant material is screened. Shatter strength is normally defined as the percentage of resultant material larger than 10 mm/6.3 mm (in case of sinter) that remains after the predetermined drops.

Tumbler/Abrasion Index The tumbler index is a measure of resistance to generate fines' handling and transportation, whereas the abrasion index is a measure of dust-generating tendency. The tumbler index of sinter signifies the cold strength of sinter representing its resistance to degradation during its transfer from sinter plant (SP) to blast furnace (BF). The TI measurement with any particular standard can reflect different meanings for different blast furnaces (i.e., the distance and drop during transportation from SP to BFs is meaningful).

These indices are measured in accordance with:

ASTM E-279-97 (reapproved 2005): Standard test method for determination of abrasion resistance of iron ore pellets and sinter by the tumbler test

IS 6495: 1984: Method for tumbler test for iron ore oxides, lump ore, sinter, and pellets: Test procedure, typically followed, is as follows.

The charge consists of 11.3 Kg of sinter having a size of 10–40 mm. The material to be tested must be dried at 105 +/– 5°C. It is then subjected to 200 revolutions in a drum at a speed of 25 rpm (for 8 min). Dimensions of the drum are as follows:

Internal diameter: 1,000 mm
Length: 500 mm
Plate thickness: 5 mm
Lifters (two): 50 mm (height) * 5 mm (breadth)

After this test, the material is screened. The weight percent of above 6.3 mm sinter is designated as tumbler index (TI).The weight percent of below 0.6 mm is reported as abrasion index.

Decrepitation Index (DI) This is a measure of resistance to thermal shock in the upper part of a BF. It is measured in accordance with

ISO 8371: 2015: Iron ores for blast furnace feedstocks- Determination of the decripitation index.

Reduction–Degradation Index (RDI) The reduction–degradation behavior of sinter is the most important parameter and has been well researched and discussed pertaining to the sinter plant and blast furnace. The phenomenon arises from the low-temperature reduction of sinter. It takes place in the upper part of the blast furnace in the temperature range of 550–600°C. During gaseous reduction, the hematite is reduced to become magnetite. The magnetite, being porous, results in a volume increase. This induces a stress and initiates a crack in the matrix, which propagates with further reduction. This causes the disintegration of the sinter mass and generates fines.

These are measured in accordance with:

***JIS M8720*-2001:** Iron ores—Determination of low-temperature reduction–disintegration.

ISO 4696-1:2015: Iron ores for blast furnace feedstocks- Determination of low temperature reduction disintegration indices by static method- Part 1: Reduction with CO, CO_2, H_2 and N_2 ***ISO 4696-2:2015:*** Iron ores for blast furnace feedstocks- Determination of low temperature reduction disintegration indices by static method- Part 2: Reduction with CO and N_2

Test procedure and presentation of the results, typically followed, are as follows:

Sample size: 10–15 mm.
Dry the sample at 105°C for 2 h in an air oven.
Weight of sample taken in the reacting vessel = 500 g.

The reacting vessel containing the sample is placed inside the heating chamber and N_2 gas is passed through the reacting vessel at 5 dm^3./min. The temperature is raised to 550°C and is stabilized at that temperature for 10 min. N_2 gas flow is stopped and a mixture of 70% N_2 and 30% CO gas is passed for 30 min. after drying the gas by passing it through a potassium hydroxide (KOH) solution, pyrogallol solution, and silica gel. The vessel is then cooled to room temperature under the flow of N_2 gas at 5 lit./min. The cold sample is then taken out and rotated in a drum (200 mm length and 130 mm ID, having 2 lifters) for 30 min at 30 rpm. It is then screened through a 3.15 mm-sized screen, and the RDI value is calculated as follows:

$$\text{RDI} = \left[\left(\text{Weight of below 3.15 mm size fraction}\right) / \left(\text{Weight of sample taken}\right)\right] * 100$$

Reducibility Reducibility of sinter is defined as the ease with which the oxygen combined with it can be removed. Obviously, a higher reducibility means a greater extent of indirect reduction in the blast furnace, resulting in lower coke rate and higher productivity. It is one of the most important criteria to assess any iron-bearing material for its utilization in a blast furnace.

This is measured in accordance with:

IS 8167-1976 Method for determination of reducibility of iron ore and sinter
IS: 11292-1985 Method for determination of relative reducibility of iron oxides, lump ores, sinter, and pellets
JIS M 8713-2000 Iron ores—Determination of the reducibility by the final degree of reduction index

Test procedure and presentation of the results, typically followed, are as follows:

Sample size: 10-15 mm.
Dry the sample at 105°C for 2 h in an air oven.
Weight of sample taken in the reacting vessel = 500 g.

The reacting vessel containing the sample is placed inside the heating chamber and N_2 gas is passed through the reacting vessel at 5 dm^3./min. The temperature is raised to 900°C and is stabilized at that temperature for 10 min. N_2 gas flow is stopped and a mixture of 70% N_2 and 30% CO gas is passed for 180 min after drying the gas by passing it through a KOH solution, pyrogallol solution, and silica gel. The vessel is then cooled to room temperature under the flow of N_2 gas at 5 dm^3/min. The cold sample is then taken out and weight is taken using a digital electronic balance. The reducibility index (RI) value is calculated as follows.

$$RI = \left[(\text{Loss in weight of the sample}) / (\text{Weight of available oxygen in the sample})\right] * 100$$

Softening and Meltdown Characteristics The softening and melting behaviour of burden materials determine the parameters of the cohesive zone and thus significantly affect blast furnace operation and quality of the produced iron. The melting of sinter usually starts with dripping of small quantity of slag, followed by simultaneous run-out of the main portion of metal and slag. In case of pellets, drops of metal appear first with subsequent simultaneous run-out of metal and slag. The formation of primary slag in the blast furnace is important, particularly for the countries, like India, Ukraine, etc. the burden material has high gangue amount and the blast furnaces operate under conditions of relatively high slag yield (400–500 kg per ton of hot metal produced).

The viscous-plastic flow of iron ore material under load results in bed shrinkage which is a complex phenomenon, owing to heterogeneousness nature of the material, superposition of thermo-rheological behaviour of particular material components namely, iron sponge, gangue, primary liquid slag etc., and the simultaneous development of physico-chemical transformations in both metal and slag phases. Four periods can be identified with respect to the viscous-plastic flow of burden material under load as follows:

- high and constant apparent viscosity in a solid state (period 1),
- fast softening followed by formation of primary liquid slag (period 2),
- plastic mass and formation of metallic sponge (period 3),
- rapid decrease of apparent viscosity due to melting of the iron sponge and separation of metal and slag to independent fluid phases (period 4).

The indices used to characterise the softening and melt-down behaviour are as follows.

T_1 - corresponds to rapid gas permeability loss in the bed in the course of the material's softening (end of period 1);

T_2 – indicates the dripping of the largest portion of liquid products (during period 4).

Various indices have been proposed in literature to determine the softening and melting behaviour of iron ore materials. The softening and meltdown characteristics of pellets are determined in accordance with:

IS 9660-1980 Guideline for determination of softening characteristics of iron ore pellets

Shatokha et al. [6] have devised laboratory experiments to identify the softening and melt-down indices from the shrinkage data during reduction of sinters and pellets produced from iron ores in Ukraine and Russia.

The sample was pre-reduced prior to melt-down experiments. The sample was placed in graphite crucible, heated to 1,000°C with heating rate of 15°C per min. and held at this temperature during 3 hours in the reduction gas flow. The flow rate of reduction gas having composition, CO:25–30 vol.% and N_2: 65–70 vol.% was maintained at 3 dm^3/min. Pre-reduced material was put in the graphite crucible placed to the alumina reactor of Tamman furnace. With the heating rate of 5°C/min, material in crucible was heated to 1,600°C, while reducing gas was introduced.

Shrinkage of the studied material and gas pressure drop were continuously recorded. The method of analysis based on calculation of apparent viscosity of iron ore material in the bed under load was applied.

ORGANIZATIONS OF STANDARDS REFERRED

E: **ASTM:** ASTM International (formerly known as American Society for Testing and Materials), 100 Barr Harbor Drive, PO. Box C 700, West Conshohocken, PA 19428-2959, USA, 2005

IS: Indian Standards, Bureau of Indian Standards, Manak Bhavan, 9 Bahadur Shah Zafar Marg, New Delhi 110002

JIS Japanese Industrial Standards, Japanese Standards Association Tokyo, Japan, www.jsa.or.jp

ISO: International Organization for Standardization, ISO Central Secretariat Chemin de Blandonnet 8, CP 401-1214 Vernier, Geneva, Switzerland

REFERENCES

1. H. A. Kortmann and O. P. Burghardt, Test methods for evaluating iron ores, pellets and sinter, in: *Proceedings of the 2nd International Symposium on Agglomeration, Agglomeration 77*, Atlanta, March 6–10, 1977, Ed. K. V. S. Sastry, Society of Mining Engineers of AIME, (1977), 219–242.
2. Sportel, H. and J. Droog, Influence of pore saturation on compressive strength of green iron ore pellets, *Ironmaking & Steelmaking*, 24, (1997), 213–223.
3. Ball, D. F., J. Dartnell, J. Davidson, A. Grieve, and R. Wild, *Agglomeration of Iron Ore*, Sec. 2, Chap. 11, Heinermann Education Books Ltd., London, (1973).
4. Taniguchi, S. and M. Ohmi; Factors influencing the crushing strength of a metallised iron ore pellet after hydrogen reduction, *Trans JIM* **1978** (19), 581–587
5. Y I Ling- yun. Y. I., Z. C. Huang, T. H. LI, and T. JIANG. Sticking of iron ore pellets in direct reduction with hydrogen and carbon monoxide: behavior and prevention **J. Cent. South Univ.** (2014), 21, 506–510
6. Shatokha, V; Velychko, O. Study of Softening and Melting Behaviour of Iron Ore Sinter and Pellets High Temp. *Mater. Proc.*, Vol. **31** (2012), pp. 215–220

Index

Note: Locators in *italics* represent figures and **bold** indicate tables in the text.

H

Q

R